NUCLEAR DYNAMICS IN THE
NUCLEONIC REGIME

Related Titles

Neutrons, Nuclei and Matter
J Byrne

Statistical Models for Nuclear Decay
A J Cole

Basic Ideas and Concepts in Nuclear Physics (2nd Edition)
K Heyde

Non-accelerator Particle Physics
H V Klapdor-Kleingrothaus and A Staudt

Nuclear Physics: Energy and Matter
J M Pearson

Nuclear Decay Modes
D N Poenaru

Nuclear Particles in Cancer Treatment
J F Fowler

Linear Accelerators for Radiation Therapy (2nd Edition)
D Greene and P C Williams

Nuclear Methods in Science and Technology
Y M Tsipenyuk

SERIES IN FUNDAMENTAL AND APPLIED NUCLEAR PHYSICS

Series Editors
R R Betts and W Greiner

NUCLEAR DYNAMICS IN THE NUCLEONIC REGIME

Dominique Durand

*Laboratoire de Physique Corpusculaire,
ISMRA and Université de Caen,
Caen, France*

Eric Suraud

*Membre de l'Institut Universitaire de France,
Laboratoire de Physique Quantique, Université Paul Sabatier,
Toulouse, France*

and

Bernard Tamain

*Laboratoire de Physique Corpusculaire,
ISMRA and Université de Caen,
Caen, France*

INSTITUTE OF PHYSICS PUBLISHING
BRISTOL AND PHILADELPHIA

British Library Cataloguing-in-Publication Data

A catalogue record for this book is available from the British Library.

ISBN 0 7503 0537 1

Library of Congress Cataloging-in-Publication Data are available

Commissioning Editor: James Revill
Publisher: Nicki Dennis
Production Editor: Simon Laurenson
Production Control: Sarah Plenty
Cover Design: Victoria Le Billon
Marketing Executive: Colin Fenton

Published by Institute of Physics Publishing, wholly owned by The Institute of Physics, London

Institute of Physics Publishing, Dirac House, Temple Back, Bristol BS1 6BE, UK

US Office: Institute of Physics Publishing, The Public Ledger Building, Suite 1035, 150 South Independence Mall West, Philadelphia, PA 19106, USA

Typeset in TEX using the IOP Bookmaker Macros
Printed in the UK by Bookcraft, Midsomer Norton, Somerset

To our families

Contents

Preface

The main goal of this book is to provide a pedagogical introduction to the physics of nuclear collisions in the so-called nucleonic regime. A few words of explanation are necessary concerning the title of this book. We define the nucleonic domain as the incident (beam) energy regime in which the sub-nucleonic degrees of freedom (quarks, gluons and hadronic resonances such as pions or kaons) do not play an important role. Although there is no well-defined frontier between this domain and the domain of hadronic matter, a natural upper limit of the nucleonic regime can be associated with collisions with an incident energy corresponding to the pion production threshold (i.e. 290 MeV/u incident energy for a nucleon on a fixed target). In other words, this book deals mainly with nuclear collisions below 100–200 MeV/u. However, some aspects of hadron–nucleus and nucleus–nucleus collisions at relativistic energies are discussed because of their strong connection with the topics addressed in this book. An introduction to the relativistic energy domain may be found, for instance, in the book by Wong [498].

The main motivation of the present work is the study of dissipative reactions in relation to the properties of nuclear matter far from equilibrium. Therefore, elastic and quasi-elastic collisions are not described. More generally, nuclear reactions studied in the context of what is traditionally called nuclear structure such as single-nucleon transfer, pick-up or break-up reactions are not discussed. Low-energy nuclear collisions close to the Coulomb barrier are mentioned but not studied in detail. In particular, the important questions of the synthesis of superheavy elements and cluster radioactivity are not addressed.

This book is in some ways a continuation of a monograph by U Schröder and J Huizenga published in the 'Treatise on Heavy Ion Science' edited by A Bromley in 1984. In this work damped reactions were discussed extensively in the energy range accessible in the 1970s and early 1980s. Since then, a new generation of heavy-ion facilities has emerged and heavier mass numbers and higher incident energy beams have become available. A large body of data has thus been accumulated thanks to the advent of powerful multidetectors. In parallel, many important developments in the theory have been undertaken mainly through the emergence of microscopic transport models.

Up to now, however, no book has discussed these new features in a thorough self-contained way; this has been the main motivation for our project. The present book is written at a level that should make it easily accessible to graduate students but it should also be useful for newcomers and researchers in the field. In particular, the first five chapters cover the general basic considerations concerning both the theoretical and experimental aspects of nuclear collisions. We believe that this material could constitute a basis for lectures on advanced nuclear physics. The last three chapters are more specialized since they discuss in detail our current understanding of subjects widely studied nowadays. They are intended to describe as completely as possible important and timely issues related to nuclear dynamics and the physics of hot nuclei. The material presented in these chapters is a matter of active discussion at conferences and workshops.

Summarizing in a few hundred pages the enormous amount of data and theoretical work about nuclear collisions is quite a difficult task. In addressing the important issues raised in this book we have endeavoured to quote as exhaustively as possible the numerous works published in the literature. However, it is unfortunately highly probable that some aspects may have been overlooked. We would therefore like to apologize in advance to those physicists whose work has not been properly reported.

Needless to say we have benefited from the help of many colleagues and friends through numerous and fruitful discussions over several years of passionate research. An alphabetical list of these many individuals would be lengthy and would reduce to mechanics irreducible personal interactions. Above all this would hide the pleasure we took in these many exchanges. The concerned people will, for sure, know that we are aware of how much we owe them. Last but not least, we would like to thank the institutions which have supported us during the realization of this work, namely our home laboratories and the Institut Universitaire de France.

Dominique Durand
Eric Suraud
Bernard Tamain
December 1999

Chapter 1

Introduction

Atomic nuclei entered physics in a 'shadowy' way. In the very last years of the 19th century radioactivity constituted the first to be identified, although indirectly, nuclear property. With Rutherford's experiments just before the First World War nuclei as such became a subject of research. Some of their basic properties were accessed in the 1930s with the identification of neutrons as constituents, together with protons, and with the pion exchange picture of nucleon–nucleon (neutron or proton) interactions. By the late 1940s both collective and single-nucleon behaviours were, to some extent, unravelled. The shell model allowed us to understand the so-called magic numbers and both fission and giant resonances had been observed, if not fully understood. The following decades saw the possibility of accelerating nuclei and smaller particles, to higher and higher energies, in larger and larger facilities, developed following the pioneering works on cyclotrons in the late 1920s. According to Heisenberg's uncertainty principle these facilities provided access to smaller and smaller structures inside the nucleus, which gave birth to a new field of research: particle physics. In turn nucleonic physics dealing with systems of nucleons, namely nuclei in their ground state or in moderately excited states, became, and still is, a major concern of nuclear physics studies.

1.1 Nuclear and nucleonic physics

Although nucleons are compounds of quarks and gluons, ground-state nuclei can *safely* be viewed as ensembles of interacting nucleons. Quarks and gluons then remain bound in nucleons and it thus makes sense to consider nucleons as the effective elementary constituents of nuclei. The basic underlying theory, QCD (quantum chromo dynamics), which describes the strong interactions between quarks and gluons, can thus also be safely hidden (outside nucleons) in an effective interaction between the nucleons. The situation here is pretty similar to the case of Van der Waals interactions in molecular physics. While the basic interaction is the Coulomb interaction with its soft $1/r^2$ dependence, the effective

1

interaction between inert atoms, such as argon, is reduced to $1/r^n$ ($n \gg 2$)
terms reflecting dipole–dipole (or higher multipoles) interactions, produced by the
reciprocal polarizations of the electronic clouds of the atoms. Physics of nuclei
in the vicinity of their ground state is thus a physics of nucleons interacting via
the nucleon–nucleon interaction. Such a picture of nuclei as sets of interacting
nucleons constitutes the basis of most studies devoted to the understanding of
structure and dynamical properties of nuclei.

Over the last decades the pile of static properties of nuclei has mounted up
and now ranges from single-nucleon characteristics to collective observables. Let
us cite spectroscopic properties such as single-particle levels, separation energies
or shape analysis as typical static quantities (although they may also be accessed
in a dynamical way). From a dynamical point of view, nucleonic motion has
also been extensively studied, in terms of single-nucleon degrees of freedom,
for example, in charge exchange reactions, or in terms of collective degrees of
freedom, as in fission or giant resonances. All these observables have allowed
us to picture nuclei as complex systems in which individual degrees of freedom
coexist with more or less collective ones. An overall concept underlying these
various findings is that of the mean field which models nuclei as sets of particles
moving 'nearly' independently from each other in a common potential well. To
a large extent, many single-particle and dynamical properties of nuclei can be
understood within this general framework. Still, the simple nucleonic mean field
is by no means the end of the story.

In order to access nuclear properties, nuclei must be excited. Depending on
the aim, the excitation may be very gentle or very strong and use various probes.
For example, electromagnetic interactions are known to allow a particularly clean
access to charge densities in nuclei. After providing the systematics of nuclear
charge profiles, electron beams, now in the multi-GeV energy range, are now
offering clues on properties of nucleons *inside* the nuclear medium. They also
allow us to study sub-nucleonic degrees of freedom inside nuclei because their
'quantum size' (e.g. as estimated from their de Broglie wavelength) is much
smaller than the nucleon radius. In such studies, however, the nucleus becomes
a laboratory for our understanding of nucleonic properties rather than a true
subject of investigation. In heavy-ion collisions, on the other hand, a nucleus is
bombarded with another nucleus and the produced nuclei, or at least the formed
(possibly short-lived) nuclear composites are the focus of study. Both their static
properties and the dynamical aspects of the reactions have to be addressed. They
can be described by using macroscopic concepts or by a microscopic description
involving the constituents of nuclei. These are nucleons as long as the considered
beam energies are not large enough to excite sub-nucleonic degrees of freedom.
Quarks and gluons then remain actually sufficiently deeply bound inside nucleons
to accommodate these violent (but still 'external') perturbations. In this book we
will consider a beam energy range (≤ 200 MeV/u), in which the latter assumption
is valid, and we shall call this energy range the *nucleonic* regime. It should,
however, be noted that this energy range does not correspond to a moderate

excitation regime for nucleons as such. Here one definitely leaves the safe realm of close-to-equilibrium physics to truly enter one of out-of-equilibrium situations.

1.2   Heavy-ion collisions in the nucleonic regime

As already noted, the last decades have seen an accumulation of an impressive corpus of nuclear properties close to equilibrium. What do the heavy-ion collisions of today bring to this picture? The answer is manifold. There is, first, the aspect of principle, linked to the question of what does 'know a physical system' mean? In any physical system time is running and, although one usually focuses first on the static properties, there is no reason, in principle, to give less credit to the dynamical properties. To start with the static properties is usually simpler and seems to make more sense. Still, it is also well known that one may understand the dynamical behaviour of a system without necessarily fully understanding its static properties. Think, for example, of the incredible impact of percolation models. Once a short range for the interaction is assumed, most of the physics is under control... and may work pretty well. Hence, as a point of principle, one should *a priori* study dynamical properties on the same footing as static ones. And heavy-ion collisions in the nucleonic domain fulfil exactly this methodological requirement for nuclei.

Beyond methodology, the study of heavy-ion collisions in the nucleonic regime is thus twofold. There is, first, an interest in understanding the time evolution of the reaction starting from a highly-out-of-equilibrium situation (two cold colliding nuclei) towards a possible thermalized system by means of dissipation. Second, reaction products are in *extreme* states in the sense that they can be either highly exotic or hot. Exotic nuclei have unusual neutron/proton ratios or a very large number of nucleons (superheavy elements). The ongoing studies focus here on the existence and structure properties but have not yet truly attacked the dynamical aspects of nuclei. Hot nuclei have excitation energies close to or even higher than their total binding energies. Studies of nuclei at finite temperature are incomplete without an explicit account of the dynamics. Investigations aiming at understanding the physics of hot nuclei thus rely heavily on the physics of the reaction mechanisms themselves, simply as the signals provided by a hot nucleus can mostly be accessed through its de-excitation which usually involves complex dynamical behaviours.

Before attacking such dynamical questions, which will turn out to constitute many discussions in this book, it is interesting to briefly discuss the idealized picture of infinite nuclear matter at various densities and temperatures. In proper thermodynamical terms, the aim of these investigations is to explore the phase diagram of nuclear matter.

1.3   Exploring the phase diagram of nuclear matter

1.3.1   Nuclear matter

Systematic measurements show that the density inside heavy nuclei, such as lead, is more or less constant, which leads us to define an ideal *infinite* system of interacting nucleons in which the Coulomb interaction has been switched off: nuclear matter. Nuclear matter is a generic system (as, for example, liquid ^{3}He and electron gas) for the theoretical description of the N-fermion quantum problem. Despite its idealization, understanding the properties of nuclear matter is a prerequisite for any consistent theory of nuclei as a finite piece of nuclear matter constitutes the core of heavy nuclei. The study of nuclear matter thus complements approaches dealing with the specific properties of individual nuclei. In turn, nuclear matter only provides a gross (synthetic) description of nuclei as it overlooks the key finite-size effects.

The nuclear matter equation of state (energy versus density and/or temperature) is one of the most important concepts in nuclear physics. It underlies much research not only in nuclear physics itself (in particular in heavy-ion physics) but also in nearby fields (for example the physics of supernovae and neutron stars). We do not aim here to give an extensive review of this fascinating topic. Several review papers, as well as many conference proceedings, have addressed the most recent developments in this field. We hence refer the reader to these texts for extensive discussions (see, for instance, [23]). Instead we would like to present superficially some aspects of the nuclear matter equation of state in connection with the dynamics of heavy-ion collisions. We shall try to show how the equation of state constitutes an essential ingredient of our understanding of heavy-ion collisions and, conversely, how these reactions might give us information on the equation of state itself.

1.3.2   The nuclear phase diagram

The phase diagram of nuclear matter gathers in the density–temperature plane the various observed or predicted phases of nuclear matter. Terrestrial nuclei possess a common central density known as the density of saturation, hereafter denoted by ρ_0 and whose value, for symmetrical nuclei ($N = Z$), is of order $\rho_0 \approx 0.17 \text{ fm}^{-3}$. The point of density ρ_0 and zero temperature is known as the saturation point of nuclear matter. High densities/temperatures may be obtained only by strongly perturbing nuclei. This occurs naturally in the cores of type II supernovae where nuclei may be heated up to temperatures of order 10 MeV[1], which are sufficient to sensibly affect the structure of nuclei. On the other hand, supernovae cores are not extremely dense, their density being only of order ρ_0.

---

[1] The temperature is generally expressed in energy units which means that, in nuclear physics, what is called temperature is in fact the product of the usual temperature (expressed in Kelvin) by the Boltzmann constant. Thus, a nuclear temperature of 1 MeV corresponds to 1.2×10^{10} K.

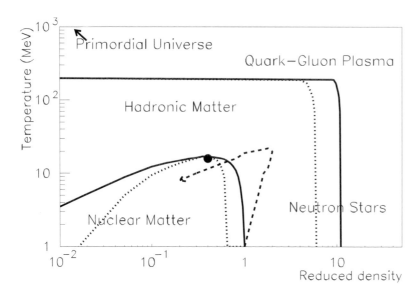

Figure 1.1. Phase diagram (density (in units of saturation density) versus temperature (in MeV)) of nuclear matter. The core of the nuclei in their ground states corresponds to $T = 0$ MeV and $\rho/\rho_0 = 1$. The boundary of the predicted QGP is indicated by the full line at high T (the broken line corresponds to the limit of the predicted coexistence region with hadrons). Very high temperatures and densities are presumably the physical conditions which prevailed in the primordial universe. Neutron stars correspond to a dense (≈ 3 times saturation density) and cold phase. Temperatures between about 20 and 200 MeV correspond to a region in which all hadrons are present: this is called hadronic matter. At low temperatures, only nucleons in their ground state are present. They can be 'confined' in drops of matter: this is the nuclear matter region. Due to the structure of the nucleon–nucleon interaction, the coexistence of a liquid and a gas phase is predicted. The full line in this region is the coexistence curve while the black point is the critical point. The region between the full and broken lines is the metastable region while the domain below the broken line is the so-called spinodale region which will be discussed in detail later. One expects to be able to explore large regions of the phase diagram through heavy-ion collisions, within varying beam energies (a schematic trajectory is indicated in the figure). In this case the collision must be described in a dynamical framework, namely as a path in the phase diagram. This makes the interpretation of results difficult. From [336].

Neutron stars, in contrast, involve very high densities (typically two to three times ρ_0) but temperatures virtually vanish, from a nuclear physics point of view, in these objects ($T \approx 10^6$ K $\approx 10^{-4}$ MeV).

At moderate temperatures, the structure of the nucleon–nucleon interaction suggests properties similar to real Van der Waals fluids. Indeed, calculations described in section 2.3.2.1 predict a liquid–gas phase transition with critical parameters (ρ_c, T_c) corresponding to the black point in figure 1.1. The exploration in the loose vicinity of this point $(\rho_c \lesssim \rho \lesssim 1.5\text{–}2\rho_0, T \lesssim T_c)$ by means of nuclear collisions is the main topic of this book.

At very high temperature/density $(\rho \approx 10\text{–}20\rho_0$ and/or $T \approx 150\text{–}200$ MeV) one expects a transition of nuclear matter to a quark gluon plasma (QGP). At such high energies quarks and gluons, usually bound in nucleons and/or mesons and baryons, become deconfined. QCD calculations provide estimates of deconfinement for energy densities of the order of a few GeV per fm^3, which is comparable to the nucleon mass. This deconfined phase should have existed during the very first instants of the universe, according to the Big Bang model. Such a state could be reached again (a few billion years later!) although probably at a somewhat higher density in ultra-relativistic heavy-ion collisions. Below the transition region towards the QGP one also encounters 'exotic' phases appearing as mixtures of nucleons and mesons (in particular pions) and/or internal excitations of nucleons (such as the Δ excitation for example). The latter regions of the phase diagram are often referred to as regions of pionic or hadronic matter. As we shall see later, these extreme conditions of temperature or density may be obtained, but only during vanishingly small times, in the course of heavy-ion collisions.

1.3.3   How to explore the nuclear matter phase diagram

What do we know about nuclear matter? A direct access to supernovae cores or neutron star matter is impossible. Furthermore, the physics of these objects is by no means a purely nuclear problem: access to the nuclear aspects is complicated by astrophysical questions. On earth, it is the central part of heavy nuclei which constitutes the best access to nuclear matter. But ground-state nuclei allow at best an exploration of the close vicinity of the saturation point. Here again difficulties show up. First one should keep in mind the intrinsic limitation due to the finiteness of nuclei which typically contain less than 250 nucleons, because of coulombic effects. To extract the universal properties of *infinite* nuclear matter from such *finite* systems is thus by no means simple. Another difficulty lies in the fact that ground-state nuclei *naturally* provide very little information on nuclear matter. In order to explore large regions of the nuclear matter phase diagram one needs to perturb nuclei significantly, which will add an extra complication to extracting the nuclear matter properties themselves.

Heavy-ion collisions turn out to constitute the best tool for investigating the properties of nuclear matter in large regions of its phase diagram. In the course of the collision matter is compressed and heated up. Depending on initial conditions, various densities and temperatures can be reached and thus a large scale exploration of the nuclear matter phase diagram is made possible. For example, the transition from hadronic matter to a QGP is actively sought in today's experiments with ultra-relativistic heavy ions, at beam energies of several tens of GeV per nucleon. As already stated, we shall focus in the following on nucleonic physics, which means, in terms of the phase diagram, regions of densities less than typically 1.5–$2\rho_0$ and temperatures typically below 10–20 MeV, as attained in heavy-ion collisions of beam energies up to about 200 MeV per nucleon.

Heavy-ion collisions hence seem to offer a unique opportunity for exploring the phase diagram of nuclear matter. However, one has to pay a heavy tribute for this possibility: namely the fact that this exploration is no longer static but dynamical. Heavy-ion collisions thus appear as paths rather than points in the phase diagram of nuclear matter. A typical heavy-ion collision lasts at most a few 10^{-20} s, often hardly long enough to allow a proper definition of the notion of temperature in such a system. The exploration of the nuclear phase diagram can hence only be understoood in a non-equilibrium context, which outlines the limitations of such studies. One has thus to remain cautious in the interpretation of nuclear collisions in the context of the nuclear matter phase diagram, because it makes sense *only* as long as the thermodynamical variables such as temperature are properly defined. Furthermore, understanding the underlying physics and linking it to the nuclear matter equation of state requires the development of specific dynamical approaches. This makes the problem more complicated but it should also be noted that it makes it richer. Heavy-ion collisions are thus not to be considered only as a tool for investigating the nuclear matter phase diagram. They also lead to far-from-equilibrium dynamical situations in finite quantum systems. Their understanding requires original techniques which may find valuable applications in several fields of physics.

1.4 A short summary of the forthcoming pages

Our text is organized in two parts. In the first part we review both theoretical and experimental basic tools for investigating nucleonic physics by means of heavy-ion collisions. In the second part we discuss how heavy-ion collisions bring us some valuable pieces of physical information. For this purpose, we consider the time evolution of a typical collision, and discuss its various stages, while trying to summarize both our theoretical and experimental understanding of the various encountered situations.

The chapter entitled 'Some basic properties of nuclei: static and statistical concepts' provides a rapid overview of basic nuclear properties. A key idea here is the importance of the mean field. We also briefly review equilibrium statistical physics for future use. We finally discuss some properties of the nuclear matter equation of state and present a statistical description of nuclear de-excitation. The following chapter 'Macroscopic and microscopic descriptions of heavy-ion collisions' is devoted to the dynamical models developed to understand nuclear collisions in the nucleonic regime. We discuss these questions in the general framework of non-equilibrium statistical physics which consists of reducing the original many-body problem to a set of relevant variables. In the nuclear case one-body descriptions represent well-founded, well-adapted and efficient reductions of the many-body problem. In particular, extended mean-field theories are often attacked via the phase space, the energy scales involved washing out detailed quantal effects. These theoretical approaches give access to the transport properties of excited nuclear matter.

The chapter 'Basic experimental tools' presents a comprehensive discussion of the experimental tools developed for analysing the dynamics of collisions in the nucleonic regime. Basically, nuclear reaction mechanisms can be described as a process during which an incident energy is shared among various degrees of freedom, leading to various fragments or particles. The underlying properties of the colliding system can only be understood if one reaches a general overview of the collisions, namely if one detects all the outgoing products among which the available energy has been shared. One thus needs to carry out experiments involving 4π detection. But the total information obtained for a single event is huge and has to be reduced in order to sort the events and to extract reliable physical quantities, hence the necessity of defining relevant and robust global variables.

Once the basic theoretical and experimental tools have been settled, and before entering more specific discussions, a general overview of reaction mechanisms is needed. This is the subject of the chapter 'Reaction mechanisms'. In the nucleonic regime the incident energy becomes larger than the Fermi energy which induces a strong evolution of dissipation mechanisms. Furthermore the reaction and thermalization times become smaller than the typical nuclear decay times, which induces changes in the decay processes. The transition from fusion and deep inelastic processes at low energy to the participant–spectator picture at high energy reflects the competition between various timescales associated with collective and intrinsic degrees of freedom. The analysis of intermediate energy reactions has, nevertheless, to rely somewhat on the pictures provided by these simpler, low- or high-energy, situations.

The following chapters are devoted to an analysis of the collisions and the extraction of the corresponding relevant physical information. The entrance channel is discussed in 'Fast processes towards thermalization', where processes involving nucleon–nucleon collisions are considered. Energetic particles or γ-rays can be produced during the early stage of the reaction. The interesting feature at this level lies in the degree of collectivity revealed by the data: particles can be created far below the corresponding nucleon–nucleon threshold and collective behaviour is observed in the transverse mean velocity of fast emitted particles. These phenomena reflect both the in-medium effects on the nucleon–nucleon collision cross-section and strong momentum fluctuations.

The chapter 'Decay modes of hot nuclei: from evaporation to vaporization' addresses the physics of the de-excitation of hot nuclei, once formed. It is possible to establish a relationship between deposited energies and collective variables such as temperature. Intermediate-energy heavy-ion collisions can be used to follow the evolution of such relations when very large excitation energies are reached. Collective motions are of special interest because they reflect the fundamental properties of nuclei. For example, the characteristics of giant resonances have been established as a function of excitation energy. In turn, the evolution of fission probability with excitation energy reveals the typical times needed to strongly deform a nucleus and the competition with thermal instabilities such as evaporation. It depends on the corresponding viscosity of the nuclear matter and its evolution at large temperature. Finally, the transition from nuclear fission to fragmentation and vaporization is discussed.

Nuclear fragmentation is the process describing the transition from a liquid-like state of nuclear matter to a vaporized gas state. It is the subject of the last chapter: 'Nuclear fragmentation and the liquid–gas phase transition'. We reach here the limits of today's research programmes. This chapter thus contains more open questions than definite answers. The physics of multifragmentation associated with the disassembly of hot nuclear systems on a short timescale is detailed. The question of the instabilities responsible for this process is one of the highly debated issues in this field. The theoretical approaches to nuclear multifragmentation are reviewed from both a dynamical and a statistical point of view with a special emphasis regarding connections with the nuclear equation of state. The relevance of the concept of a phase transition in finite systems is also discussed. Experimental characterizations of nuclear multifragmentation are then developed in terms of timescales and collective motion and the experimental signatures of a liquid–gas phase transition are detailed.

In the 'Epilogue', we draw conclusions and discuss some possible avenues and perspectives for the future of the field.

Chapter 2

Some basic properties of nuclei: static and statistical concepts

In the nucleonic regime nuclei behave as sets of interacting nucleons. This picture is not only valid in the case of ground-state nuclei but it also holds in the dynamical situations we shall encounter in the following. The standard concepts introduced for describing ground-state properties of nuclei thus constitute, to a large extent and provided with some extensions (which will be discussed in the next chapter), the basic tools needed for understanding nucleonic physics. In this chapter we aim hence at remembering some basic concepts used in the description of ground-state nuclei and, by extension, the nucleonic regime.

A key idea, on which much theoretical machinery is founded, is the concept of the nuclear mean field, which basically relies on the fact that nucleons move quasi-independently from one another inside a nucleus. This approximation requires some words of caution as well as some explanations. In the context of nuclear collisions in the nucleonic regime, it is precisely this independence of nucleons which is gradually degraded with increasing beam energy. It is thus of prime importance for the forthcoming discussions to clearly define its range of applicability. This will indeed constitute a 'theme' along this chapter and we shall specifically discuss this aspect in a true dynamical context in chapter 3.

Although the mean field will underlie many of our discussions, one should not forget the elementary nucleon–nucleon interaction from which it is built. We shall thus also briefly review the gross properties of this interaction, particularly its renormalization in the nuclear medium because of the Pauli principle.

In the course of a heavy-ion collision a possibly hot composite is frequently formed and access to its properties requires the introduction of the concepts of statistical physics. We shall thus also briefly describe some of the basic tools of equilibrium statistical physics, leaving the out-of-equilibrium aspects for chapter 3. These tools will allow us to investigate in some detail the equation of state of infinite nuclear matter. Finally, the last part of the chapter will be devoted to a presentation of the statistical model, which constitutes the basic tool for investigating the decay of hot nuclei, following the pioneering work of N Bohr.

Most of the topics covered in this chapter are discussed in standard nuclear physics textbooks. We thus refer the reader to these more exhaustive references, a list of which can be found in the bibliography section. Hence, we have particularly used [68,69,155,398] and [376,426]. Accordingly, only a few original references for some seminal papers or very recent developments are given in this chapter. The discussions are often concise, the main goal here being consistency rather than completeness.

2.1 Nuclei as sets of interacting nucleons

2.1.1 Nuclei and nucleons

2.1.1.1 Nuclei made of nucleons

Nuclei are composed of nucleons (neutrons and protons) which interact via nuclear and Coulomb interactions. While Coulomb repulsion between protons tends to blow the whole system apart, the attractive part of the nuclear interaction binds nucleons together until a balance between the two competing effects is found. Although nuclei are ultimately constituted of quarks, the latter are bound in the nucleons so that this simple picture holds, at least as long as one does not deposit too much energy into the system (basically an energy density of the order of the mass energy of a nucleon (1 GeV) in its own volume, namely a fraction of fm^3). In the nucleonic domain of energy we consider in this book, quarks remain safely bound inside nucleons, even when the nucleus is 'strongly' perturbed. The effective elementary constituents of nuclei thus remain the nucleons, even in dynamical situations. We shall, nevertheless, also have to consider some other particles such as pions or photons which may be produced in the course of heavy-ion collisions. However, there are only very few such particles produced in a given collision so that they can generally be considered in a perturbative way, namely without accounting for possible feedback effects on the nucleons themselves. For the sake of completeness we, nevertheless, give in table 2.1 some basic properties (masses, charges, spin, etc) of the particles we shall encounter in the course of the forthcoming discussions.

The picture of nuclei as sets of interacting nucleons can, to some extent, be visualized in electron scattering experiments. Over the years electron scattering has allowed access, with a high degree of accuracy, to the charge density of nuclei, from which one recovers the proton density. These results are illustrated in figure 2.1 in which the proton densities of some nuclei of various masses are plotted. Compact measurement of the extension of the nucleonic cloud is, in turn, provided by the systematics of nuclear radii. For nuclei of mass A typically larger than 15–20, nuclear radii scale as

$$R \simeq r_0 A^{1/3} \qquad (2.1)$$

Table 2.1. Properties of some 'particles' relevant for the discussions in this book. Column 1 gives the name of the particle, column 2 the usual symbol for denoting it, column 3 its mass (mc^2, in MeV), column 4 its charge (in elementary charge unit ($e = 1.6 \times 10^{-19} C$)). In the fifth column the spin (in Planck's constant $\hbar$ unit) is indicated and column 6 gives the lifetime in seconds (s). Lifetimes denoted ∞ correspond to 'stable' particles. From 1990 *Phys. Lett.* B **239** 1.

Particle	Symbol	Mass (mc^2, MeV)	Charge (e)	Spin ($\hbar$)	Lifetime (s)
Proton	p	938.3	+1	1/2	∞ (?)
Neutron	n	939.6	0	1/2	900
Delta	Δ	≈ 1232	$-1, 0, 1, 2$	3/2	6.0×10^{-24}
Pion	π^0	135	0	0	$\approx 8.0 \times 10^{-17}$
Pion	$\pi^\pm$	140	± 1	0	$\approx 2.0 \times 10^{-8}$
Kaon	$K^\pm$	494	± 1	0	$\approx 1.2 \times 10^{-8}$
Eta	η	549	0	0	
Electron	e	0.511	-1	1/2	∞
Photon	γ	$\approx 0 (<3 \times 10^{-33})$	0	1	∞

with $r_0 \simeq 1.12$ fm. This scaling relation actually constitutes an expression of the well-known saturation property of nuclear matter (see section 2.2.2.1).

2.1.1.2   Energy scales

At this early stage of our discussion it is interesting to remember a few basic energy scales in order to better define the so-called nucleonic energy domain. This will also allow us to recall the energy scales associated with typical nuclear properties such as pairing or giant resonances. We restrict ourselves here to 'static' excitation energies, overlooking on purpose any dynamical effects which will be extensively discussed in the following.

In order to make the discussion more quantitative we consider a medium size nucleus of mass of order 100. The typical energy scale associated with pairing is given by the gap size which is of order 1 MeV in such a nucleus, when pairing effects are observed. This thus corresponds to an excitation energy per nucleon E^*/A of order 0.01 MeV/u. Giant resonance energies typically lie in the 15–20 MeV range for such a nucleus, which amounts to about $\epsilon^* = E^*/A \sim 0.15$–0.20 MeV/$u$. The 'beyond MeV/$u$' excitation energy range is attained in the course of heavy-ion collisions, typical of the nucleonic regime. For ϵ^* beyond about 2–3 MeV/u (a typical energy range discussed in this book) a wealth of phenomena is observed among which fragmentation is one of the most intriguing and complex. We shall discuss it at length in the following (see, in particular, chapters 7 and 8) but must bear in mind that, in this case, one has to account for possibly strong dynamical effects.

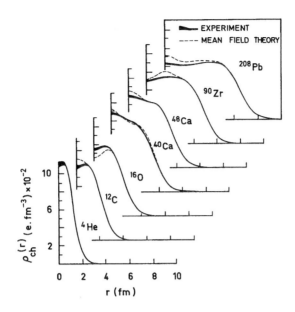

Figure 2.1. Charge density profiles as obtained by electron scattering experiments. Results obtained from mean-field calculations (section 2.2.3) are also indicated for comparison. From [196].

Beyond about $\epsilon^* \sim 8$ MeV/u one enters the domain of intrinsically unstable systems, as the excitation energy per nucleon becomes of the order of magnitude of the binding energy of nucleons in nuclei. Note again that this threshold is 'static', hence probably somewhat overestimated. In terms of excitation energy per nucleon, the upper limit of the nucleonic regime is attained for $\epsilon^* \sim 100$ MeV/u, beyond which internal excitations of nucleons start to play an important role: this also traces the entrance into the so-called *hadronic* domain. In the hadronic domain pions are produced in large amounts. This pionic component may even become dominant around $\epsilon^* \sim 1$ GeV/u. Finally, the transition to the QGP is expected for typical excitation energies of a few tens of GeV/u. However, these energy ranges lie well beyond the domain of this book and thus we shall not discuss them further.

2.1.2 Some basic nuclear models

2.1.2.1 *Liquid drop model*

The simplest, and still extremely powerful, nuclear model is the so-called liquid drop model (LDM) [334, 398], in which the nucleus is pictured as an

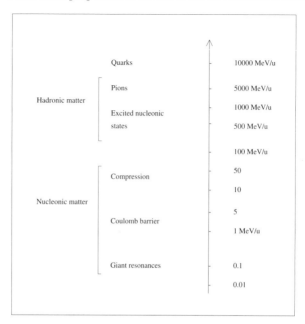

Figure 2.2. Nuclear excitation energy scale (in MeV per nucleon). The scheme shows various 'phases' of matter (left column), the corresponding typical physical phenomena (second column) and the corresponding energies.

homogeneous droplet of nuclear fluid. Such a model is *macroscopic* in the sense that it provides a global description of the system. Most of the physics of the LDM is contained in the evaluation of the energy of the drop. This energy is composed of various terms representing complementing contributions. For a nucleus of mass A, with Z protons and N neutrons, the LDM energy per nucleon can be written in the Bethe–Weiszäcker form as

$$E/A = -a_{\mathrm{v}} + a_{\mathrm{s}} A^{-1/3} + a_{\mathrm{C}} Z^2 A^{-4/3} + \cdots \tag{2.2}$$

where $a_{\mathrm{v}} \simeq 16$ MeV represents the volume energy, $a_{\mathrm{s}} \simeq 18$ MeV the surface term and $a_{\mathrm{C}} \simeq 0.7$ MeV the Coulomb contribution. Note that several other terms may be added to this simple expression (2.2), in order to account for asymmetry ($N \neq Z$), pairing effects. But for the forthcoming discussions we shall content ourselves with the previously defined terms. The LDM is extremely useful in many situations. Beyond the mere (and highly successful) estimates of nuclear binding energies, it also constitutes a useful tool, for example, in the overall description of fission.

2.1.2.2 Fermi gas model

The next step in the description of nuclei consists in accounting for nucleonic motion *inside* the nucleus. This requires an account of the confinement mechanism of nucleons inside the nucleus, namely a description, in LDM terms, of the boundary of the drop. The simplest approach here is to assume an infinite wall, and freely moving nucleons inside this potential well. This will constitute the so-called Fermi gas model. Before briefly discussing its properties, however, one should spend some time on the latter assumption concerning free motion of nucleons. First, the term 'free' is somewhat exaggerated in the sense that nucleons, being fermions, are subject to the Pauli exclusion principle which prevents them from occupying the same quantum state. Second, the assumption of lack of interaction between the nucleons may seem irrelevant at first sight. A way to quantify this effect is to evaluate the nucleon's mean free path, which, mainly because of the Pauli principle, is very large—typically comparable to the size of the system itself (see section 2.1.3.1). It thus makes sense to 'forget' elementary interactions between nucleons, at least in ground-state nuclei. Note, nevertheless, that interactions are not totally absent from the picture. Nuclei are self-bound objects and the potential 'well' confining the nucleons itself results from the interactions between nucleons, but in an average way. For the time being, we simply omit a description of how to evaluate this potential from the elementary interactions and we take a model potential instead. We shall see later how the nuclear potential well can indeed be constructed from elementary interactions (section 2.2.3).

In the Fermi gas model [68, 186, 376] there is, strictly speaking, no model potential but the system is supposed to be confined in a finite volume V. Nucleonic wavefunctions are then assumed to be plane waves, each one characterized by its momentum $\boldsymbol{k}$. Single-particle energies are purely kinetic and hence not quantized but the Pauli principle enters into the occupation numbers of the levels. A Fermi gas is characterized by its Fermi energy ϵ_F which is the energy of the last occupied (highest energy) level. Denoting by A the mass of the nucleus, one can relate ϵ_F to the density $\rho = A/V$. At zero temperature (a *degenerate* Fermi gas) the number of particles of energy $\epsilon = (\hbar^2/2m)\boldsymbol{k}^2$ below ϵ_0 is given by

$$N(\epsilon_0) = \frac{gV}{(2\pi)^3} \int_{|\boldsymbol{k}| \leq k_0 = \sqrt{2m\epsilon_0}/\hbar} \mathrm{d}^3 k. \qquad (2.3)$$

In this expression g is the degeneracy of level k: in a spin degenerate case $g = 2$; when isospin degeneracy is furthermore assumed, within assimilating neutrons and protons, $g = 4$. By definition $N(\epsilon_F) = A$, which leads to the relation

$$\rho = \frac{A}{V} = \frac{g}{(2\pi)^3} \int_0^{k_F} 4\pi k^2 \, \mathrm{d}k = \frac{g}{6\pi^2} k_F^3 \qquad (2.4)$$

which provides the usual relation between density and Fermi energy or Fermi momentum k_F ($\epsilon_F = (\hbar^2/2m)k_F^2$). In nuclei, a typical value of the Fermi energy is $\epsilon_F \simeq 40$ MeV. In symmetrical matter ($N = Z$), this corresponds to an average density $\rho_0 \simeq 0.17$ fm^{-3} and a Fermi momentum $k_F \simeq 1.4$ fm^{-1} (see also section 2.2.2 on the properties of infinite nuclear matter).

As energy is purely kinetic in the Fermi gas model, it can easily be evaluated as

$$E_{FG} = \frac{gV}{(2\pi)^3} \int_0^{k_F} \frac{\hbar^2}{2m} k^2 \, d^3k = \frac{gV}{10\pi^2} \frac{\hbar^2}{2m} k_F^5. \qquad (2.5)$$

This expression can be rewritten in terms of the density ρ and expressed in a simple way by inserting the Fermi energy ϵ_F and the total number of nucleons A

$$E_{FG} = A \frac{3}{5} \frac{\hbar^2}{2m} \left(\frac{6\pi^2 \rho}{g} \right)^{2/3} = \frac{3}{5} A \epsilon_F. \qquad (2.6)$$

The latter relations for kinetic energy express the fact that in a Fermi gas particles constantly move. This 'Fermi motion' reflects the Pauli principle which states that two fermions cannot have the same set of quantum numbers or, in the Fermi gas model, that two fermions move with respect to each other. This Fermi motion can thus be associated with a pressure

$$P_{FG} = -\frac{\partial E_{FG}}{\partial V} = \rho^2 \frac{\partial (E_{FG}/A)}{\partial \rho} = \frac{2}{5} \frac{\hbar^2}{2m} \left(\frac{6\pi^2}{g} \right)^{2/3} \rho^{5/3} = \frac{2}{5} \rho \epsilon_F. \qquad (2.7)$$

We shall see later how this pressure enters into nuclear matter calculations (section 2.2.2.2). Fermi motion inside a nucleus also implies that two nucleons inside a nucleus have a non-vanishing relative velocity. As nucleons may have a variety of velocities (from zero to Fermi velocity), the relative velocity may itself take any value between zero and twice the Fermi velocity. More precisely, one can compute the distribution of relative momenta $k = \|k\|$ which simply reads

$$\mathcal{P}(k) = \int_0^{k_F} d^3k_1 \int_0^{k_F} d^3k_2 \, \delta(\|k_1 - k_2\|/2 - k) \qquad (2.8)$$

(mind that the relative momentum is associated with the reduced mass of the two nucleons, hence the 1/2 factor). After some straightforward algebra, this leads to the simple form [376]

$$\mathcal{P}(k) \, d^3k = \frac{6}{\pi k_F^3} \left[1 - \frac{3}{2} \frac{k}{k_F} + \frac{1}{2} \left(\frac{k}{k_F} \right)^3 \right] dk \qquad (2.9)$$

which exhibits a bell shape between $k = 0$ and $k = k_F$, with its peak around $\langle k \rangle \sim 0.6 k_F$. This means that a first guess at the average relative momentum in nuclei is about $0.6 k_F$.

2.1.2.3 Shell model

The basic defect of the Fermi gas picture of nuclei lies in the quasi-absence of finite-size effects. The latter are, to some extent, better accounted for in the LDM approach. Still, nuclei are neither liquid drops nor a Fermi gas, although both approaches may provide useful orders of magnitudes in many situations. From the quantal viewpoint the lack of finite-size effects in the Fermi gas model reflects itself in the fact that the single-particle nucleon spectrum is structureless: a nucleon can take any momentum below Fermi momentum. Data, in contrast, show that nucleon spectra in nuclei are highly structured in shells. The well-known shell effects, responsible, for example, for the shape of nuclei, and the associated sequence of magic numbers of neutrons or protons (corresponding to neutron or proton shell closure) thus cannot be accounted for in a Fermi gas picture. The shell model, as proposed in the late 1940s, brought the first satisfactory answer to this problem [240, 310, 311].

At simplest level, the shell model can be built up starting from an *ad hoc* central potential in which nucleons evolve independently from each other. In light nuclei a harmonic potential provides a simple and accurate starting point. For heavier nuclei, Woods–Saxon type potentials are better suited:

$$V(r) = \frac{V_0}{1 + \exp\left(\frac{r-R}{a}\right)} \qquad (2.10)$$

where R is the nuclear radius ($R \sim r_0 A^{1/3}$, equation (2.1)), $2a$ the surface diffusivity ($2a \sim 1$–1.5 fm) and V_0 the potential depth at centre $V_0 \sim -50$ MeV. The nuclear potential is furthermore complemented by a Coulomb term acting on protons. In order to recover the proper sequence of magic numbers, the external potential, in which nucleons evolve, has furthermore to contain a spin–orbit term

$$V_{LS} = V_{LS}^0 \boldsymbol{L} \cdot \boldsymbol{S} \qquad (2.11)$$

where $\boldsymbol{L}$ is the orbital momentum and $\boldsymbol{S}$ the spin of the nucleon.

Shell effects, as accounted for in the shell model, will only play a minor role in the forthcoming discussions, although they will be mentioned at some places. It is thus sufficient here to keep in mind orders of magnitude of the values of shell spacing, which strongly depends on the nuclear mass. In the case of the harmonic oscillator, shell spacing reduces to the harmonic frequency

$$\hbar\omega \simeq 41. A^{-1/3} \text{ (MeV)} \qquad (2.12)$$

which provides values between about 7 MeV in heavy nuclei to 15 MeV in light ones.

2.1.3 Independent particle motion in nuclei

As already suggested, independent particle motion is a key feature of nuclei in their ground state. This somewhat surprising property (in view of the short-range singularity exhibited by the nucleon–nucleon interaction (section 2.2.1) and of the 'high' density of nuclei) has long been established, both on experimental and theoretical grounds. The success of the shell model description is actually a typical manifestation of this property. The latter shell model picture is robustly supported by much experimental evidence, ranging from galactic abundancies to the systematics of nuclear level density parameters (section 2.4.2) or separation energies. It is hence certain that nucleons do evolve more or less independently from each other in a common potential, at least in ground-state nuclei. Still, the latter potential can only be understood as stemming from elementary nucleon–nucleon interactions, as nuclei are self-bound systems, with no external confining agent (as, for example, is the case for valence electrons in molecules). The picture emerging from such simple considerations is thus simply one of nuclei bound by a common 'soft' mean field (section 2.2.3) resulting from an average taken over (non-soft) elementary nucleon–nucleon interactions (section 2.2.1). This picture turns out to be quite realistic for ground-state nuclei. As is obvious, this simple picture may have to be revisited in the dynamical context of the heavy-ion collisions we aim to describe. But it is useful to recall some basic features of the independent particle picture in order for a better grasp of how the picture can be altered. This is what is discussed in this section.

2.1.3.1   The mean free path of nucleons

The relevance of an independent particle picture can be quantified by the value of the mean free path λ of nucleons in the nuclear medium. Still, the calculation of λ remains difficult because the elementary nucleon–nucleon interaction has to be renormalized by medium effects (section 2.2.1.2). The results of a microscopic calculation of λ are reported in figure 2.3, where one should note the strong energy dependence of λ [141].

It is also interesting to discuss the value of λ in a simplistic model, inspired by kinetic gas theory. If ρ is the nucleon density and σ the elementary nucleon–nucleon cross-section one can then evaluate the mean free path as $\lambda \simeq 1/(\sigma\rho)$. This relation emphasizes the dependence of λ on ρ and σ and thus points to the strong energy dependence of λ (through σ, figure 2.5). In a nucleus $\rho \sim \rho_0 \simeq 0.17$ fm^{-3}, but estimating σ is more delicate. In a Fermi gas the relative momentum between two nucleons is centred around $0.6k_F$ (section 2.1.2.2), which corresponds to a relative energy of order 40% of the Fermi energy, hence typically around 20 MeV. At such low relative energy the free nucleon–nucleon cross-section is huge, of the order several hundreds of mb (figure 2.5), which would lead to a vanishingly small value of λ ($\lambda \simeq 1/(\sigma\rho) \simeq \frac{1}{10} \ll 1$ fm). But the Pauli principle blocks these interactions at low relative energy. The fermionic nature of the nucleons thus leads to a significant reduction in the in-medium nucleon–nucleon cross-section (section 2.2.1.2). When taking into account this effect, one obtains values of λ of order 5–10 fm, which is comparable to the size of the nucleus itself, which justifies mean-field approaches in the ground state.

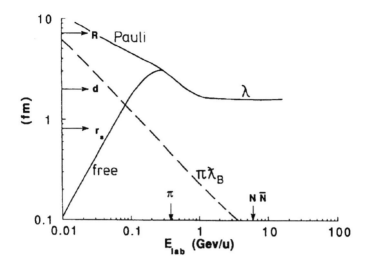

Figure 2.3. Evolution of the mean free path λ (in fm) as a function of beam energy (in GeV/u). Two evaluations of λ are plotted with (full line, Pauli) and without (full line, free) Pauli correlations. In the 'free' case one can see that the mean free path tends to zero at zero energy, which reflects the fact that the free nucleon–nucleon cross-section tends towards infinity at zero energy (figure 2.5). In fact, inside a nucleus (corresponding to the 'Pauli' case), Fermi correlations forbid collisions and make the mean free path very large, comparable to the size of the nucleus itself (R), which justifies mean-field approaches at low energy. In the energy range we consider here, one also notes that λ strongly depends on energy and becomes comparable to the average distance between nucleons (d), so that the mean field becomes insufficient. The vertical arrows indicate the π and N$\bar{\text{N}}$ thresholds in a nucleon–nucleon free collision while the broken line ($\pi\lambda_B$) is the de Broglie wavelength associated with a nucleon with an incident energy E_{lab}. From [141].

The latter conclusion deserves some comment. An independent particle or a mean-field model does not eliminate, *a priori*, all the interactions between the nucleons. These are the so-called residual interactions, i.e. they are beyond the average effect of the interactions between nucleons contained in the mean field (section 2.2.3). The justification for the independent particle model hence lies in the smallness of the residual interaction, rather than on a true lack of interaction. Furthermore, these arguments hold for nuclei close to their ground states. However, λ does strongly depend on energy and may become very small at high beam energy (figure 2.3), so one may face situations in which the independent particle model does not make sense as such. This is, in fact, a typical situation for heavy-ion collisions in the energy range we consider and we shall thus discuss this aspect at length in chapter 3.

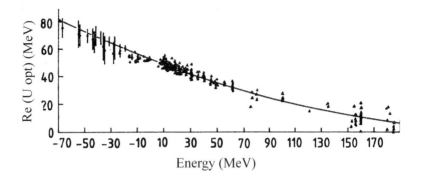

Figure 2.4. Compilation of experimental values of the depth of real part $U_0(\epsilon)$ of the optical potential (in MeV). It is the opposite of the physical (negative, in this energy range) value of $U_0(\epsilon)$, which is plotted here. Note that the single-particle energy ϵ is not defined relative to the bottom of the potential well as in section 2.1.2.2 but in such a way that $\epsilon = 0$ means a zero kinetic energy outside the nucleus. Results obtained with various nuclei have been superimposed here. Isospin and Coulomb corrections have also been taken into account. The full line represents a fit to the experimental data. One should note that the optical potential strongly depends on energy. At incident energies of order 200–300 MeV, it even changes sign and becomes positive, namely repulsive. From [32].

2.1.3.2   *Optical potential*

Nucleon–nucleus scattering provides an almost direct probe of the potential experienced by nucleons inside a nucleus. It has thus been studied in a systematic way for many years. Beyond mere elastic processes, one also observes absorption/emission channels. Systematic measurements of such scattering cross-sections have shown that these results can be interpreted in terms of an optical potential, namely a model in which nucleons in the nucleus move in a complex potential $V(r) = U(r) + iW(r)$. In analogy to standard optics, the real part of the potential accounts for elastic scattering (diffusion), while the imaginary part is associated with absorption. The real part is thus roughly associated with 'mean-field' effects, while the imaginary part accounts for residual elementary interactions beyond the mean field and is thus directly linked to the mean free path.

A comparison with experimental data shows that the real part of the optical potential $U(r)$ can be fitted by a Woods–Saxon potential [426]

$$U(r) = U_0 \frac{1}{1 + \exp((r - R)/a_U)}. \tag{2.13}$$

At low energy the imaginary part $W(r)$ can also be fitted to Woods–Saxon-like shapes, although in a slightly more complicated way,

$$W(r) = \left(W_0 - 4W_1 a_W \frac{\partial}{\partial r}\right) \frac{1}{1 + \exp((r - R)/a_W)}. \qquad (2.14)$$

In these expressions, equations (2.13) and (2.14), R is the nuclear radius, $a_U \sim$ 0.65 fm and $a_W \sim 0.5$ fm is the surface width, which mainly depend on the nucleus, while U_0, W_0 and W_1 are intensity parameters, which depend on the incident nucleon energy ϵ.

A compilation of experimental data for the depth of the real part of the optical potential $U_0(\epsilon)$ is shown in figure 2.4. It is interesting to note that U_0 strongly depends on the energy ϵ, and goes to zero around $\epsilon \sim 200$ MeV, an energy range typically attained in the nucleonic regime. Beyond $\epsilon \sim 300$ MeV, $U_0(\epsilon)$ even becomes repulsive and the usual notion of an attractive potential well disappears in such a dynamical context. The strong energy dependence of the real part of the optical potential $U_0(\epsilon)$ also points to the limitations of a 'static' mean-field picture.

The imaginary part W of the optical potential, in turn, accounts for absorption of nucleons in the nuclear medium. It should be noted here that absorption may correspond to various processes. Depending on the energy, it may imply a loss of flux in the elastic channel (N − N → N − N) to the benefit of inelastic channels (N − N → N − Δ, for example) and not necessarily a true absorption. In any case, absorption is directly linked to the mean free path λ of nucleons inside nuclei. Indeed, let us assume, for the sake of simplicity, that U and W do not depend on position r, and let us consider a nucleon of momentum k entering the nucleus with wavefunction $\phi(r) \propto \exp(-i k \cdot r)$ at instant $t = 0$ [426]. The time evolution of the nucleon wavefunction follows the Schrödinger equation $i \partial \phi / \partial t = h \phi$ with $h = (\hbar^2/2m)k^2 + U + iW$, which leads to a solution of the form

$$\phi(r, t) \propto \exp(-i k \cdot r) \exp\left[-i \cdot \left(\frac{\hbar^2}{2m} k^2 + U\right) t\right] \exp(-Wt). \qquad (2.15)$$

The probability density associated with ϕ thus decreases exponentially in time as $\exp(-2Wt)$. If one pictures the mean free path as the probability that a nucleon 'survives' after having travelled a distance λ corresponding to a time interval Δt such that $\exp(-2W \Delta t) = 1/e$, one can identify $\lambda \simeq (\hbar k/m)1/2W$ (with W strongly depending on energy ϵ).

2.1.3.3 A critical view on the independence of nucleons

As discussed in the preceding sections the independence of nucleons inside the nucleus is (only) a (good) approximation to reality. Even if nucleons are, to a large extent, independent to each other in ground-state nuclei, which is, in particular, reflected by their large mean free path, definitively, one cannot totally overlook the role of the 'residual' interaction beyond the average common mean field, as examplified by the optical potential results. It is thus interesting to try to quantify more precisely the amount of independence of nucleons in ground-state nuclei, possibly with the help of dedicated approximations.

For example, a simple and efficient way of accounting for the energy dependence of the optical potential $V(\epsilon)$ is to introduce an effective mass $m^*(\epsilon)$:

$$\frac{m^*(\epsilon)}{m} = 1 - \frac{\mathrm{d}V(\epsilon)}{\mathrm{d}\epsilon}. \tag{2.16}$$

A nucleon, once 'dressed' with such an effective mass, can then be viewed as moving in an energy independent potential. Experimentally, $m^*/m \sim 0.7$ for single-particle energies of order $20 < \epsilon < 80$ MeV, while in the vicinity of the Fermi energy ($\epsilon \sim -8$ MeV) it is sizably larger ($m^*/m \simeq 1.2$) [301, 302].

In fact, even when not explicitly accounting for the energy dependence of the optical potential, the notion of effective mass shows up very naturally in the nuclear context. As soon as one faces a finite range interaction, exchange terms introduce non-local (or momentum dependent) components into the picture (section 2.2.1.3). Recall that the indistinguishability of particles in quantum mechanics was imposed to account for our 'ignorance' of who is who in an elementary scattering process. One has thus to complement the direct term ('$1 + 2 \to 1 + 2$') by the exchange one ('$1 + 2 \to 2 + 1$') to account for this effect. If the range of the interaction is not too large, the momentum dependence is small (by virtue of Fourier transform properties) and one may content oneself with a small momentum expansion (see, for example, section 2.2.1.3), which, for symmetry reasons, provides a leading term which is quadratic in the momentum and thus effectively renormalizes the mass of the particles. Indeed one obtains

$$V(\boldsymbol{r}, \boldsymbol{p}) \simeq V_0(\boldsymbol{r}) + V_1(\boldsymbol{r})\boldsymbol{p}^2 + \cdots \tag{2.17}$$

so that the single-particle energy can be rewritten as

$$\epsilon = \frac{\boldsymbol{p}^2}{2m} + V(\boldsymbol{r}, \boldsymbol{p}) \simeq \frac{\boldsymbol{p}^2}{2m^*} + V_0(\boldsymbol{r}) \quad \text{with} \quad \frac{1}{2m^*} = \frac{1}{2m} + V_1(\boldsymbol{r}) \tag{2.18}$$

again leading to the introduction of an effective mass.

To use an effective mass for nucleons is usually, at least with moderate excitation, a good ansatz and it will be routinely used in the following. Still, the concept of an effective mass does not exhaust all the pending questions concerning the independence of nucleons, even in ground-state nuclei (dynamical correlations will be discussed in chapter 3). A central issue here concerns the key role of the Pauli principle. Indeed, even if one considers an independent particle model of nuclei, the Pauli principle has to be accounted for and single-particle levels populated accordingly. The example of the Fermi gas here is generic as by nature it is a model in which *only* the Pauli principle applies (and, as we shall see later, the Fermi gas model often provides an instructive zero level picture).

One sometimes speaks here of the 'Pauli correlation'. This term has to be taken with some caution as there is no interaction in the game and thus it does not correspond to a correlation in the sense of *classical* statistical physics. Still, one can compute the correlation function of, for example, a Fermi gas representing a ground-state nucleus, which allows us to quantify the effect. The calculation is well known and we refer the reader to standard textbooks for details [68]. The result is, nevertheless, instructive. One introduces the two-body density matrix (section 2.3.1.1):

$$\rho_{12}(\mathbf{r}_1, \mathbf{r}_2) = \sum_{j<k} [\delta(\mathbf{r}_1 - \mathbf{r}_j)\delta(\mathbf{r}_2 - \mathbf{r}_k) + \delta(\mathbf{r}_1 - \mathbf{r}_k)\delta(\mathbf{r}_2 - \mathbf{r}_j)] \quad (2.19)$$

and computes its average over a ground-state (spin–isospin degenerate, factor $\frac{1}{4}$ in equation (2.20)) Fermi gas which leads, after some algebra, to

$$\langle \rho_{12} \rangle = \rho^2 [1 - \tfrac{1}{4}C^2(k_F||\mathbf{r}_1 - \mathbf{r}_2||)] \quad (2.20)$$

where ρ is the Fermi gas density and where C is given by

$$C(x) = \frac{3}{x^2}\left(\frac{\sin x}{x} - \cos x\right). \quad (2.21)$$

The function C characterizes the correlations of nucleons due to the Pauli principle (hence it would be zero in a classical gas). It takes the value one at zero distance ($x = 0$) which expresses the impossibility to put two fermions at the same point with finite momenta (strong correlation). At large relative distance $x \gtrsim 5$, C vanishes, which reflects the loss of the Pauli correlation for two fermions far away from one another. Taking $x = 5$ for the loss of Pauli correlations and $k_F \sim 1.4\,\mathrm{fm}^{-1}$ leads to a typical 'healing distance' of order 3 fm in a ground-state nucleus. It is a 'huge' quantity, comparable to the nuclear radius, which reflects the key role played by these Pauli correlations in ground-state nuclei.

As a final remark let us return, but from a complementary point of view, to a basic underlying hypothesis of the Fermi gas model, namely the total absence of interactions between nucleons. We have seen the limitations of this hypothesis. In the Fermi gas picture the fact that nucleons are not strictly independent of one another reflects itself in the fact that even at zero temperature occupation numbers are not strictly zero or one. Levels below the Fermi level are thus slightly depleted while levels above are slightly populated. The depletion of occupied levels has been measured experimentally. It typically amounts to 15%, which, nevertheless, remains a sufficiently small value to justify the independent particle model [302].

2.2 From the nucleon–nucleon interaction to nuclear matter and finite nuclei

In order to go beyond the simple (still often very useful) models of section 2.1.2, one has to include in the picture the elementary interaction between nucleons.

One can then explore how the average potential well, in which nucleons evolve, can be built up from this elementary stone and thus gain a more microscopic picture of nuclei as constructed from nucleons. These questions are discussed in this section.

2.2.1   On the nucleon–nucleon interaction

The 'free' (in vacuum) nucleon–nucleon interaction is experimentally well known but it cannot be used as such for building a mean field inside a nucleus. In the nuclear medium, elementary scattering between two nucleons is 'affected' by neighbouring nucleons because of the Pauli principle effects. These *in-medium* effects have to be estimated by means of suitable approximations such as Brückner theory. This leads us to consider *effective*, in-medium, interactions, which will then serve as a basis for constructing a proper mean field inside a nucleus. Although formally appealing, Brückner's theory is unable to provide a quantitative account of basic nuclear properties, such as saturation or binding energies of simple nuclei [302, 459]. As no definite theoretical framework does fulfil these requirements, one is often bound to use phenomenological effective interactions such as the Skyrme and Gogny interactions.

2.2.1.1   *Basics of the 'free' nucleon–nucleon interaction*

Experimental data on the nucleon–nucleon interaction are numerous, ranging from total scattering cross-sections to detailed quantities such as phase shifts, which give access to the characteristics of the interaction [376]. Although providing very global accounts of the nucleon–nucleon interaction, total scattering cross-sections are very telling and useful for the forthcoming discussions. They are plotted in figure 2.5 for the various nucleon–nucleon combinations. In the case of proton-proton interactions, the Coulomb part of the interaction has been subtracted, so that only the nuclear components are active here. Note also that below the π production threshold (290 MeV for $N + N \rightarrow D + \pi$) nucleon–nucleon scattering cross-sections are purely elastic, while beyond 290 MeV both elastic and inelastic channels are plotted.

Figure 2.5 calls for several comments. First, there is a striking similarity between the various isospin channels, which reflects the so-called charge independence of the nuclear interaction. Second, the low-energy behaviour is, almost perfectly, inversely proportional to the energy ($\sigma \sim 1/E$). This, however, is true only in vacuum. Inside a nucleus the Pauli principle strongly affects nucleon–nucleon interactions (section 2.2.1.2), in particular at low relative energy, so that the resulting 'in-medium' cross-section is largely suppressed at low energy (see also figure 2.7). As a result, it is not totally foolish to keep in mind a gross value of the in-medium nucleon–nucleon cross-section of order 40 mb, which is more or less independent of the energy. This 'average' value is often used in qualitative discussions and in dynamical simulations of heavy-ion collisions

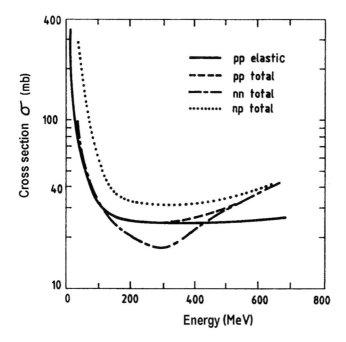

Figure 2.5. Compilation of total nucleon–nucleon scattering cross-sections (in mb) as a function of beam energy (in MeV). Beyond 290 MeV the inelastic channel of the p–p scattering is also indicated. Note the strong energy dependence of the scattering cross-sections at low relative energy and their energy independence at high relative energy. Note also that these are 'free' cross-sections, which would be significantly renormalized in the medium, by Pauli effects (figure 2.7). From [376].

as well. Finally, it is worth briefly mentioning angular effects, although they have been integrated over in figure 2.5. Indeed, at low energy (up to typically π threshold) the nucleon–nucleon differential cross-section is essentially isotropic. At higher energy, in turn, it becomes forward peaked. In most of the discussions relevant for the nucleonic regime this latter anisotropy effect can, nevertheless, be safely overlooked.

It is not our aim to discuss here all the works which have been devoted to the nucleon–nucleon interaction. We shall thus only recall a few gross properties. The shape of the interaction crucially depends on the spins of the two interacting nucleons (figure 2.6). Still, we can content ourselves with noting that the dominant part of the interaction is central (i.e. depending only on the relative distance of the two nucleons), strongly repulsive at short range (≤ 0.4 fm, hard core) and attractive at intermediate range (~ 1–1.2 fm). Note, also, that this dominant repulsive/attractive shape of the interaction, analogous to a typical Van

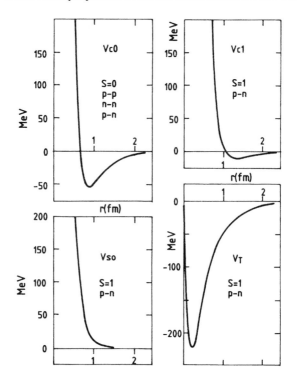

Figure 2.6. Main channels of the so-called Paris nucleon–nucleon potential. The explicit form of the interaction potential (in MeV) depends on the angular momenta of the two nucleons. For a coupling of the spin to zero (typical of low-energy n–n and p–p interactions), only the central potential V_{C0} plays a role. For a coupling to one the central component V_{C1} has to be complemented by tensor V_T and spin–orbit V_{SO} terms. From [134].

der Waals interaction [255], is responsible for the saturation mechanism of nuclear matter as well as for the possibility of liquid–gas coexistence in nuclear matter (see also figure 2.8 and section 2.2.2).

The nucleon–nucleon interaction cannot yet be derived from first principles (QCD). Nucleon–nucleon potentials are thus, at least partly, phenomenological and contain a, possibly large, number of parameters. Starting from a given functional form for the potential (fulfilling, in particular, the appropriate symmetry properties), the parameters are fitted to deuteron properties and available phase shifts. Note that this fitting procedure does not necessarily ensure a proper reproduction of many-body properties, such as saturation. Conversely, saturation may allow nucleon–nucleon interactions which otherwise lead to the same phase shifts to be triggered.

The long range part of the nucleon–nucleon interaction has long been known to correspond to pion exchange. This has led to the so-called OPEP (one pion exchange potential) models of the 1950s and 1960s. The OBEP (one boson exchange potential) approaches, developed from the 1970s on, have led to a substantial progress in terms of the foundations of the interaction [98, 376]. One assumes that nucleons interact via the exchange of mesons, the π corresponding to the attractive long range part, the ρ and ω to the shorter range part etc. Complex, multi-meson contributions are furthermore simulated by effective mesons, such as the σ, which leads to an overall simple form for the interaction, which is finally fitted to a few experimental phase shifts. OBEP approaches, although phenomenological, have thus provided an extremely fruitful framework and a sound basis for our understanding of the free nucleon–nucleon interaction, particularly in the long and medium ranges. Short range effects (hard core) have yet to be better understood and properly linked to quark degrees of freedom.

2.2.1.2 Brief outline of Brückner G-matrix theory

We have seen that the picture of quasi-independent nucleons, evolving in a common potential well, appears as an extremely relevant approach to the nuclear many-body problem. It may, however, look contradictory to the fact that the underlying two-body interaction, the nucleon–nucleon interaction, is singular at short distance (hard core). This is all the more true as this hard core part of the interaction seems to be necessary to explain the observed saturation property of nuclear matter (section 2.2.2). How should these two aspects be reconciled? In other words, how is it possible to justify a description of an ensemble of particles, strongly correlated at short relative distance, as a set of 'independent' particles? Most of the answer to this question is contained in the Pauli principle. In vacuum, the scattering of two nucleons would be described by means of scattering theory in terms of a T matrix [420]. However, inside a nucleus, the presence of neighbouring nucleons does not allow virtual intermediate scattering inside a Fermi sea and thus introduces high momentum components ($k > k_F$) in the wavefunction of the two interacting nucleons. This means that this wavefunction will be renormalized at short relative distance. This opens up two possibilities: either to use the bare interaction with a wavefunction which is correlated at short relative distance; or to use a renormalized interaction with a non-correlated wavefunction. In the latter case, Pauli correlations thus lead to an effective interaction, which differs from the bare interaction, in which the hard core is softened by the neighbouring nucleons [48, 101–103, 213, 437]. This modification of the nucleon–nucleon interaction *in the medium* does *a posteriori* justify independent particle models.

The former argument can be made more formal with the help of the so-called 'independent pair' approximation [376]. Let us take two interacting nucleons 1 and 2, and let us assume that we can 'isolate' the (1, 2) pair from the rest of the system, which is not perturbed by the elementary process of interaction between

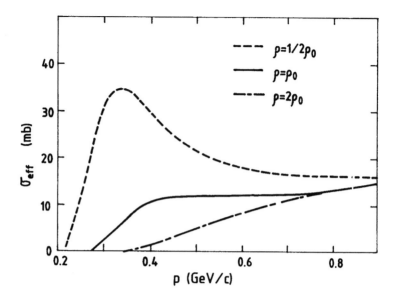

Figure 2.7. In-medium nucleon–nucleon cross-sections (in mb) at various nuclear densities, as a function of the relative momentum p (in GeV/c, p_F = 270 MeV/c at saturation density). Note the strong medium effects and their density dependence. At small relative momentum we find a vanishing cross-section, which would correspond to an infinite mean free path. From [85].

1 and 2 (which amounts to neglecting three...-body effects). The wavefunction of the total system may then be written as

$$\Psi(1,2,3,4,\ldots) = \mathcal{A}_{1\ldots A} \prod \{\psi_{kl}(1,2)\phi_m(3)\phi_n(4)\ldots\} \qquad (2.22)$$

where $\mathcal{A}_{1\ldots A}$ labels the antisymmetrization operator over the A nucleons, ψ_{kl} is the *correlated* wavefunction of the two interacting nucleons and the wavefunctions of the spectator nucleons $3, 4, \ldots$ are labelled as ϕ_i. We furthermore assume that the system may be properly described by a one-body Hamiltonian h, whose eigenfunctions are the ϕ_i's. The Brückner matrix (or effective interaction) G is then defined as

$$V|\psi_{kl}\rangle = G\mathcal{A}_{kl}(|\phi_k\rangle|\phi_l\rangle) \qquad (2.23)$$

where $\mathcal{A}_{kl}$ is the antisymmetrization operator for nucleons k and l. The G matrix thus allows us to decorrelate the wavefunction of the system, at the price of a renormalization of the interaction. It can furthermore be linked to the bare (free) interaction V, through the so-called Bethe–Goldstone equation [376]

$$G(\omega) = V + V \frac{Q}{\omega - h} G \qquad (2.24)$$

where Q denotes the projection operator on empty states (namely with momenta k above the Fermi momentum) and where ω represents the energy of the pair of interacting nucleons. Equation (2.24) is a standard scattering equation, similar to the one followed by the T matrix, the major difference lying in the Pauli blocking operator Q. The practical evaluation of G in terms of V is not easy, but a degree of freedom lies in the choice of the splitting of the interactions between h and the residual scattering process V [302]. This allows us to devise efficient approximation schemes.

Brückner theory has recently been extended to the relativistic domain (Dirac–Brückner) [14, 459]. It is well known that, as the Fermi velocity is of order $v_F \approx 0.3c$, a typical velocity of nucleons inside a nucleus is of order $v \approx 0.2c$, which should justify the use of relativistic descriptions. Relativistic theories of nuclear matter are based on the Dirac rather than on the Schrödinger equation. The results obtained in these approaches, in spite of the huge technical difficulties, bring precious insights into the nature of the saturation mechanism (section 2.2.2.2). These calculations thus provide an appealing alternative to the standard, non-relativistic, approaches of the nuclear many-body problem. They provide, in particular, information on in-medium effects. An example of the effective in-medium nucleon–nucleon cross-section, computed in the Dirac–Brückner formalism, is shown in figure 2.7. One can see from this figure that the in-medium cross-section vanishes at low energy, because of Pauli blocking, which explains the high value of the nucleon mean free path in a ground-state nucleus. One should also note the strong influence of the medium density, which reflects the fact that, in nuclear matter, the Fermi momentum is directly related to density ($\rho \sim k_F^3$). Finally, one should remember that the values of the cross-sections still depend, to some extent, on the details entering the models used.

2.2.1.3 *Phenomenological effective interactions: Skyrme and Gogny forces*

Although formally 'simple' and satisfying from the qualitative point of view, the Bethe–Goldstone equation (2.24), and its relativistic counterpart, are tough to solve in nuclear matter and even more in finite nuclei. Furthermore, the results obtained in these approaches (for example for saturation or nuclear binding energies) are not always fully satisfying (section 2.2.2.2 and [302]). It thus makes sense to consider phenomenological effective interactions which allow us to reproduce experimental data. One then starts from a sound *a priori* functional form of the effective interaction, actually inspired from Brückner theory outcomes and with proper symmetries, and the parameters entering the force are then fitted to basic nuclear properties (saturation, binding energies and/or radii of simple nuclei).

The nuclear interaction is known to be short range (at least of smaller range than the average internucleon distance), and this fact is confirmed in the medium, by Brückner calculations [376, 398]. The simplest approximation thus consists in building an effective interaction of zero range, namely proportional to $\delta(\boldsymbol{r})$ ($\boldsymbol{r}$ being the relative position). This turns out to provide a surprisingly good account of many nuclear properties. Still, a more realistic approach has to include finite range effects, as, for example, suggested by simple Yukawa forces ($V(\boldsymbol{r}) \propto \mathrm{e}^{-\mu r}/r$, $r = \|\boldsymbol{r}\|$). This finite-range effect may actually be simulated by a momentum dependence of the interaction. Starting from a central effective interaction $V(\boldsymbol{r})$, the momentum representation reads

$$\langle \boldsymbol{p}|V|\boldsymbol{p}'\rangle = \frac{1}{(2\pi)^3} \int \mathrm{e}^{-\mathrm{i}(\boldsymbol{p}-\boldsymbol{p}')\cdot\boldsymbol{r}/\hbar} V(\boldsymbol{r}) \, \mathrm{d}^3 r. \qquad (2.25)$$

A zero-range interaction thus leads to a constant in momentum space, while a finite-range force leads to an explicit momentum dependence. The simplest, rotational-invariant, momentum-dependent, interaction thus reads

$$\langle \boldsymbol{p}|V|\boldsymbol{p}'\rangle = V_0 + V_1(\boldsymbol{p}^2 + \boldsymbol{p}'^2) + V_2 \boldsymbol{p} \cdot \boldsymbol{p}' \qquad (2.26)$$

or, in real space, ($\boldsymbol{p}$ is now the momentum operator)

$$V(\boldsymbol{r}) = V_0 \delta(\boldsymbol{r}) + V_1(\boldsymbol{p}^2 \delta(\boldsymbol{r}) + \delta(\boldsymbol{r})\boldsymbol{p}^2) + V_2 \boldsymbol{p}\delta(\boldsymbol{r})\boldsymbol{p}. \qquad (2.27)$$

Effective forces do also, in general, depend on the nuclear density, which directly enters the $\boldsymbol{Q}$ projector in the Brückner matrix $\boldsymbol{G}$ (equation (2.24)). Many effective forces have been developed over the years, following this brief outline. We focus here on two types of forces which are particularly suited to moderate energies: Skyrme and Gogny interactions.

In 1956 T H R Skyrme [431] proposed a phenomenological effective interaction with two-body $\boldsymbol{V}(1,2)$ and three-body $\boldsymbol{V}(1,2,3)$ terms. The two-body term is treated in an approximation of short range (or small relative momentum $\boldsymbol{k} = -\mathrm{i}(\boldsymbol{\nabla}_1 - \boldsymbol{\nabla}_2)/2$):

$$\boldsymbol{V}(1,2) = t_0(1 + x_0 \boldsymbol{P}_\sigma)\delta(\boldsymbol{r}_1 - \boldsymbol{r}_2) + \tfrac{1}{2}t_1(\delta(\boldsymbol{r}_1 - \boldsymbol{r}_2)\boldsymbol{k}^2 + \boldsymbol{k}^2\delta(\boldsymbol{r}_1 - \boldsymbol{r}_2))$$
$$+ t_2\boldsymbol{k}\delta(\boldsymbol{r}_1 - \boldsymbol{r}_2)\boldsymbol{k} + \mathrm{i}W_0(\boldsymbol{\sigma}_1 + \boldsymbol{\sigma}_2)\boldsymbol{k} \wedge \delta(\boldsymbol{r}_1 - \boldsymbol{r}_2)\boldsymbol{k} \qquad (2.28)$$

where $\boldsymbol{P}_\sigma$ is the spin-exchange operator between spins $\boldsymbol{\sigma}_1$ and $\boldsymbol{\sigma}_2$. The three-body term is explicitly taken with zero range $\boldsymbol{V}(1,2,3) = t_3\delta(\boldsymbol{r}_1 - \boldsymbol{r}_2)\delta(\boldsymbol{r}_2 - \boldsymbol{r}_3)$. The $\boldsymbol{V}(1,2)$ and $\boldsymbol{V}(1,2,3)$ potentials altogether contain six parameters (t_0, t_1, t_2, t_3, W_0, x_0), which are fitted on nuclear matter saturation and on binding energies and radii of a few nuclei. The first realistic calculations with Skyrme forces were performed in the early 1970s in the Hartree–Fock approximation (section 2.2.3) [378, 477].

In a Skyrme interaction, the t_0 term corresponds to a zero range contribution with a spin exchange term P_σ, and t_1 and t_2 simulate finite-range effects. The W_0

term represents a two-body spin–orbit interaction. The three-body term deserves more comment. First, it should be noted that in a spin-saturated system it is equivalent to a density-dependent two-body interaction

$$V_{123} \rightarrow V_\rho(1,2) = \frac{1}{6}t_3(1 + P_\sigma)\delta(r_1 - r_2)\rho\left(\frac{r_1 + r_2}{2}\right). \tag{2.29}$$

One can thus view this term as a phenomenological representation of the density dependence of the Brückner matrix (section 2.2.1.2). This interpretation is presumably more satisfying than the one in terms of a three-body interaction, although it is well known that explicit three-body interactions are expected to play a role in nuclei [205]. Still, the density dependence of the interaction term in (2.29) raises some difficulties with respect to the incompressibility modulus K_∞ of nuclear matter (it leads to too high values of order 350–400 MeV, section 2.2.2). Modern Skyrme parametrizations hence involve a fractional density dependence, namely a term of the form $\rho^{1+\gamma}$ in equation (2.29), with typically $\frac{1}{6} \le \gamma \le \frac{1}{3}$, which leads to acceptable values of K_∞ of order 200–250 MeV (section 2.2.2.1) [31, 274]. A few extra terms may furthermore be included, leading to typical parametrizations containing about 10 parameters.

From equations (2.28,2.29) for the nucleon–nucleon potential, it is possible to build the Hartree–Fock mean field U (section 2.2.3), which takes a particularly simple form for Skyrme forces. The t_0 and t_3 zero-range terms lead to volume contributions, expressed as powers of the density. The t_1 and t_2 finite-range terms give contributions proportional to gradients of the density, which represent surface effects and lead to a density-dependent renormalization of the kinetic energy (equation (2.34), section 2.2.2.2). Still, in many dynamical calculations linked to heavy-ion collisions, these t_1 and t_2 terms, as well as spin–orbit contributions, have been omitted for practical reasons. The one-body potential U then reduces, for a spin and isospin ($N = Z$) saturated system, without effective mass ($m^*/m = 1$, equation (2.36) and sections 2.1.3.3 and 2.2.2.2) to

$$U = \frac{3}{4}t_0\rho + \frac{2+\gamma}{16}t_3\rho^{1+\gamma}. \tag{2.30}$$

This oversimplified form is often called a (t_0, t_3) potential.

Skyrme forces have been widely used, and with success, in many nuclear problems [341,378], for structure as well as for dynamical questions. A reason for this success lies in their simplicity due to the Dirac functions in equations (2.28) and (2.29), and to the fact that it is possible to derive the functional form of these forces from a Brückner matrix approach [340, 389]. While Brückner theory does not directly provide a realistic effective interaction, it thus, nevertheless, allows the building of efficient effective interactions, with a minimum number of parameters.

The major defect of Skyrme forces lies precisely in what makes them so useful, namely the ignoring of a true finite range. The momentum-dependent terms do indeed, to some extent, represent finite-range effects, but this only holds in the limit of small relative momenta, typically less than k_F. Difficulties with Skyrme forces hence appear, in particular, in dynamical situations, in which high relative momenta may be attained in the course of collisions. An alternative to Skyrme forces is provided by Gogny interactions [210], in which the t_0, t_1 and t_2 Dirac terms are replaced by Gaussians. The total potential then takes the form

$$
\begin{aligned}
\boldsymbol{V}(1,2) = & \sum_{i=1}^{2} \mathrm{e}^{-(\boldsymbol{r}_1 - \boldsymbol{r}_2)^2/\mu_i^2} (W_i + B_i \boldsymbol{P}_\sigma - H_i \boldsymbol{P}_\tau - M_i \boldsymbol{P}_\sigma \boldsymbol{P}_\tau) \\
& + \mathrm{i} W_0 (\boldsymbol{\sigma}_1 + \boldsymbol{\sigma}_2)(\boldsymbol{k} \wedge \delta(\boldsymbol{r}_1 - \boldsymbol{r}_2)\boldsymbol{k}) \\
& + t_3 (1 + \boldsymbol{P}_\sigma)\delta(\boldsymbol{r}_1 - \boldsymbol{r}_2)\rho^{1/3} \left(\frac{\boldsymbol{r}_1 + \boldsymbol{r}_2}{2} \right)
\end{aligned} \tag{2.31}
$$

where $\boldsymbol{P}_\sigma$ and $\boldsymbol{P}_\tau$ are, respectively, the spin and isospin exchange operators and $\{\mu_i, W_i, B_i, H_i, M_i, \ i = 1, 2\}$ and W_0 and t_3 are the parameters of the force. This force is less simple than Skyrme interactions, but it can be used in dynamical situations involving relative momenta typically up to $k \simeq 2 \ \mathrm{fm}^{-1}$. For higher relative momenta there is no simple standard parametrization of effective forces. One has then to rely more or less on simplified Dirac–Brückner calculations or on phenomenological parametrizations of the momentum dependence, but this mainly concerns beam energies above the range we are considering here.

2.2.2   The nuclear equation of state at zero temperature

When building a proper theoretical framework on the basis of the nucleon–nucleon interaction, the first step is to access the properties of infinite nuclear matter properly. We briefly review these aspects here.

2.2.2.1   The saturation point

The single point of the phase diagram or, more generally, of the equation of state of nuclear matter which is known experimentally is the saturation point, characterized by its density ρ_0 and energy per nucleon E/A_0 [302]

$$
\rho_0 \simeq 0.17 \pm 0.02 \ \mathrm{fm}^{-3}
$$
$$
E/A_0 \simeq -16 \pm 1 \ \mathrm{MeV}.
$$

The values of ρ_0 and E/A_0 are almost pure experimental values: electron scattering experiments have long shown that the central density of heavy nuclei is almost constant, independent of the mass (figure 2.1 and [196]). This fixes the value of ρ_0. The volume term in the Bethe–Weiszäcker mass formula (2.2) [47], in turn, represents the binding energy of nuclear matter at saturation density: $a_v \approx 16 \ \mathrm{MeV} = -E/A_0$.

The saturation point corresponds to the equilibrium point (at zero temperature) of nuclear matter, hence characterized by vanishing pressure $P =$

$\rho^2 \partial (E/A)/\partial \rho = 0$ and positive curvature $\partial P/\partial \rho \geq 0$. This is all that is known about the equation of state with a high degree of experimental confidence. The question thus becomes whether one would be able to explore *at least* the vicinity of the saturation point, which would provide the shape of the equation of state at zero temperature close to the saturation point. The first quantity of interest in this respect is the curvature of the equation of state at saturation. This quantity may be obtained through systematic studies of the monopole giant resonance [57, 58], which corresponds to radial density oscillations. Representing the nucleus as a finite piece of nuclear matter, the monopole resonance hence provides an exploration of the vicinity of the saturation point. One characterizes the curvature of the equation of state *at saturation* by the incompressibility modulus

$$K_\infty = k_{\mathrm{F}}^2 \frac{\partial^2 (E/A)}{\partial k_{\mathrm{F}}^2}. \tag{2.32}$$

In this expression k_{F} is the Fermi momentum at saturation ($\rho_0 \propto k_{\mathrm{F}}^3$, equation (2.4)). Knowing K_∞ allows us to parametrize the nuclear matter equation of state as a function of ρ *in the vicinity of* ρ_0 as

$$E/A(\rho) \approx E/A(\rho_0) + \frac{K_\infty}{18} \frac{(\rho - \rho_0)^2}{\rho_0^2} = E/A_0 + \frac{K_\infty}{18} \frac{(\rho - \rho_0)^2}{\rho_0^2} \tag{2.33}$$

which may be useful for qualitative discussions.

The first experimental values of K_∞ were obtained from the giant monopole resonance in the late 1970s. However, as K_∞ represents the curvature of the equation of state, it may be expected to play an important role in many other phenomena. Several attempts have been made in this direction, trying to relate K_∞ to many nuclear as well as astrophysical properties. This gave raise to a debate, still not fully closed, on the value of K_∞ [58, 207]. To cut a long story short, it does not seem today that any physical situation might be better suited for determining the incompressibility modulus than the actual measurements of the monopole vibration frequency. In particular, it was often proposed that astrophysical observations such as the maximum mass of neutron stars could directly point to a well-defined value K_∞. This is presumably far too optimistic for the very simple reason that neutron stars involve very high densities (≈ 2–$3\rho_0$) while K_∞ is a property of nuclear matter *at* saturation (e.g. at $\rho = \rho_0$). Heavy-ion collisions were also often propounded as a laboratory for measuring K_∞. Once again this statement is probably too optimistic. The reason is partly different here. Of course the vicinity of the saturation point may be explored in intermediate energy heavy-ion collisions. However, it is very often in the course of a highly out-of-equilibrium process and so we are unable to explore accurately the curvature of the equation of state *at* saturation, the latter exploration requiring more 'gentle' and small amplitude motion. Strongly out-of-equilibrium phases in nuclear collisions also lead to relatively high/low values of the density, again departing from a proper analysis of the saturation point. Evaluating K_∞ directly from heavy-ion collisions at intermediate energies hence appears to be difficult because of dynamical effects. For the time being, therefore, we shall primarily trust values of K_∞ obtained 'directly' from the giant monopole resonance and stick to the value:

$$K_\infty \approx 220 \pm 50 \, \text{MeV}$$

which allows us to accommodate recent discussions on the value of K_∞ [58].

Before closing this discussion on the value of K_∞ let us make a last remark. The debate on the value of K_∞ opens a wider discussion on the stiffness of the nuclear matter equation of state. Indeed, as previously outlined, one is often bound to explore regions of very high (or low) densities, either in dynamical or astrophysical contexts. For example it seems to be established that the stiffness of the nuclear matter equation of state might influence the physics of neutron stars [132]. However, one is considering here a more general property (at high density) of the equation of state. It is quite possible, although not necessary, that a stiff equation of state will indeed have a high incompressibility modulus. This then concerns the high-density behaviour, *not* the saturation point. This distinction, or rather the lack of this distinction, is often a source of confusion.

2.2.2.2 *Theories of nuclear matter*

From the theoretical side, in spite of our satisfying knowledge of the bare nucleon–nucleon interaction, sizeable difficulties still remain if one aims at producing a realistic *ab initio* model of nuclear matter. There is thus no totally satisfying theoretical scheme that allows us to recover *quantitatively*, from an elementary nucleon–nucleon interaction, the experimental saturation point. The difficulties encountered in the theoretical descriptions of nuclear matter are basically of two origins: (i) the existence of many-body effects beyond two-body ones (light nuclei seem to favour the existence of at least three-body interactions [205]); and (ii) the importance of medium effects on the elementary nucleon–nucleon interaction.

Basic results obtained from standard theories of nuclear matter can be summarized as follows. Non-relativistic theories have been developed since the mid 1950s and basically rely on Brückner theory [302]. These calculations, involving only a two-body nucleon–nucleon interaction, do not allow a satisfying reproduction of the saturation point, whatever approximate scheme is used for solving the problem [302]. One can even show that saturation points are bound to lie on the so-called Coester line, which misses the experimental saturation point [127]. The inclusion of a three-body interaction might help solve this problem of recovering the proper saturation [192]. While remaining at the level of two-body interactions, relativistic approaches may also provide a solution to this saturation problem [459]. The resulting theory is known as the Dirac–Brückner theory of nuclear matter and allows us to exit the Coester line and to move closer to the experimental saturation point. This conclusion should, however, be softened by the fact that the latter effect diminishes when the Δ excitation of the nucleons is inserted in the description [459]. One should also keep in mind the extreme complexity of these calculations and hence the fact that several approximations have to be inserted in the numerical treatment.

We shall hence content ourselves, in the following, with (partly) phenomenological approaches. Conversely, one may hope that studies of the properties of nuclear matter could reduce, or even suppress, this speculative component. In any case, phenomenological models, such as the ones based on effective forces (section 2.2.1.3), allow relevant qualitative discussions.

2.2.2.3 A phenomenological equation of state

As an example we consider the widely used Skyrme interaction for nuclear matter at zero temperature [431, 477]. In such an infinite system the Fermi gas model is exact, as wavefunctions are indeed plane waves (section 2.1.2.2). For a standard Skyrme interaction (section 2.2.1.3) the total energy density, of spin and isospin degenerate matter, can then be written as

$$\mathcal{E}(\rho) = \frac{\hbar^2}{2m}\tau + \frac{3}{8}t_0\rho^2 + \frac{1}{16}t_3\rho^{\gamma+2} + \frac{1}{16}(3t_1 + 5t_2)\rho\tau \qquad (2.34)$$

where τ is the kinetic energy density, which in the Fermi gas takes the simple form

$$\tau = \tau_{\mathrm{FG}} = \frac{2m}{\hbar^2}\frac{E_{\mathrm{FG}}}{A} = \frac{3}{5}\left(\frac{3\pi^2}{2}\right)^{2/3}\rho^{5/3} \qquad (2.35)$$

with the notation of section 2.1.2.2. It is interesting to note that the finite range terms t_1 and t_2 play a role even in this infinite system where surface terms vanish. By construction, Skyrme forces accommodate weak momentum dependence only, and it is thus no surprise that the latter dependence shows up as a term proportional to $\boldsymbol{k}^2$ (equation (2.28)). But such a term has the same functional form as the standard kinetic term in the Hamiltonian and hence effectively renormalizes the nucleon mass. One then directly obtains a density-dependent effective mass [477]

$$\frac{\hbar^2}{2m^*} = \frac{\hbar^2}{2m} + \frac{1}{16}(3t_1 + 5t_2)\rho \qquad (2.36)$$

which allows us to recast the energy density in a simple form. From the energy per nucleon one can then compute the pressure as

$$
\begin{aligned}
P &= \rho^2\frac{\partial(E/A)}{\partial\rho} = \rho^2\frac{\partial(\mathcal{E}/\rho)}{\partial\rho} \\
&= \frac{2}{3}\frac{\hbar^2}{2m}\tau + \frac{3}{8}t_0\rho^2 + \frac{5}{16}\left(\frac{3\pi^2}{2}\right)^{2/3}(3t_1 + 5t_2)\rho^{5/3} + \frac{1}{16}t_3(\gamma + 1)\rho^{\gamma+2}
\end{aligned}
$$

$$(2.37)$$

which provides a phenomenological equation of state of symmetric nuclear matter at zero temperature. Note that the saturation is then simply recovered by setting both the pressure and its derivative with respect to density, equal to zero. Finally, the incompressibility modulus at saturation ($\rho = \rho_0$) is obtained via a second derivative [398]:

$$
\begin{aligned}
K_\infty = k_{\rm F}{}^2 \frac{\partial^2 (E/A)}{\partial k_{\rm F}{}^2} &= 9\rho_0^2 \frac{\partial^2 (\mathcal{E}/\rho)}{\partial \rho^2}\bigg|_{\rho=\rho_0} \\
&= -\frac{6}{5}\frac{\hbar^2}{2m}\left(\frac{3\pi^2}{2}\right)^{2/3} \rho_0^{2/3} + \frac{3}{8}\left(\frac{3\pi^2}{2}\right)^{2/3}(3t_1 + 5t_2)\rho_0^{5/3} \\
&\quad + \tfrac{9}{16}t_3\gamma(\gamma + 1)\rho_0^{\gamma+1}.
\end{aligned}
\tag{2.38}
$$

As is clear from this set of formulae, the Skyrme force has sufficient flexibility to accommodate experimental results on saturation (density, energy and compressibility) while leaving open a wide choice of parameters. In the simplest, still widely used, case $m^*/m = 1$, only three parameters are left free (t_0, t_3 and γ), which can thus be directly expressed in terms of the three previous constraints on saturation. The value of K_∞ then directly fixes γ, once given saturation density and energy. For example, assuming a 'canonical' value of 220 MeV for the incompressibility modulus, gives $\gamma \simeq \frac{1}{4}$ which leads to $t_0 = -2096$ MeV $\cdot$ fm^{+3} and $t_3 = 13\,872$ MeV $\cdot$ fm$^{+15/4}$.

2.2.3 The Hartree–Fock model

As already mentioned, a microscopic foundation of an independent particle picture of nuclei has to account for the interactions between nucleons, which, after all, are responsible for the binding. The idea is thus to extract from these many interactions an average potential which will allow us to consider nucleons as independent from each other *inside* this self-consistent field. The aim of the Hartree–Fock theory is to fulfil precisely this requirement. Of course, the major interest of such an approach is that it establishes a link between the *average* mean field and elementary nucleon–nucleon interactions. This aspect is especially important for our purpose of exploring dynamical situations in which a self-adaptative time-dependent mean field is obviously unavoidable.

The basic idea of the Hartree–Fock approach lies in approximating the A-body ket $|\Psi(1, 2, \ldots, A)\rangle$ by an 'optimal' antisymmetrized product of one-body wavefunctions $|\phi_i(i)\rangle$ of the form (Slater determinant)

$$
|\Psi_{\rm HF}(1, 2, \ldots, A)\rangle = \mathcal{A}_{1\ldots A}\left(\prod_{i=1}^{A}|\phi_i\rangle\right)
\tag{2.39}
$$

where $\mathcal{A}_{1\ldots A}$ labels the A-body antisymmetrization operator, which ensures a proper account of the fermionic nature of the nucleons. The Hartree approximation (without Fock!) would simply consist of replacing $\mathcal{A}_{1\ldots A}$ by the unit operator. But the Hartree approximation is unrealistic in nuclear physics where the Pauli principle is known to play a considerable role.

We start the derivation with the A-body Hamiltonian, built here, for the sake of simplicity, with only a two-body interaction V_{ij} and neglecting the spin degree of freedom [426]. In the $|r\rangle$ representation it thus reads

$$H = \sum_{i=1}^{A} K_i + \sum_{i<j} V_{ij} = -\sum_{i=1}^{A} \frac{\hbar^2}{2m}\Delta_i + \sum_{i<j} V_{ij}(r_i - r_j) \qquad (2.40)$$

in the simple case of a two-body potential depending on relative position (K_i labels here the one-body kinetic energy operator). We furthermore introduce the one-body density matrix ρ, defined as (section 2.3.1.1):

$$\rho = \sum_{i=1}^{A} |\phi_i\rangle\langle\phi_i| \qquad (2.41)$$

which in the $|r\rangle$ representation is simply expressed from the one-body wavefunctions ϕ_i as

$$\langle r'|\rho|r\rangle = \rho(r,r') = \sum_{i=1}^{A} \phi_i^*(r')\phi_i(r) \qquad (2.42)$$

where the star denotes the complex conjugate.

The Hartree–Fock equations can now be obtained by minimizing the total energy of the system $\langle \Psi_{\mathrm{HF}}|H|\Psi_{\mathrm{HF}}\rangle$, in the space of Slater determinants, with respect to the $|\phi_i\rangle$, and under the constraint that each one-body wavefunction $|\phi_i\rangle$ is normalized to unity. One hence introduces A Lagrange parameters ϵ_i accounting for these normalizations, which leads us to write the variational principle as

$$\frac{\delta}{\delta\phi_i^*}\left[\langle \Psi_{\mathrm{HF}}|H|\Psi_{\mathrm{HF}}\rangle - \sum_{j=1}^{A} \epsilon_j \int \mathrm{d}^3 r\, \phi_j^*(r)\phi_j(r) \right] = 0. \qquad (2.43)$$

This leads, after some algebra, to the static Hartree–Fock equations for the one-body wavefunctions ϕ_i:

$$\epsilon_i\phi_i(r) = -\frac{\hbar^2}{2m}\Delta\phi_i(r) + U^{\mathrm{dir}}(r)\phi_i(r) - \int \mathrm{d}^3 r'\, U^{\mathrm{exc}}(r,r')\phi_i(r')$$

$$= \int \mathrm{d}^3 r'\, h(r,r')\phi_i(r') = \langle r|h|\phi_i\rangle. \qquad (2.44)$$

In this expression

$$h = K + U \qquad (2.45)$$

is, by definition, the one-body Hartree–Fock Hamiltonian, composed from the kinetic energy operator K and from the one-body mean-field potential U, itself composed from the direct

$$U^{\mathrm{dir}}(r) = \int \mathrm{d}^3 r'\, V(r - r')\rho(r',r') \qquad (2.46)$$

and exchange

$$U^{\text{exc}}(\boldsymbol{r}, \boldsymbol{r}') = V(\boldsymbol{r} - \boldsymbol{r}')\rho(\boldsymbol{r}, \boldsymbol{r}') \tag{2.47}$$

potentials. The Lagrange parameters ϵ_i hence represent the energies of the single-particle levels (of wavefunctions ϕ_i). Equations (2.44)–(2.47) clearly express the self-consistent nature of the Hartree–Fock procedure, namely the fact that the potential part of $\boldsymbol{h}$ itself depends on the solutions ϕ_i of the Schrödinger-like equations (2.44), through the one-body density matrix ρ. This self-consistency also appears in the fact that the total energy of the system E_{HF} does not reduce to the mere sum of the single-particle energies ϵ_i. Indeed, in order to avoid double counting of the potential energy terms, the total energy reads

$$E_{\text{HF}} = \langle \Psi_{\text{HF}} | \boldsymbol{H} | \Psi_{\text{HF}} \rangle = \sum_{i=1}^{A} \langle \boldsymbol{K}_i \rangle + \frac{1}{2} \sum_{i=1}^{A} \langle \boldsymbol{U}_i \rangle = \frac{1}{2} \sum_{i=1}^{A} (K_i + \epsilon_i) \tag{2.48}$$

with obvious notation ($\epsilon_i = \langle \boldsymbol{K}_i \rangle + \langle \boldsymbol{U}_i \rangle = K_i + \langle \boldsymbol{U}_i \rangle$). Note also that the exchange potential $\boldsymbol{U}^{\text{exc}}$, which is, in general, complex to evaluate, is highly simplified in the case of a zero-range interaction. This is one of the reasons for the success of Skyrme interactions in Hartree–Fock calculations (section 2.2.1.3).

It may finally be useful, for forthcoming discussions, to recast the HF equation in a compact operator form involving the one-body density matrix ρ. Start from equation (2.44), multiply it by $\phi_i^*(\boldsymbol{r}')$ and sum over single-particle levels i, which leads to

$$\sum_i \epsilon_i \phi_i(\boldsymbol{r})\phi_i^*(\boldsymbol{r}') = \int \mathrm{d}^3 r'' \, h(\boldsymbol{r}, \boldsymbol{r}'') \sum_i \phi_i(\boldsymbol{r}'')\phi_i^*(\boldsymbol{r}')$$

$$= \int \mathrm{d}^3 r'' \, h(\boldsymbol{r}, \boldsymbol{r}'')\rho(\boldsymbol{r}'', \boldsymbol{r}') = \langle r' | \boldsymbol{h}\rho | r \rangle. \tag{2.49}$$

Starting from the complex conjugate of equation (2.44) at point $\boldsymbol{r}'$, and multiplying it by $\phi_i(\boldsymbol{r})$ similarly leads to

$$\sum_i \epsilon_i \phi_i(\boldsymbol{r})\phi_i^*(\boldsymbol{r}') = \langle r' | \rho \boldsymbol{h} | r \rangle \tag{2.50}$$

so that finally

$$\langle r' | \boldsymbol{h}\rho | r \rangle - \langle r' | \rho \boldsymbol{h} | r \rangle = \langle r' | [\boldsymbol{h}, \rho] | r \rangle = 0 \quad or \quad [\boldsymbol{h}[\rho], \rho] = 0. \tag{2.51}$$

Note that the derivation of this relation from equations (2.44)–(2.47) follows the same lines as the one which leads from the standard Schrödinger equation to the Liouville–von Neumann equation (2.54).

2.3 Nuclei as statistical physics systems

In the following we shall often face situations in which statistical physics concepts have to enter our description. Following essentially [28] and [75] we thus recall in this section a few basic relations of equilibrium statistical physics for future use and we discuss how nuclei can be described with such tools. In addition to [28], the reader may also consult standard statistical physics textbooks such as [249,386] or [122,275]. Out-of-equilibrium statistical physics will be discussed in chapter 3.

2.3.1 Basics of equilibrium statistical physics

2.3.1.1 Notion of density matrix

How to describe a physical system with A particles? If its quantum state $|\Psi\rangle$ is perfectly known, the average value of any observable Q is simply obtained as $\langle Q \rangle = \langle \Psi | Q | \Psi \rangle$. This 'well-defined' problem corresponds to what is called a *pure state* in statistical physics. It occurs, however, extremely rarely as it requires complete knowledege of the state of the system, which is virtually impossible.

One is thus led to consider 'poorly known' systems, which one can describe by ensembles of accessible states $\{|\Psi_\lambda\rangle\}$ and corresponding probabilities q_λ that a given state $|\Psi_\lambda\rangle$ does indeed describe the state of the system. The density matrix (or density operator) D associated with this *statistical mixture* $\{q_\lambda, |\Psi_\lambda\rangle; \lambda = 1, N\}$ (in the simple illustrative example of a discrete finite ensemble) is then defined as

$$D = \sum_\lambda |\Psi_\lambda\rangle q_\lambda \langle\Psi_\lambda| \qquad (2.52)$$

from which constitutive relations follow, namely hermiticity, unit trace ($\mathrm{Tr}\,D = 1$) and semi-definite positiveness of D. From the density matrix D of an A-body system, one can also define 'reduced' density matrices by partial traces (integrations) over some components of the system. For example the one-body density matrix ρ is obtained by tracing over all but one particle. It contains, of course, much less information than D (see chapter 3).

Once introduced to the density matrix D, the average value of an operator Q can then simply be expressed as

$$\langle Q \rangle = \sum_\lambda q_\lambda \langle\Psi_\lambda | Q | \Psi_\lambda\rangle = \mathrm{Tr}(QD). \qquad (2.53)$$

It is to be noted that all the information about the system is actually contained in D. Two different statistical mixtures with the same density matrix are thus indistinguishable by any measurement. One can also show that the density matrix D of a system with Hamiltonian operator H evolves in time according to the Liouville–von Neumann equation

$$i\hbar \frac{\partial D}{\partial t} = [H, D] \qquad (2.54)$$

which can be derived from the Schrödinger equation (see the derivation of equation (2.51)).

The density matrix D does contain, through the probabilities q_λ's, much more information on a system than the one contained in the mere $|\Psi_\lambda\rangle$'s. Details about the thermodynamical conditions are precisely contained in the q_λ's, and hence in D. Depending on the situation, the q_λ's (or D), will thus take specific forms, reflecting the physical conditions under study. Most of the time, the state of the system is only partially known, but in order to provide the most realistic description, it is necessary to account for the information at hand. Two cases may occur:

(i) The information known about the system is exact. In this case, one directly includes it in the definition of the space containing the $|\Psi_\lambda\rangle$'s themselves, and it thus becomes a *variable*.

(ii) The information is only known as an average $Q^0 = \text{Tr}(QD)$. In this case, it is introduced in the description as a *constraint* on the density operator.

2.3.1.2 *Equilibrium density matrixes: the Boltzmann–Gibbs distribution*

The equilibrium state of a system is associated with the density matrix D which is least biased in terms of probabilities. The latter criterion is based on the missing information which is quantified in terms of the entropy

$$S = -k\,\text{Tr}(D\log D) \qquad (2.55)$$

where k is the Boltzmann constant. For a system about which nothing is known, the least biased choice (maximum entropy) is obviously the most democratic one, in which the probability associated with any of the N $|\Psi_\lambda\rangle$'s is simply given by $q_\lambda = 1/N$ $(\lambda = 1, N)$. One then recovers the well-known relation $S = k\log N$. This variational principle of least biased choice is extended to the general case of a system characterized by a set of 'natural' variables and constraints (including the constitutive relation $\text{Tr}(D) = 1$)

$$\langle Q_i \rangle = \text{Tr}(Q_iD) = Q_i^{\,0}, \quad i = 1, p \qquad (2.56)$$

by seeking for a maximum of the entropy $S(D)$ under the set of constraints given by equation (2.56). The constraints are included by means of Lagrange multipliers $\{\lambda_i, i = 1, p\}$ and one thus obtains the general equilibrium solution in the form of the so-called Boltzmann–Gibbs distribution

$$D = D_{\text{eq}} = \frac{1}{Z}\exp\left(-\sum_{i=1}^{p}\lambda_iQ_i\right) \qquad (2.57)$$

where the *partition function Z* ensures the normalization of D ($\text{Tr}\,D = 1$). The partition function thus reads

$$Z = Z(\{\lambda_i\}) = \text{Tr}\left(\exp\left(-\sum_{i=1}^{p}\lambda_iQ_i\right)\right) \qquad (2.58)$$

so that $\frac{\partial}{\partial \lambda_i} \log Z(\{\lambda_i\}) = -\langle \boldsymbol{Q}_i \rangle = -Q_i^0$. The Lagrange parameters $\{\lambda_i, i = 1, p\}$ are finally determined by imposing the condition that the Boltzmann–Gibbs distribution does indeed fulfil the initial constraints (equation (2.56)). Note also that the Lagrange parameters $\{\lambda_i, i = 1, p\}$ are the natural variables of the partition function Z, and the average values $\langle \boldsymbol{Q}_i \rangle$, the ones of the statistical entropy S. The pairs $\{\langle \boldsymbol{Q}_i \rangle, \lambda_i; i = 1, p\}$ form pairs of *conjugate variables*.

2.3.1.3 Thermodynamical potentials and statistical ensembles

In the following we shall use three standard statistical descriptions: the *microcanonical* ensemble, the *canonical* one and the *grand canonical* one. We briefly discuss here the physical situations they correspond to and define the corresponding thermodynamical potentials.

A thermodynamical potential is a function of the variables characterizing a system which is extremal at equilibrium. The partial derivatives furthermore have simple physical interpretations. The thermodynamical potentials thus provide the most transparent links between statistical and thermodynamical descriptions. In this respect, the entropy, as a function of the internal energy U, the number of particles A and possibly other variables x_α's (volume, external field, etc), constitutes a good example of a thermodynamical potential. It is associated with the *microcanonical* ensemble. Its differential reads:

$$\mathrm{d}S = \frac{1}{T} \mathrm{d}U - \frac{\mu}{T} \mathrm{d}A - \frac{1}{T} \sum_\alpha X_\alpha \, \mathrm{d}x_\alpha \qquad (2.59)$$

which enlightens the pairs of conjugate variables: $x_\alpha \leftrightarrow -X_\alpha/T$, internal energy $U \leftrightarrow$ inverse temperature $1/T$, number of constituents $A \leftrightarrow -\mu/T$ (μ chemical potential).

In the standard *microcanonical* ensemble, the system is described by its number of constituents A, its volume V and internal energy U. The associated thermodynamical potential is the *entropy* $S = S(U, V, A)$. The microcanonical description is typical of isolated systems. Temperature is therefore not fixed exactly, but may fluctuate. For mainly historical reasons, thermodynamics focuses on the *internal energy* $U(S, V, A)$, which is the inverse function of the entropy $S(U, V, A)$, and also a thermodynamical potential.

For describing closed systems in contact with a thermostat, one, in turn, takes the temperature as a variable and one obtains a new thermodynamical potential, the *free energy*

$$F = F(T, V, A) = U - TS \qquad (2.60)$$

the natural variables of which are the ones of the *canonical* ensemble: the temperature T, the number of particles A and the volume V.

Finally, one can choose as variable the chemical potential μ and not the number of particles A. One then introduces the *grand potential* (or Gibbs potential)

$$\Omega = \Omega(T, V, \mu) = U - TS - \mu A \qquad (2.61)$$

Table 2.2. Elementary properties of the three usual statistical ensembles. The 'natural' variables are given in column 2. In column 3 is given the associated thermodynamical potential and conjugate variables can be found from the differential of the potential (column 4). The pairs of conjugate variables (x, X) are linked: $X = \pm \partial \mathcal{F}/\partial x$, where the x's are the natural variables and $\mathcal{F}$ the thermodynamical potential. From [28].

Statistical ensemble	'Natural' variables	Potential	Differential of the potential
Microcanonical	S, V, A	U	$dU = T\,dS - P\,dV + \mu\,dA$
Canonical	T, V, A	$F = U - TS$	$dF = -S\,dT - P\,dV + \mu\,dA$
Grand canonical	T, V, μ	$\Omega = U - TS - \mu N$	$d\Omega = -S\,dT - P\,dV - N\,d\mu$

the variables of which are the ones of the *grand canonical* ensemble, temperature T, volume V and chemical potential μ. A typical situation in which the grand canonical description is used is the case of an open system in contact with a thermostat. For the sake of completeness, the characteristics of these three statistical ensembles are gathered together in table 2.2.

As is clear from this rapid discussion, the statistical ensembles, and the corresponding thermodynamical potentials, are defined in terms of the variables chosen to describe the physical system under consideration. This is exactly the line that we have followed since the introduction of the concept of density matrices. There is obviously a strong link between the two, statistical and thermodynamical, pictures. Actually both the canonical and grand canonical thermodynamical potentials can be directly calculated from the corresponding partition functions:

$$F = -T \ln(Z_{\mathrm{c}}) \quad \text{and} \quad \Omega = -T \ln(Z_{\mathrm{gc}}) \tag{2.62}$$

with obvious notation. These relations thus point to the intrinsic connection between the statistical and thermodynamical descriptions of a given system. Note finally that in the microcanonical case the relation is even more direct as the internal energy *is* by construction a *variable* of the corresponding statistical description.

Thermodynamical potentials involve variables in pairs of conjugate variables, such as pressure, volume, etc. Experience furthermore shows that, for large systems, such as the ones considered in thermodynamics, one of the variables of a couple of conjugate variables is intensive and the other one is extensive. An extensive variable (energy, entropy, etc) is proportional to the volume V when V goes to infinity, while intensive variables such as temperature or chemical potential are independent of the volume and become equal to each other at equilibrium, when two systems are set in contact. In general the constraints of type $\langle Q_i \rangle$ are extensive and the corresponding Lagrange multiplier $(T, \mu, \ldots)$ intensive.

2.3.1.4 *Fermions and bosons in nuclear physics*

Indistinguishability of quantal particles leads to the introduction of two types of statistics, one for fermions (with half-integer spins) and one for bosons (with integer spins) [28, 277]. In nuclear physics, nucleons are spin-$\frac{1}{2}$ fermions, while π's, ρ's, etc (vectors of the nucleon–nucleon interaction) are bosons, and may play an important role in dense nuclear matter. In the nucleonic regime we shall mainly deal with fermions.

The fermionic nature of particles is revealed in the way they occupy levels. Thus A fermions occupy the first A lowest energy levels of a single-particle energy spectrum, up to the Fermi level (energy ϵ_F, see, for example, section 2.1.2.2). If one denotes by $\{|\phi_i\rangle, i = 1, A\}$ the corresponding eigenstate kets, the A-body state can then be written as an antisymmetrized product $|\Psi\rangle = \mathcal{A}_{1 \ldots A}(\prod_{i=1,A} |\phi_i\rangle)$ (Slater determinant). This corresponds to a ground-state zero-temperature nucleus. In terms of the occupation numbers of levels, one sees that zero temperature fermions have occupation numbers $n_i = 1$ for levels below Fermi level ($\epsilon_i \leq \epsilon_F$) and $n_i = 0$ otherwise. At finite temperature T occupation numbers may be calculated, for example, in the grand canonical ensemble. This leads to the well-known result:

$$n_i = \frac{1}{1 + \exp((\epsilon_i - \epsilon_F)/T)}. \tag{2.63}$$

Note that at finite temperature all levels $\{\epsilon_i, i = 1, \ldots\}$ of the single-particle spectrum are occupied with a non-zero probability. A complete description of the system hence requires knowledge of all eigenstates $\{|\phi_i \rangle, i = 1, \ldots\}$. This, in turn, allows us to evaluate the one-body density matrix ρ as

$$\rho = \sum_{i=1}^{\infty} n_i |\phi_i\rangle\langle\phi_i| \tag{2.64}$$

which displays a form typical of density matrices (2.52) with probabilities q_λ's given by the occupation numbers n_i.

In heavy-ion collisions in the nucleonic regime one may attain a wide range of temperatures. Typical values of the temperatures reached provide an estimate of the more or less degenerate nature of the system. For example, at $T = 4$ MeV $\sim \epsilon_F/10$ one obtains $n(0.8\epsilon_F) \sim 0.88$ and $n(1.2\epsilon_F) \sim 0.12$ and for $T = 8$ MeV $\sim \epsilon_F/5$, $n(0.8\epsilon_F) \sim 0.73$ and $n(1.2\epsilon_F) \sim 0.27$. These occupation numbers show that statistics is extremely robust. Hence, one cannot, *a priori*, overlook it in the case of heavy-ion collisions in the nucleonic regime.

It is also interesting to consider the zero temperature limit of Fermi occupation numbers as we shall also often be interested in physical situations for which the temperature is not so large ($T/\epsilon_F \sim 0.1$–0.2). In the zero temperature limit, one can expand Fermi occupation numbers in powers of temperature as

$$\frac{1}{1 + e^{(\epsilon - \epsilon_F(T))/T}} \simeq \theta(\epsilon - \epsilon_F(T)) - \frac{\pi^2}{6}T^2\delta'(\epsilon - \epsilon_F(T)) + \cdots \qquad (2.65)$$

where δ' is the derivative of the Dirac distribution and θ is the Heaviside step distribution. Note that the Fermi energy $\epsilon_F = \epsilon_F(T)$ is *a priori* a temperature-dependent object. Expression (2.65) turns out to be very useful for deriving approximate expressions at moderate temperatures. The strictly zero-temperature term is also quite interesting by itself, as it directly provides the standard Fermi gas model (section 2.1.2.2). Indeed, let us rewrite the integral relation (2.4) in terms of energies $\epsilon = \hbar^2 k^2/2m$ rather than momenta k; it then reads:

$$\rho = \frac{g}{(2\pi)^3}\int_0^{k_F} 4\pi k^2 \, \mathrm{d}k = \frac{g}{4\pi^2}\left(\frac{\hbar^2}{2m}\right)^{-3/2}\int_0^{\epsilon_F} \sqrt{\epsilon}\,\mathrm{d}\epsilon$$

$$= \int_0^{\epsilon_F} \omega(\epsilon)\,\mathrm{d}\epsilon = \int_0^{\infty} n_{T=0}(\epsilon)\omega(\epsilon)\,\mathrm{d}\epsilon \qquad (2.66)$$

where $\omega(\epsilon) = \mathrm{d}N(\epsilon)/\mathrm{d}\epsilon$ (equation (2.94)) is the single-particle level density (expressed here in the particular case of the Fermi gas, but see also section 2.4.2) and where $n_{T=0}(\epsilon) = \theta(\epsilon_F - \epsilon)$. Conversely, by replacing $n_{T=0}(\epsilon)$ by $n(\epsilon)$ (equation (2.63)) one directly obtains the expression of the density in a Fermi gas at finite temperature.

2.3.2 Nuclear systems at finite temperature

2.3.2.1 The nuclear equation of state at finite temperature

The equation of state of a system reflects the underlying elementary interactions between its constituents. The nucleon–nucleon interaction contains a dominant repulsive/attractive central term (section 2.2.1), which has a form analogous to the one of an intermolecular Lennard-Jones type potential. In spite of the different natures of the system, this suggests that the equation of state of infinite nuclear matter has a form similar to a Van der Waals equation of state [255]. This analogy is only formal: nucleons are fermions (not classical particles) and hence subject to the Pauli principle; it is furthermore extremely difficult, in contrast with many real fluids, to establish a proper link between the elementary interaction and the equation of state (section 2.2.2.2). Still, this analogy allows us to understand the general form of the isotherms of the nuclear equation of state (figure 2.8).

This analogy between equations of state furthermore suggests the possibility of liquid–gas phase coexistence. Such a coexistence presumably takes place during the latest stages of the collapse of type II supernovae cores, during which an equilibrium between nuclear clusters and a nucleon gas occurs. In heavy-ion collisions some dynamical behaviours, such as the fragmentation of a hot

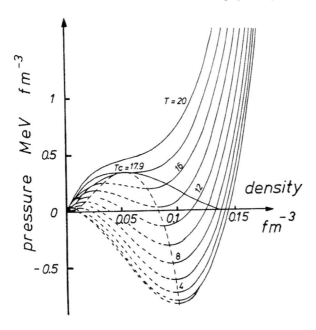

Figure 2.8. Isotherms, in the pressure (in MeV · fm^{-3}) versus density (in fm^{-3}) plane, of a phenomenological nuclear matter equation of state, obtained by means of a Skyrme interaction (section 2.2.1.3). The saturation point corresponds to the zero pressure point at finite density. The full line is the liquid–gas coexistence line. Temperatures are given in MeV. The critical point corresponds to $T_c = 17.9$ MeV. Finally, the broken lines correspond to the mechanically unstable spinodale region (section 8.2.1.3). From [410].

and compressed nucleus into a set of several lighter nuclei, also suggest a link with phase transitions (this is the subject of chapter 8). The liquid phase could then correspond to the nuclei, while the gaseous phase would represent the free nucleons produced in the collision (section 8.1.2). This 'phase coexistence' has been, and still is, actively sought in many experiments and calculations (section 8.4.3). Finite-size effects, nevertheless, make delicate the signature of such a phenomenon (section 8.4.3.5). The critical temperature for the 'liquid–gas' transition in nuclear matter should be of order $T_c \sim 16$–18 MeV and the corresponding critical density $\rho_c \sim 0.05$–0.06 fm^{-3}. It should also be noted that there exists a low-density region, which is mechanically unstable (spinodale region), namely in which fluctuations are amplified, instead of being damped as in nuclear matter close to saturation density [43, 363]. The spinodal region typically corresponds to densities below two-thirds of the saturation density.

As an example we show in figure 2.8 the equation of state for symmetric nuclear matter, as obtained from a Skyrme interaction (section 2.2.1.3). The various trends qualitatively discussed earlier are obtained in this phenomenological calculation. It is interesting to complement this figure by simple analytical estimates from a typical simple (and widely used) t_0, t_3 Skyrme interaction (equation (2.30) and section 2.2.2.3). The equation of state of nuclear matter at finite temperature is then simply obtained from the zero temperature case (equation (2.37)) by omitting the t_1, t_2 term and replacing the zero-temperature kinetic term, corresponding to a zero-temperature Fermi gas equation (2.7), by its temperature-dependent counterpart.

While at zero temperature (section 2.2.2.3) the natural potential is the energy from which the pressure is deduced by derivation (equation (2.37)), the finite temperature case requires us to consider the canonical or grand canonical ensemble and the associated thermodynamical potential (section 2.3.1.3). In the simple case of a Fermi gas, the grand canonical ensemble provides a simple frame in which to work out general expressions. Starting from the well-known thermodynamical expression $\Omega = -PV = -2/3E$ (energy E reduces to kinetic energy E_K^T in the non-interacting Fermi gas) [186] leads directly to the general expression (spin–isospin degenerate Fermi gas)

$$
\begin{aligned}
P_\mathrm{K}^T &= \frac{2}{3}\frac{E_\mathrm{K}^T}{V} = \int_0^\infty n(\epsilon)\epsilon\omega(\epsilon)\,\mathrm{d}\epsilon \\
&= \frac{2}{3\pi^2}\left(\frac{\hbar^2}{2m}\right)^{-3/2}\int_0^\infty \sqrt{\epsilon\epsilon}\frac{1}{1+\exp\left(\frac{\epsilon-\epsilon_\mathrm{F}}{T}\right)}\,\mathrm{d}\epsilon \\
&= \frac{2}{3\pi^2}\left(\frac{\hbar^2}{2m}\right)^{-3/2} T^{5/2} J_{3/2}(\epsilon_\mathrm{F}/T)
\end{aligned}
\tag{2.67}
$$

which defines the $J_{3/2}$ Fermi integral. At finite temperature the kinetic energy E_K^T, as well as the pressure P_K^T, are, nevertheless, not simple analytical functions of the density ρ and one usually resorts to numerical calculations. Still, the moderate and high temperature cases can be worked out analytically. At low temperature ($T \ll \epsilon_\mathrm{F}$) the kinetic pressure can thus be expressed as (the spin–isospin degenerate case)

$$
P_\mathrm{K}^T(\rho) \underset{T\to 0}{\simeq} \frac{2}{5}\rho\left[\epsilon_\mathrm{F} + \frac{5\pi^2}{12}\frac{T^2}{\epsilon_\mathrm{F}}\right] \quad \text{with } \epsilon_\mathrm{F} = \frac{\hbar^2}{2m}\left(\frac{3\pi^2\rho}{2}\right)^{2/3}
\tag{2.68}
$$

while a correction to the oversimplified perfect gas pressure provides a reasonable ansatz at high temperature:

$$
P_\mathrm{K}^T(\rho) \approx \rho T + \frac{1}{2^{9/2}}\left(\frac{2\pi\hbar^2}{m}\right)^{3/2}\frac{\rho^2}{\sqrt{T}}.
\tag{2.69}
$$

The total finite temperature equation of state thus finally reads

$$
P(\rho, T) = P_\mathrm{K}^T(\rho) + \frac{3}{8}t_0\rho^2 + \frac{1}{16}t_3\rho^{\gamma+2}
\tag{2.70}
$$

with the appropriate expression for $P_K^T(\rho)$.

The critical point is then characterized by the two equations

$$\left(\frac{\partial P}{\partial \rho}_{|T}\right)_{|\rho=\rho_c, T=T_c} = 0 \quad \text{and} \quad \left(\frac{\partial^2 P}{\partial \rho^2}_{|T,T}\right)_{|\rho=\rho_c, T=T_c} = 0 \quad (2.71)$$

which define the critical density ρ_c and temperature T_c (and hence the critical pressure $P_c = P(\rho_c, T_c)$). The isothermal spinodale line corresponds to $\partial P/\partial \rho_{|T} \leq 0$. The largest root of the latter inequality will be, in the following, labelled as the spinodale density ρ_s. As an example, taking the numerical values of t_0, t_3 and γ of section 2.2.2.2, one obtains a 'reasonable' critical point and zero temperature spinodale density: $\rho_c \simeq 0.056$ fm$^{-3} \simeq 0.33\rho_0$, $T_c \simeq 16.5$ MeV and $\rho_s(T = 0) \simeq 0.11$ fm$^{-3} \simeq 0.65\rho_0$.

2.3.2.2 *The nuclear equation of state in the astrophysical context*

To form nuclear matter at finite temperature is not only the privilege of heavy-ion collisions. Finite temperature nuclear matter is actually formed 'naturally' during some specific phases of stellar evolution. While most phases of nucleosynthesis only require vanishingly small temperatures (at the nuclear level), the late phases of the evolution of massive stars allow access to huge temperatures of the order of several MeV, together with densities of the order of the saturation density. This is a typical situation encountered in pre-supernova cores, during the phase of stellar collapse preceding the supernova explosion. This collapse phase is extremely short in terms of stellar timescales (some milliseconds at most) but is crucial for the future evolution of the system, particularly for the fate of the explosion and the possible ensuing formation of a neutron star. This question has thus been the focus of a lot of attention, in particular during the 1980s, both from the astrophysical and the nuclear point of view. It should be remembered that the nuclear system under consideration is, nevertheless, quite different from the ones formed in the course of heavy-ion collisions. First, the system is huge and the whole process is quasi-static at the nuclear timescale. The concept of temperature is thus thermodynamically well defined. Second, the composition of the system is admittedly somewhat unusual: besides nuclear phases, it contains electrons (which ensure global electroneutrality) and neutrinos. Detailed calculations of the equation of state of supernovae cores have thus acquired a high degree of sophistication, at least from the nuclear point of view and here is not the place to give credit to all the works performed in this direction. We shall thus only mention a few general references and reviews [49, 99, 100, 261, 330] in which most developments (from the nuclear side) were achieved in the 1980s. We also finally recall one of the most exotic results obtained, i.e. the existence of phases of various nuclear geometries during the collapse (nuclei, rods, tubes, planes, holes, etc) [279, 384, 497].

The explosion of a type II supernova is expected to leave behind a compact, neutron-rich residue: a neutron star. Neutron stars are also fascinating objects for nuclear physicists because they have huge densities. While supernovae cores are moderately dense but hot, neutron stars are cold but extremely dense (typically two to three times the saturation density) and they are almost exclusively composed of neutrons (unbound neutron matter!). Understanding the equation of state of dense neutron matter is thus still the focus of much effort. A key problem here lies in the high densities which require a relativistic treatment and a proper account of pionic degrees of freedom. Neutron star matter is thus even more on the margin of our topic than supernovae cores. And we shall again only give a few general references [49, 261, 424], not forgetting the pedagogical seminal paper of Walecka [423, 485] which has played a key role in the development of relativistic treatments of nuclear physics. We also finally mention the calculation by [140] in which a direct application of Brückner theory to neutron star matter is performed.

2.3.2.3   Statistical physics of finite nuclear systems

Terrestrial nuclei are relatively small ($A \lesssim 250$). They can be studied in a statistical physics framework, but extrapolation to the thermodynamical limit ('infinite' system) is by no means trivial. On the other hand, by its very nature, nuclear matter can be treated at the thermodynamical limit, and quantities such as temperature (or pressure) recover their usual thermodynamical meaning. This is not the case in an isolated nucleus, in which the temperature, for example, is defined only in terms of the statistical ensemble chosen for describing the system. For example, in the microcanonical ensemble, energy is supposed to be known exactly, and one can thus only access an 'average' value of the temperature (see equation (2.82) and related discussions). Conversely, in the canonical ensemble, energy is only fixed on average while temperature is known exactly. The choice of a particular ensemble depends on the physical conditions, and on the knowledge one has of these conditions. Different statistical ensembles will thus correspond to different values of the temperature. One then speaks of fluctuations between statistical ensembles. Fluctuations around average values can be evaluated directly from the partition function Z, using the relation

$$\langle Q_i Q_j \rangle - \langle Q_i \rangle \langle Q_j \rangle = \frac{\partial^2}{\partial \lambda_i \partial \lambda_j} \log Z. \tag{2.72}$$

Let us take the example of energy fluctuations in the canonical ensemble. We note H the Hamiltonian of the system and $U = \langle H \rangle$ its internal energy. The energy fluctuations are then given by $\langle (H - U)^2 \rangle = \partial^2 (\ln(Z_{\text{can}})/\partial \beta^2$. One can show that $\ln(Z_{\text{can}})$, as $\langle (H - U)^2 \rangle$ are proportional to the volume V of the system. The statistical fluctuations of the energy are thus proportional to $V^{1/2}$ and small compared to the energy itself, in the limit of large volumes; conversely, the smaller the system is, the larger the fluctuations are. The smallness of the fluctuations in the limit of large volumes allows us to understand the equivalence between ensembles, in the limit of large systems: it then becomes equivalent to inserting a constraint, such as energy or particle number, exactly or an average.

Table 2.3. Evolution of $\Delta T/T$ as a function of the mass number A and temperature T following equation (2.75).

	A			
T (MeV)	20	40	100	200
1	0.45	0.31	0.20	0.14
8	0.16	0.11	0.07	0.05
16	0.11	0.08	0.05	0.04

In contrast, in finite systems such as nuclei, it is *a priori* important to distinguish the various statistical ensembles.

In a nucleus, it is energy which is well defined and conserved in time. The microcanonical ensemble should thus *a priori* provide the natural framework for description. Temperature is not a natural variable in this ensemble and is thus bound to fluctuate. Let us take an independent particle model as an example. The excitation energy is linked to the average temperature by the approximate relation $E^* \simeq aT^2$ (equation (2.93)). The variance of the excitation energy distribution assuming a system in equilibrium at temperature T then reads:

$$\sigma^2 = 2aT^3 \tag{2.73}$$

so that

$$\frac{\sigma}{E^*} = \frac{\sqrt{2}}{(aT)^{\frac{1}{2}}} \tag{2.74}$$

which leads to

$$\frac{\Delta T}{T} = \frac{1}{\sqrt{2aT}}. \tag{2.75}$$

Table 2.3 gives typical values of the $\Delta T/T$ ratio with the level density parameter taken as $a \simeq A/8$ (section 2.4.2) where A is the mass number of the considered nucleus.

Hence, the use of the concept of temperature in nuclei is not simple. Only in very heavy nuclei at very high temperature are temperature fluctuations small enough for the canonical ensemble to provide a reasonable description. One is then 'close' to the thermodynamical limit and fluctuations between statistical ensembles start to be washed out. It should also be noted that, in contrast to other situations, the notion of temperature in a nucleus does not imply the existence of a thermostat: the nucleus constitutes the whole system and thus provides itself the thermostat. This is another way of revealing finite-size effects and the difficulty in choosing a statistical ensemble. To some extent, these difficulties are, nevertheless, overlooked in dynamical approaches.

2.4   The statistical model

Experimentally speaking, it is mainly in terms of their de-excitation that the properties of excited nuclei may be accessed (chapter 7). There are two big classes of nuclear excitations, depending on the type of states they lead to. The first class concerns excitations leading to specific states: this is typically the case of giant resonances. In turn, these states may decay either in a specific way, following proper selection rules, or in a statistical way. In the second class of de-excitations, a large set of states become populated in a non-specific way. This is the situation typically encountered in hot nuclei, close to equilibrium (moderate temperatures up to $T \leq$ 4–5 MeV and little compression). In such cases, the time evolution of the system may be followed in a statistical framework once the accessible states, together with their relative weights, have been properly identified [66, 175, 193, 486].

2.4.1   Basics of the statistical model

2.4.1.1   *The key role of the density of states*

Let us consider an excited nucleus of mass A, excitation energy E^* and, possibly further characterized by its charge Z and/or angular momentum J. The point is to evaluate towards which states the system will preferentially decay. The transition probability from an initial state i to a final state f is given by the Fermi golden rule:

$$\frac{\mathrm{d}N_{i\to f}}{\mathrm{d}t} \propto |M_{i\to f}|^2 \rho_f \qquad (2.76)$$

in which $M_{i\to f}$ is the transition matrix and ρ_f is the final density of states (section 2.4.2). The basic assumption of statistical theory is to consider that all transition matrices are equal so that the probability of observing a given state is solely governed by its density of states.

The nucleus is an isolated system which may be properly described in a microcanonical approach (section 2.3). Let us apply the Fermi golden rule in a case where the final state 'f' corresponds to the emission by a parent nucleus 'i' (initial) of a particle 'b' of spin s, emitted with a kinetic energy between ϵ and $\epsilon + \mathrm{d}\epsilon$ (see figure 2.9). The corresponding emission (evaporation) probability per unit of time for the process $i \to b + f$ may be written as:

$$P_b(\epsilon)\,\mathrm{d}\epsilon = C_0 \rho_f(E_f^*)\,\mathrm{d}E_f^*(2s+1)\frac{4\pi p^2\,\mathrm{d}pV}{h^3} \qquad (2.77)$$

This quantity may be expressed as the product of three terms. C_0 is a normalization constant discussed later. The term $(\rho_f(E_f^*)\,\mathrm{d}E_f^*)$ is the number of states available for the excited (E_f^*) daughter nucleus: this is the product of the density of states $\rho_f(E_f^*)$ (section 2.4.2) and of an energy interval $\mathrm{d}E_f^*$. The last term $((2s+1)4\pi p^2\,\mathrm{d}pV/h^3)$ is the number of states of the emitted particle with

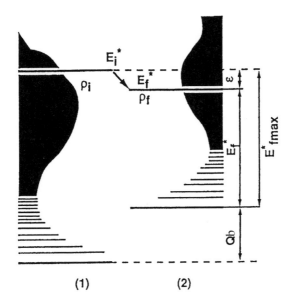

Figure 2.9. A schematic picture of the decay of an initial excited nucleus towards a final nucleus and a particle with a kinetic energy ϵ. The initial excitation energy E^* is shared between various terms: E_f^*, the excitation energy of the final nucleus, Q_b (respectively ϵ), the binding energy (respectively kinetic energy) of the emitted particle b. It is assumed here that this particle is not excited.

a linear momentum between p and $p + \mathrm{d}p$; V is the volume of an imaginary box where the decay takes place. The proportionality constant C_0 may be obtained from the detailed balance principle, which states that microscopic phenomena are time reversal invariant. This means that the decay probability is linked to the reverse fusion reaction rate (equation (2.77)):

$$\frac{P_b(\epsilon)\,\mathrm{d}\epsilon}{n_f} = \frac{P_{\text{fus}}}{n_i} \tag{2.78}$$

where P_{fus} is the fusion rate (i.e. the number of fusions per unit of time) between the particle b and the final nucleus f leading to the initial nucleus i, and n_i the corresponding number of available states.

The number of available states n_i may again be expressed through the density of states ρ_i (section 2.4.2) of the initial nucleus

$$n_i = \rho_i(E^*)\,\mathrm{d}E^* \tag{2.79}$$

and the fusion rate can be expressed as

$$P_{\text{fus}} = \frac{v\sigma_c(\epsilon)}{V} \tag{2.80}$$

where $\sigma_c(\epsilon)$ is the capture (fusion) cross-section of the particle b by the final nucleus f and v the particle velocity. When combining these expressions and noting that (i) $v = d\epsilon/dp$ (non-relativistic case) and (ii) $dE^* = dE_f^*$ (resulting from energy conservation), one finally obtains

$$P_b(\epsilon)\,d\epsilon = \frac{\rho_f(E_f^*)}{\rho_i(E^*)}(2s+1)\frac{4\pi p^2}{h^3}\sigma_c(\epsilon)\,d\epsilon. \tag{2.81}$$

A de-excitation channel will thus be all the more favoured if the number of accessible states ($\propto \rho_f$) is large, which emphasizes the key role played by the density of states (section 2.4.2).

2.4.1.2 Ingredients of the statistical model

To go beyond equation (2.81) it is necessary to express the ingredients of this formula, namely the densities of states and the inverse capture cross-section. The number of states available for a nucleus with an excitation energy between E^* and $E^* + \Delta E^*$ may be connected to the corresponding entropy S of the system (with Boltzmann constant $k = 1$)

$$S = \ln(\rho(E^*)\Delta E^*) \tag{2.82}$$

and to its temperature, defined in the microcanonical approach as

$$\beta = \frac{1}{T} = \frac{dS}{dE^*} \approx \frac{\Delta \ln \rho(E^*)}{\Delta E^*}. \tag{2.83}$$

The density of states thus exhibits an exponential evolution with the excitation energy:

$$\rho(E^*) \propto e^{E^*/T} \tag{2.84}$$

which emphasizes the sensitivity of statistical models to this quantity.
   The capture cross-section in equation (2.81) may be written:

$$\sigma_c(\epsilon) = \sum_{l=0}^{\infty}(2l+1)\pi\lambda^2 T_l(\epsilon). \tag{2.85}$$

If the transmission coefficients T_l are set to unity, one obtains:

$$\sigma_c(\epsilon) = \pi R^2 \left(1 - \frac{B_b^{\text{Coul}}}{\epsilon}\right) \quad \text{for } \epsilon \geq B_b^{\text{Coul}} \tag{2.86}$$

and

$$\sigma_c(\epsilon) = 0 \quad \text{for } \epsilon \leq B_b^{\text{Coul}} \tag{2.87}$$

where B_b^{Coul} is the Coulomb barrier associated with the emission of particle b. From all these equations, one finally obtains

$$P_b(\epsilon) = \frac{\epsilon - B_b^{\mathrm{Coul}}}{T^2} e^{-(\epsilon - B_b^{\mathrm{Coul}})/T} \quad \text{for } \epsilon \geq B_b^{\mathrm{Coul}}. \qquad (2.88)$$

In this expression $P_b(\epsilon)$ has been normalized to unity. The energy E_f^* is expressed as a function of ϵ: $E_f^* = E_{f\max}^* - \epsilon$ (see figure 2.9). The exponential term of equation (2.88) comes from equation (2.84) used to express $\rho(E_f^*)$ as a function of ϵ. This means that the temperature T is the temperature of the final nucleus. It may be considered as independent of ϵ only for significant excitations. In other words, this statistical formalism is not valid for nuclei close to their ground state. This limitation is also required to derive simple expressions for the density of states (section 2.4.2).

The emission probability $P_b(\epsilon)$ exhibits a maxwellian shape which is typical of the decay of an equilibrated nucleus. We will see in chapter 4 that it may be used to measure the temperature from the evaporated particles' kinetic energy spectra. It is worthwhile to remark that equation (2.88) has been obtained without the need of any nuclear model: the only ingredients which have been used are the microcanonical description of isolated systems, the density of states, the entropy and the temperature. Nuclear physics begins when one wants to be more precise, i.e. when one wants to compare the relative probabilities of two decay channels or to link the excitation energy with the temperature, because such steps imply the calculation of the absolute value of the nuclear density of states.

The competition between various channels (this means the emission probability for different particles) may be obtained from the integration of equation (2.88) before normalization. The total emission probability of a given particle b is essentially governed by the final density of states of the daughter nucleus:

$$P_b \propto \rho(E^* - Q_b - B_b^{\mathrm{Coul}}). \qquad (2.89)$$

In other words, for similar Q values, particles for which the Coulomb barrier is low, are preferentially emitted. This is why neutrons are highly favoured in the decay of heavy nuclei.

2.4.1.3 Partial widths and lifetimes

In the statistical model, the initial (parent) nucleus is not in a stationary state so that one can only access de-excitation probabilities. Hence, the aim here is to evaluate the lifetime τ of the system, through the various accessible channels. In this respect a statistical approach provides, to some extent, a pseudo-dynamical picture of the de-excitation process (section 2.4.4). The total lifetime τ of the system is defined as

$$\frac{1}{\tau} = \sum_f \frac{\Gamma_f}{h} \qquad (2.90)$$

where the summation runs over all the possible de-excitation channels f, the de-excitation rate, in a given channel f (or partial width of the channel) reading $1/\tau_f = \Gamma_f/\hbar$. And the total de-excitation probability of the system reads $e^{-t/\tau}$, while the one in the specific f channel reduces to $P_f = \Gamma_f/(\sum_j \Gamma_j)$.

Once all the partial widths Γ_f are known, it becomes possible to simulate the statistical de-excitation of an excited nucleus of excitation energy E^*. This is done by sampling. The relative weight $P_f = \Gamma_f/(\sum_j \Gamma_j)$ of any channel is known. At a given step of de-excitation, it is hence sufficient to sample the accessible states, according to the probabilty law given by the P_f's. Provided there is sufficiently large sampling, one will obtain a statistical ensemble in which each channel will effectively have the weight P_f. But this represents only one step of the whole process of de-excitation. After this step nuclei are, in general, themselves excited and can de-excite again in a statistical way. One can thus iterate the former step, up until the final nuclei with insufficient energy for further de-excitation are reached. To follow these events along several steps, and while respecting the relative weights of the various de-excitation channels, requires huge statistical samplings of several millions of events. As already noted, these de-excitation sequences simulate, to some extent, a dynamical process. But this dynamics is not real, as it makes sense only provided the various de-excitation channels correspond to physical processes with compatible durations (see section 2.4.4).

One has finally to stress the fact that, in the statistical description of the system, it is implicitly assumed that the time τ (equation (2.90)) is long compared with all the other involved timescales: the time needed for the nucleus to explore the whole phase space (thermalization time) or the time needed for a specific process. The time needed to emit (evaporate) a particle b (once the system has 'decided' to emit this particle) is one of these typical timescales. The duration time of a fission process, i.e. the time needed to deform a system beyond the corresponding saddle-point, is another example. It turns out that neglecting such typical times does not play any role at low excitation energy because the value of τ is rather large in this case. This is no longer the case at larger excitations and so the pure statistical picture becomes more and more questionable. The fission process, because it is the slowest, is the first one which is influenced by such transient effects and an improved description has to include them in a dynamical description of the fission barrier passing: such a description is extensively discussed in chapter 7. In this section we limit ourselves to the pure statistical decay mechanisms where all the previous timescales are neglected.

2.4.2   Density of states

For a given energy E and particle number A, the density of states may formally be defined as

$$\rho(E, A) = \sum_{i,\nu} \delta(A - A_\nu)\delta(E - E_i(A_\nu)) \qquad (2.91)$$

where the summation runs over all the states with A_ν particles and E_i energy. One usually uses an approximate expression when the nucleus can be described in a model of independent particles with single-particle energy levels ϵ_k, and for moderate excitation energies ($E^* = E - E_{gs}$)

$$\rho(E, A) \simeq \frac{1}{\sqrt{48E^*}} e^{2\sqrt{aE^*}} \qquad (2.92)$$

where 'a' is the so-called level density parameter. At the same level of approximation, one can link the excitation energy to the temperature

$$E^* \simeq aT^2 \quad \text{with } a = \frac{\pi^2}{6} \omega(\epsilon_F) \qquad (2.93)$$

where the zero-temperature single-particle level density $\omega(\epsilon)$ reads:

$$\omega(\epsilon) = \sum_k \delta(\epsilon - \epsilon_k). \qquad (2.94)$$

Note that the single-particle level density $\omega(\epsilon)$ counts here the number of single-particle levels per unit energy, while the density of states $\rho(E, A)$ counts the number of accessible states as a function of the total energy of the nucleus. Finally, one should keep in mind that equations (2.92) and (2.93) only hold for moderate temperatures ($T \leq 3$–4 MeV). At very low temperatures these simple expressions are also insufficient to account for the various patterns exhibited by discrete excited states.

The Fermi energy ϵ_F plays a dominant role in the calculation of a (equation (2.93)) and thus in the expression of $\rho(E, A)$ (equation (2.92)). This reflects the fact that at low temperature the first levels to become empty are the least bound ones, namely the ones close to the Fermi energy. Elementary excitations which promote nucleons lying just below the Fermi level to just above it then become dominant.

The density of states is a difficult object to evaluate. Starting from an independent particle picture, which allows noticeable simplifications, the problem then occurs in the evaluation of the single-particle level density, and especially the level density parameter 'a'. Realistic mean-field calculations do not allow the recovery of the experimental value of 'a'. One has to take into account effects beyond the mean field, in order to reproduce the data. These calculations are, nevertheless, quite involved in realistic cases. It is sufficient, here, to keep in mind the average experimental value $a \simeq A/8$ (up to shell effects).

2.4.3 The neutron clock

The dissipative collisions we are interested in here, lead, as we shall see in detail later, to more or less thermalized objects, notwithstanding simultaneous mechanical excitations. The simplest way for cooling down a 'hot' nucleus is

particle emission. It is thus of importance to analyse this phenomenon and to see
on which timescale it takes place. This question was tackled long ago in the early
days of the introduction of statistical concepts in nuclear physics by Bohr [66]
and Weisskopf [486]. We shall not recount here these early analyses but rather
briefly show, which is equivalent, how to characterize nucleon emission in the
statistical model. For simplicity we restrict our discussion to neutron emission,
which is obviously favoured compared to proton emission because of the absence
of a Coulomb barrier.

We start from the probability for neutron emission as it may be written from
equation (2.81):

$$P_n(\epsilon)\, d\epsilon = \frac{\rho_f(E^* - S_n - \epsilon)}{\rho_i(E^*)} \frac{8\pi p^2}{h^3} \sigma_c(\epsilon)\, d\epsilon. \tag{2.95}$$

For simplicity we again make the simplest approximation, namely to assume E^*
to be large compared to both the neutron separation energy S_n and ϵ, which may
again make sense in a big hot nucleus. Following equation (2.92), the ratio of the
final and initial density of states then reduces to

$$\frac{\rho_f(E^* - S_n - \epsilon)}{\rho_i(E^*)} = \frac{E^*}{E^* - S_n - \epsilon} e^{2\left(\sqrt{a(E^* - S_n - \epsilon)} - \sqrt{aE^*}\right)} e^{-S_n/T} e^{-\epsilon/T} \tag{2.96}$$

which provides the usual maxwellian form of the neutron emission spectra
and which now allows a simple integration over ϵ, yielding a factor quadratic
in temperature. Considering that the neutron capture cross-section σ may be
approximated by the geometrical cross-section ($\sigma = \pi R^2$), one finally obtains
for the neutron emission lifetime:

$$\tau_n \simeq \frac{h^3}{8\pi} \frac{1}{2m} \frac{1}{T^2} \frac{1}{\pi R^2} e^{S_n/T}. \tag{2.97}$$

In this expression R is the nucleus radius. For $R \simeq 5$ fm and $S_n = 8$ MeV
one approximately obtains $\tau_n \simeq (1.7/T^2) \exp{(8/T)} 10^{-21}$ s (with temperature
in MeV), which provides a reasonable approximation to more sophisticated
estimates [449].

All calculations of neutron lifetimes τ_n point to the strong dependence
of τ_n on temperature. Neutrons thus constitute a thermal clock (or a 'time
thermometer'). This property has, for example, been used to evaluate fission
time (section 7.2.3.2). Between $T = 1$ MeV and $T = 5$ MeV the value of τ_n
decreases by four orders of magnitude. Up to $T \simeq 2$ MeV τ_n is large compared
to any characteristic time of the system, even fission time (section 7.2.3.2). When
the temperature reaches 2–3 MeV, τ_n becomes comparable to fission time but
still remains large compared to other nucleonic timescales (see also figure 3.1).
Difficulties occur for temperatures of the order 5 MeV for which τ_n becomes
comparable to the thermalization time and elementary process times, which
makes the notion of a hot nucleus itself questionable. One should, nevertheless,

be cautious here, because τ_n only represents an average value and because the present picture somewhat overlooks the underlying dynamical effects.

2.4.4 Limitations of the statistical model

The statistical model is widely used in data analysis. Starting from the final de-excitation residues, it allows us, in principle, access to the characteristics of the initial excited nucleus. In spite of its successes the statistical model suffers from some defects. A first problem lies in the evolution of the various components entering the model (barrier heights and level density, mainly) as a function of temperature (or excitation energy). Fission barrier heights, and thus corresponding fission probabilities, do depend on temperature. Barrier heights should decrease with temperature, which would significantly increase expected fission rates. Fission barriers are, for example, known to disappear at temperatures of the order $T \simeq 4$ MeV. And it is likely that such a barrier collapse also occurs for other emission channels. This is *a priori* an important effect which should be accounted for. The level density parameter also depends on temperature because of the disappearance of correlations, an effect which becomes particularly sensitive beyond about $T \simeq 4$ MeV (section 7.2.1). As the level density becomes exponential (equation (2.92)) one may expect an important effect on the density of states.

These problems remain, to a large extent, technical in nature. But the statistical model also raises fundamental conceptual problems. The original idea underlying this approach is Bohr's picture of the compound nucleus, namely a nucleus hit, for example, by a neutron which gradually transfers its energy to the nucleus in the course of a series of interactions with the nucleons inside the nucleus. The statistical description precisely stems from this 'random' sequence of interactions. In this case the statistical model applies perfectly. In hot nuclei formed in heavy-ion collisions the situation may be quite different. The formation of hot/excited nuclei in these nuclear collisions is usually accompanied by strong dynamical effects, which may affect, and even dominate the de-excitation process. It is, for instance, the case when compression occurs in a central violent nucleus–nucleus collision. Such effects cannot be taken into account in a statistical picture of a quasi-stationary state as a starting point for a statistical de-excitation. Violent collisions are, nevertheless, not the only way to form hot nuclei. Collisions with light ions, for example, presumably allow large energy deposit, with smaller mechanical effects than in heavy-ion collisions. But even this would not solve the 'intrinsic' problems of timescales. The neutron emission lifetime, for example, becomes vanishingly small with increasing temperature (section 2.4.3). In particular, it becomes small compared to the time needed to achieve a whole fission process or to achieve a thermalization of the available energy in the nucleus. This competition between evaporation and fission can be accounted for by explicitly including the fission time in statistical models: this approach is discussed in chapter 7. The lack of thermalization of the initial nucleus is a major problem since it concerns a fundamental hypothesis of the model. In such cases, the statistical approach is simply irrelevant and dynamical approaches have to be developed. In fact, the weights P_f, as previously defined, correspond to *stationary* regimes. These stationary regimes do not settle instantaneously. If the neutron emission time τ_n becomes small compared to the time needed to establish a stationary regime for the nuclear system, the statistical description loses its meaning. This is precisely what occurs in hot nuclei when the temperature reaches about 3–4 MeV and this indeed fixes the limits of applicability of the statistical model.

2.5   Conclusion of the chapter

In this chapter we have recalled some basic properties of ground-state nuclei, basing most of our discussions on the idea that nucleons are, to a large extent, independent from one another in such nuclei. This idea of the independence of nucleons is the basis of many approaches. In particular, it provides a basic justification for mean-field theories, although these are constructed from an originally singular elementary interaction. The Pauli principle plays a crucial role here in removing undesirable short range effects. We have thus discussed at length the impact of the Pauli principle.

The physics of excited nuclei as formed in heavy-ion collisions in the nucleonic regime has to rely on the concepts of statistical physics. Overlooking the dynamical aspects, which will be addressed in chapter 3, we have focused here on a brief overview of the basic tools of statistical physics at equilibrium. This also allowed us to address in some detail the physics of the equation of state of infinite nuclear matter, both at zero and finite temperature. We have discussed, in particular, the low-density region of the phase diagram where a liquid–gas phase transition is conceivable and where a mechanically unstable region is expected. Finally we have given a critical discussion of the statistical model of nuclear decay, which provides a basic tool for investigating the de-excitation of hot nuclei as formed in the course of heavy-ion collisions.

Chapter 3

Macroscopic and microscopic descriptions of heavy-ion collisions

The general approach to non-equilibrium statistical physics consists in reducing the original many-body problem to a (restricted) set of relevant variables. This procedure is all the easier and all the more justified if the timescales can be decoupled. The relevant variables are then associated with slow dynamics while (too!) rapid degrees of freedom are treated in an approximate way, buried in a stochastic force, as in the generic example of Brownian motion. The decoupling of timescales, which allows this elegant reduction of a complex dynamics into a, if not simple, at least ordered picture, is only marginally true in the case of the nuclear collisions we aim to describe. Still, this general decoupling scheme constitutes one of the firmest bases of our understanding of nuclear dynamics in the nucleonic regime, via the key role played by one-body degrees of freedom, as already widely illustrated in chapter 2 for nuclei close to equilibrium.

In the nuclear case one-body descriptions indeed represent both well-founded and efficient reductions of the many-body problem. The dominance of mean field or extended mean-field effects is now well established in the nucleonic regime of nuclear dynamics. We shall thus spend some space on deriving the basic kinetic equations in use in this field of physics, starting from standard methods for reducing the many-body problem via hierarchies of density matrices. The resulting extended mean-field theories are, nevertheless, often attacked via phase space, the energy scales involved washing out detailed quantal effects. This finally provides the widely used BUU-like approaches, which have constituted one of the basic working tools for analysing experimental data in the field since the mid 1980s.

Still, BUU is by no means the end of the story, as the situations encountered in heavy-ion collisions in the nucleonic regime cannot be fully addressed at such a gross level of reduction of the dynamics. This opens up the way to two kinds of approach. The first class of methods again emphasizes the key role of one-body degrees of freedom: BUU is here complemented by a stochastic component in the spirit of Brownian motion. This leads to involved stochastic differential equations in phase space, which have not yet been extensively applied to realistic collision cases. In the second class of approaches the problem is attacked by molecular dynamics methods, inherited from molecular physics. Such techniques allow a proper and cheap handling of many-body effects, at any order, but at the price of an almost complete loss of quantal effects which raises basic problems in the nuclear context. Re-establishing the Pauli principle turns out to constitute a difficult issue which, to a large extent, again points to the key role played by the mean field in the descriptions of heavy-ion collisions in the nucleonic domain.

3.1   Collision dynamics: collective effects

3.1.1   Microscopic and macroscopic scales

3.1.1.1   Some basic scales

It is useful to remember here some orders of magnitude of distances, energies and timescales, which will frequently enter our forthcoming discussions of collision processes. In the energy range we consider ($E_{\mathrm{lab}}/A \leq$ 200–300 MeV/A), the relevant energies are the nuclear binding energy $E_{\mathrm{B}}/A \sim$ 8 MeV, the Fermi energy $\epsilon_{\mathrm{F}} \sim$ 40 MeV and the pion production threshold $E_{\pi} \simeq$ 290 MeV (the real pion produced in a nucleon–nucleon interaction), although the latter does not correspond to a dominant channel. The typical distance between nucleons inside a nucleus is about 2 fm, while the range of the elementary nucleon–nucleon interaction is again of order 1 fm. Two 'dynamical' length scales also play an important role: the mean free path λ, which provides a measure of the independence of nucleons with respect to each other; and the de Broglie wavelength λ_{B}, which allows us to quantify the more or less quantal nature of the system (figure 2.3). The large values of λ and λ_{B} at low energy suggest the relevance of quantal mean-field approaches. At high energies, in contrast, the small values of λ and λ_{B} allow classical descriptions based on elementary nucleon–nucleon processes rather than on the mean field. The intermediate energy range we consider, in turn, calls for hybrid descriptions (mean field + elementary processes), possibly in a semi-classical approximation (section 3.2).

3.1.1.2   The entrance channel

Energies and distances allow us to evaluate dynamical times, which are typical of the dynamics of the colliding nuclei, in particular, in the entrance channel. In a ground-state nucleus the Fermi energy defines a typical time for a nucleon to cross the nucleus as $\tau_{\mathrm{tr}} \approx 2R/v_{\mathrm{F}} \sim$ 30–40 fm/c. During a heavy-ion collision, the most trivial timescale is the overlap duration. But it is a difficult quantity to evaluate, as it cannot be reduced to a mere geometrical effect. One has to take into account the interactions between nucleons of the target and projectile, which slows down the penetration of the latter into the former. Altogether, one may estimate this overlap time (which of course strongly depends on beam energy) to some tens of fm/c. Interactions between nucleons during overlap furthermore heat up the system. It turns out that one can very roughly identify the thermalization time with the overlap time. Finally, the collision, beyond thermal effects, also leads to sizeable mechanical effects, in particular, in terms of compression/dilatation. The characteristic timescale here is given by the period of the corresponding giant modes (in particular the monopole mode), which leads to a typical value of order $\tau_{monop} \sim 70$ fm/c.

A comparison of the previous timescales shows that, in the entrance channel, the most relevant timescales are of comparable orders of magnitude, as can also be seen from figure 3.1. This means that during this early phase of the collision no timescale is dominant, which implies that no observable *a priori* provides a 'compact' account of the whole process. As a consequence, a description of the entrance channel of such collisions has to be based on microscopic descriptions of the dynamics, in order to account for 'any' timescale.

It should be noted here that the impossibility of identifying a few relevant variables *a priori* does not mean that such variables do not exist. In this respect, it would not be contradictory to build (or extract) from a microscopic description a set of more 'global' relevant variables, characterizing the dynamics in a more compact way. It is actually exactly the way experimentalists try to address this question. From the huge amount of data furnished by modern 4π detectors (chapter 4), a proper access to a few global variables is obviously a way to try to account, in a simple fashion, for a complex situation.

3.1.1.3 *Collective dynamics*

This lack of a hierarchy of relevant timescales/observables is, nevertheless, not necessarily the rule. For example, at low beam energy, close to the Coulomb barrier, it is well known that the dynamics can, to some extent, be understood by a few global quantities characterizing the geometry of the collision, and complemented by the appropriate more microscopic components [487]. The situation is similar in fission [1, 195], but this time, to a large extent, it is independent of the entrance channel. Fission is a slow motion, taking place on a very long timescale compared to any of the previously discussed timescales ($\tau_{fission} \sim 10^{-21}$–$10^{-20}$ s, section 7.2.3.2). In such a case, the whole dynamical process can be reached by means of a few characteristic global variables, such as the elongation of the system. Such a variable 'q' evolves at a slow pace and corresponds to a heavy inertia (the nucleus mass more or less). The equation of motion of q has, nevertheless, to account for other timescales, such as the ones characterizing the ('lighter') degrees of freedom, which, in turn, show up at a rapid timescale (section 3.1.2).

More generally speaking, when a system exhibits such a hierarchy of timescales (or masses), it becomes possible to *a priori* identify a few 'relevant' variables on which the dynamics will be built up, even if the effect of other

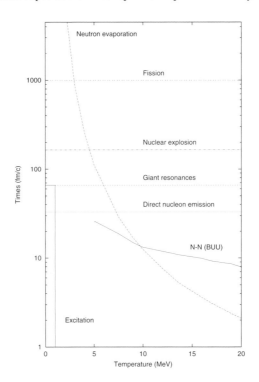

Figure 3.1. Estimation of various characteristic times (in fm/c) in nuclear reactions in the Fermi energy range as a function of the nuclear temperature T (in MeV). The dotted line corresponds to the neutron evaporation time τ_n (equation (2.97)). Other timescales are indicated on this figure: the line 'nuclear explosion' is associated with the characteristic timescale for the expansion and break-up of a very excited nucleus (see section 8.4.1.1), while timescales for giant resonances are typically between 50 and 100 fm/c (see section 7.2.2) and the timescale for fission lies in the 1000 fm/c range (see section 7.2.3). Above $T \simeq 5$ MeV, neutron emission time, as well as the decay of giant resonances, becomes comparable to the time needed for a hot nucleus to form (called here 'Excitation', around 60–70 fm/c) and becomes comparable with the timescale for direct (pre-equilibrium) particle emission (see chapter 6). This raises conceptual difficulties, and suggests that dynamical effects are likely to play a key role here.

variables has to be accounted for. As these relevant variables have a global nature, one speaks, in such cases, of macroscopic approaches. We shall briefly discuss them in section 3.1.2, bearing in mind, in particular, the interesting recent application to nuclear fission in hot nuclei (see also section 7.2.3).

3.1.2 Macroscopic approaches

3.1.2.1 A generic case : Brownian motion

In Brownian motion, a heavy (hence slow) particle M of mass m and velocity v (the Brownian particle) is supposed to move inside a gas of light (hence rapid) particles, characterized by its temperature T. Interactions between the gas particles and the Brownian particle slow down the latter. One may characterize this slowing down by a friction γ and hence write down an equation of motion for M of the form:

$$m\frac{dv}{dt} = -\gamma v. \tag{3.1}$$

But this simple picture does not fit experimental facts. In order to describe the 'erratic' aspect of Brownian trajectories [203], Langevin's [278] idea is to complement equation (3.1) by a stochastic term, which simulates some details of the collisions between M and the gas, beyond the average description provided by the friction γ. Equation (3.1) thus becomes

$$m\frac{dv}{dt} = -\gamma v + F(t) \tag{3.2}$$

where the complementary term F is a Gaussian stochastic force (Langevin force), characterized by its average value and its correlation function, respectively,

$$\langle F(t)\rangle = 0 \quad \text{and} \quad \langle F(t)F(t')\rangle = \alpha^2\delta(t-t') \tag{3.3}$$

where α represents the intensity of F. The meaning of the average values $\langle\cdot\rangle$ deserves some comment. When passing from a deterministic (equation (3.1)) to a stochastic description (equation (3.2)) one replaces the velocity v of the Brownian particle by an ensemble of possible velocities $v \longrightarrow \{v_\lambda, \lambda = 1,\ldots\}$. Each element of this ensemble evolves according to its own 'history', as a result of the values taken by the force F. The brackets $\langle\cdot\rangle$ in equation (3.3) represent an average over this ensemble of samplings of F, hence on the ensemble of the velocities, at a given instant. Note also that the correlation function (3.3) is Markovian: two successive values of F are independent of each other.

With only assumptions equation (3.3), and the fact that the gas is supposed to be at equilibrium at temperature T, one can show that α is not a free parameter of the problem, but reads $\alpha^2 = 2\gamma T$. This result, known as Einstein's relation, expresses the fluctuation dissipation theorem [122]. It establishes an important relation between the intensity of the dissipation (γ) and the intensity of the fluctuations (α): the more dissipative a process involving a variable q is, the larger the fluctuations of q are around its average value.

3.1.2.2 Langevin equations

In a Langevin approach *non-relevant* degrees of freedom are assumed to produce an average dissipative behaviour (friction γ in equation (3.1)). One then considers that the neglected part of the interactions between the Brownian particle and the molecules of the bath varies sufficiently fast that it can be represented by a *noise*, characterized by average dissipative features. Such a description is clearly applicable to many various physical problems [203, 475], much beyond the original Brownian motion. As soon as it is possible to identify well separated time (or mass) scales one may hope to extract an average 'slow' motion and simulate the neglected part of the dynamics by a Langevin term. Indeed, this type of description has been used in many fields of physics and chemistry, and this constitutes one of the major interests of the Langevin picture.

Equation (3.2) represents a simple prototype of Langevin equations, as the deterministic (dissipative) part of the equation is linear in velocity. More generally, one calls 'linear Langevin equations' stochastic differential equations of the form (written in one dimension for the sake of simplicity)

$$M_q \frac{\mathrm{d}q}{\mathrm{d}t} = -\gamma q + F(t) \qquad (3.4)$$

where M_q is the mass associated with the q degree of freedom and γ the friction, assumed constant, and acting on q motion. The nature of the variable q is not specified here. It is a *relevant* variable, chosen for representing the dynamics of a complex process in which the 'slow' motion of q can be separated from the 'rapid' motion associated with the other degrees of freedom. The 'relevant' variable q thus interacts with a 'bath' (usually assumed of thermal nature) of *non-relevant* degrees of freedom. The effect of this 'bath' is twofold : a 'slow' average effect represented by γ, and a rapidly varying effect represented by the fluctuating force $F(t)$. A key ingredient of the description thus turns out to be the random force F, which is characterized by its first two moments

$$\langle F(t) \rangle = 0 \quad \text{and} \quad \langle F(t)F(t') \rangle = \alpha^2 \delta(t - t') \qquad (3.5)$$

where the brackets $\langle \cdot \rangle$ label an average over the ensemble of realizations of the stochastic process. One can then obtain Einstein's relation linking α to friction γ and bath temperature T. Equation (3.5) reflects the Markovian nature of the stochastic force. One usually calls such a force a white noise. It is a reasonable assumption as long as little is known about the coupling between q and the bath, and it is furthermore compatible with the instantaneous effect of the friction term $(-\gamma q)$ (fluctuation–dissipation theorem). It should, nevertheless, be noted that equation (3.5) only specifies the first two moments of F, which is not enough to fully characterize F. Generally, one chooses a Gaussian form for F, which is then fully defined by its first two moments. This then provides a complete description of the Langevin equation.

Among all the hypotheses leading to a Langevin equation, it is the Markov approximation which is probably the most questionable one. It forces the stochastic term to act instantaneously, compared to the average slowing down due to friction—in many situations this hypothesis is too restrictive. One can then consider generalizations of linear Langevin equation (3.4), as

$$M_q \frac{dq}{dt} = - \int_0^t \gamma(t - t')q(t')\, dt' + F(t) \tag{3.6}$$

where a non-constant, generalized friction $\gamma(t - t')$ is introduced. The stochastic force then follows the two relations

$$\langle F(t) \rangle = 0 \quad \text{and} \quad \langle F(t)F(t') \rangle = \phi(t - t') \tag{3.7}$$

characteristic of a so-called 'coloured noise' (compared to the Markovian white noise, equation (3.5)). Einstein's relation is generalized according to

$$\phi(t - t') = 2\gamma(t - t')T \tag{3.8}$$

which again expresses the fluctuation–dissipation theorem. One usually speaks here of *memory effects*, to reflect the fact that the stochastic force keeps some sort of memory of the preceding dynamics.

Even more general forms of Langevin equations can be thought of. It is, for example, possible to consider so-called 'non-linear' Langevin equations in which the noise term itself contains the stochastic variable 'q'. In an ensemble average of such an equation the stochastic term does not trivially vanish, as before. In this respect, one starts far away from the original (elementary) line of reasoning leading to a basic linear Langevin equation. The separation between average behaviour and fluctuations is no longer trivial. Correlatively, the interpretation to be given to such an equation requires some caution. For example one faces the well-known problem of the appearance of spurious white noise, an effect which reflects the fact that on a time interval Δt between t and $t + \Delta t$, one does not exactly know how (when) to define the intensity of the random force, as it itself depends on the stochastic variable q.

3.1.2.3 Langevin or Fokker–Planck descriptions?

An alternative description is provided by the Fokker–Planck equation [203, 399, 475]. We again consider the Langevin equation (3.4) with Gaussian white noise, and we introduce the distribution $W(q, t)$ of the realizations $\{q(t)\}$ of the stochastic process. One can then show that $W(q, t)$ follows the Fokker–Planck partial differential equation

$$\frac{\partial W}{\partial t} = \frac{\gamma}{M_q} \frac{\partial}{\partial q}(qW) + \gamma T \frac{\partial^2 W}{\partial q^2} \tag{3.9}$$

in which the friction γ, temperature T of the 'thermal bath' and mass M_q of the q variable enter. The Fokker–Planck equation (3.9) is a diffusion equation in the space of variations of the relevant variable q. The product $D = \gamma T$ in front of the second-order term $(\partial^2/\partial q^2)$ is called the diffusion coefficient, while the coefficient of the first-order term $(\partial/\partial q)$ is known as the drift term (here γ/M_q).

The Fokker–Planck equation, which describes the temporal evolution of $W(q,t)$, is deterministic, as it neither contains any stochastic term nor is it applicable to a stochastic variable. Still, W is an object of probabilistic nature, representing the distribution of realizations of the stochastic variable q. The Fokker–Planck equation thus allows us to evaluate average values as

$$\langle A \rangle_{\mathrm{FP}} = \int A(q,t) W(q,t) \, \mathrm{d}q \qquad (3.10)$$

while in the Langevin approach, the same average value would be evaluated as

$$\langle A \rangle_{\mathrm{L}} = \frac{1}{N_{\mathrm{ev}}} \sum_{i=1}^{N_{\mathrm{ev}}} A(q_i, t) \qquad (3.11)$$

where N_{ev} is the number of samplings q_i of the stochastic process (in other words $q_i = q_i(\omega)$ is the ith realization of the stochastic variable ω characterizing the stochastic force F).

For Gaussian white noise, the Langevin and Fokker–Planck equations are equivalent. If the noise is coloured, when there are memory effects, one can extend the Fokker–Planck equation by replacing the right-hand side of equation (3.9) by a time integral as in the Langevin case (equation (3.6)). If the noise is non-Gaussian, one can also generalize the Fokker–Planck equation in order to explicitly account for higher moments of the stochastic force. This yields an equation with high-order partial derivatives.

Although the Fokker–Planck equation is deterministic a practical difficulty is linked to the fact that it is a partial differential equation, which may become very complex as soon as one does not consider the simplest Gaussian Markov case and as soon as several variables are to be taken into account. These limitations are significantly less important in the Langevin case. Furthermore, the Langevin picture is more transparent, and it offers an event-by-event description of the process, which may possibly allow us to identify 'singular' events. Still, the Langevin description may also raise technical difficulties, as outlined earlier, particularly in the non-linear case. It is also often limited by the huge statistical ensembles to be computed. In the following we shall use the Langevin picture to describe the stochastic extensions of kinetic equations in the nuclear context (section 3.2.5) and a Fokker–Planck approach to describe fission of hot nuclei (section 7.2.3).

3.1.3 On reduced theories

The idea of reducing a complex dynamical process to a (small) set of relevant variables [27–29, 133, 505] is frequently used in nuclear physics, for example, at the level of one-body reductions, but also in more 'reduced' cases (involving, for example, collective variables). The ensuing basic problem of loss of information by reduction of the dynamics is, nevertheless, not specific to nuclear physics. It is, in fact, a central question of statistical physics to seek the relevant variables which will allow a reduction of the complete description to the one provided by the sole 'relevant' variables. The term 'relevant' deserves some comments here. In general, the number of variables characterizing a system is huge, and it is thus often impossible to describe this system in terms of these basic variables from a theoretical as well as from an experimental viewpoint. Even worse, a 'complete' description might even hide the dominant features of the dynamics inside a useless amount of detail. To reduce the description thus appears to be both a formal and practical necessity. However, to reduce means to select, and to select *the* variables which one estimates to be the ones most suited to describe the process under consideration. The choice of these variables does not follow any absolute rule. Conserved quantities, such as mass, energy, etc, are usually part of the chosen relevant variables but the choice of complementary variables is usually wide open. It is directly dictated by the physical situation under consideration, and by the degree of coarseness aimed at in the description.

In the nuclear applications we consider here, we shall mainly have to deal with two levels of reduction of the nuclear many-body problem: one-body reduction and collective variables. Several theoretical arguments do support reductions of the nuclear many-body problem to one-body approaches, as discussed at length in many places (sections 2.2.3 and 3.2, in particular). The simplicity and transparence of such approaches presumably also constitutes an important argument. We shall see later that the one-body reductions also provide a basis for extending of simple mean-field theories (section 3.2.1.1). More reduced descriptions can also be constructed. In the case of collision dynamics, the hydrodynamical models provide a typical example of a description intermediate between microscopic and macroscopic levels. The relevant degrees of freedom are taken here as being purely local quantities, such as the local density and local current (section 3.1.4). At an even more macroscopic (hence reduced) level of description, the liquid drop model (section 2.1.2), in which the nucleus state is specified only by its mass, charge, energy, provides another example of a successful reduced description, at least at a static level. In spite of the gross character of such a description some properties, for example, of rotating nuclei, can, nevertheless, be attained. Fission, in the same spirit, can be attacked by a proper choice of one or two 'collective' (relevant) variables, such as an elongation parameter or multipole moments. Together with a phenomenological potential barrier and an adequate friction parameter, the dynamics of fission can then be discussed in some detail [1, 195] (see section 7.2.3.1).

Reduced descriptions can hence provide excellent ansatze of a dynamics with a large number of degrees of freedom. Still, the consequences of such a reduction may, in some cases, turn out to be 'dramatic'. Fluctuations do provide a measure of this possible degradation of the description (see also section 2.3.2.3). To reduce the dynamics, or to project it onto a small number of degrees of freedom, in fact means to write down a set of closed equations for these relevant variables, without reference to the other degrees of freedom. The reduction of the

BBGKY hierarchy (section 3.2.2) to a kinetic equation provides a typical example of such a situation for the one-body density matrix ρ. Indeed, the kinetic equation contains only ρ, and no other reduced density matrix. As a consequence, the reduced description is not only incomplete but also approximate, in the sense that it contains only part of the whole dynamics, through its restriction to the relevant variables. How to cure the possible defects of these approximations is indeed a central question in nuclear dynamics and it will be discussed later on some examples (sections 3.2.5, 3.2.6 and 3.3).

3.1.4   Hydrodynamical models at very high beam energy

Nuclear fluid dynamics (NFD) models, although basically dedicated to a beam energy range beyond the scope of the present discussions, have, nevertheless, played such an influential role in the description of heavy-ion collisions that they deserve a special mention. By nature, they furthermore lie in between the macroscopic models we have just discussed and the truly microscopic approaches which will constitute the core of the forthcoming discussions. As such they thus require a specific account, although the one given here is highly incomplete. As we do not aim at a detailed review of the topic we shall refer the reader to a few general references such as [361, 444], where more extensive discussions of the topic, in particular, in relation to the key question of access to the properties of the nuclear matter equation of state can be found.

Examining the required criteria for the validity of NFD actually fixes immediately the energy regime for which it is valid. In an NFD picture the description of the dynamics is reduced to a few local quantities such as, in particular, (and primarily) the local density and current. For such an approximation to be valid local equilibrium is required, which is only marginally true in realistic cases of nuclear collisions. Still, in the case of sufficiently large numbers of partipant nucleons and at sufficiently large energies, the nucleon mean free path is sufficiently small (figure 2.3) to allow for such an assumption of ('instantaneous') local equilibrium to be valid. The hydrodynamical approach is then essentially applicable. But we have to keep in mind the restrictions attached to the model, in terms of energy, mass (heavy nuclei) and kinematics (central collisions to provide large overlap).

The derivation of NFD equations may be achieved from various viewpoints, either directly from the quantal world, following the pioneering path of Madelung, [300] or from standard classical (or semi-classical) kinetic theory [249]. Whatever way (and overlooking interpretational differences in terms of content) the basic equations of NFD can be written as (in the non-relativistic limit and with an implicit summation over repeated indices)

$$\frac{\partial \varrho}{\partial t} + \frac{\partial}{\partial x_i}(\varrho v_i) = 0 \tag{3.12}$$

$$\frac{\partial}{\partial t}(\varrho v_i) + \frac{\partial}{\partial x_j}(\varrho v_i v_j) = \frac{\partial}{\partial x_j}P_{ij} - \varrho\frac{\partial\Psi}{\partial x_i} \qquad (3.13)$$

$$\frac{\partial\epsilon}{\partial t} + \frac{\partial}{\partial x_i}(\epsilon v_i) = \frac{\partial}{\partial x_j}(v_i P_{ij}) - \frac{\partial q_i}{\partial x_i} - \varrho v_i\frac{\partial\Psi}{\partial x_i} \qquad (3.14)$$

where ϱ, ϱv_i and $\epsilon = \varrho(mv^2/2 + W)$ are, respectively, the local densities of baryon number, momentum and energy. The local velocity components are denoted by v_i and $q_i = -\kappa\partial T/\partial x_i$ are the components of the vector of heat current according to Fourier's law (κ here is the coefficient of thermal conductivity). The Coulomb (and possibly finite-range Yukawa) potentials are gathered in the field Ψ. Finally, the components of the pressure tensor are expressed as

$$P_{ij} = -p\delta_{ij} + \eta\left(\frac{\partial v_i}{\partial x_j} + \frac{\partial v_j}{\partial x_i} - \frac{2}{3}\delta_{ij}\frac{\partial v_k}{\partial x_k}\right) + \xi\delta_{ij}\frac{\partial v_k}{\partial x_k} \qquad (3.15)$$

where η and ξ are, respectively, the coefficients of bulk and shear viscosity and p is the local pressure. Note that when neglecting heat conduction ($\kappa = 0$) one recovers the well-known Navier–Stokes equations, which for a non-viscous fluid ($\eta = \xi = 0$) reduces to the Euler equations.

The first NFD calculations were performed at the end of the 1970s, at the simplest level of Euler equations [343,443]. They allowed the exploration of some aspects of nuclear collisions at energies of a few hundreds of MeV/A beam energy. They provided several results in relation, in particular, to the important question of sidewards flow (section 6.2). In the following developments, both viscosity and thermal conductivity were added into the picture. Still, even at that higher level of sophistication, NFD models rely on the assumption of local equilibrium, which is hardly justified, particularly during the first phases of the collisions (basically during the highly anisotropic and dissipative overlap phase). This led to the so-called three-fluid pictures [136] in which the system was split into three NFD pieces corresponding, respectively, to target, projectile and overlap zones (see also the participant–spectator models, section 5.2), thus introducing a necessary amount of anisotropy into the picture. Conversely, NFD pictures cannot be used up to asymptotic times. During the late stages of the collisions the expanding system reaches a density below which it starts to break up into free particles. The treatment and the implementation of the freeze-out stage (see section 8.3.2) also turns out to be an important issue in this context.

The major defect of NFD approaches, in particular, in the direction of low beam energies, lies in the underlying assumption of instantaneous equilibrium, associated with an assumed short mean free path. For this fundamental reason, NFD models are not well suited to the energy domain we aim to discuss here. Still, one should not overlook their importance, both in terms of applications at higher energies and in terms of their influence on the theoretical developments at lower beam energies. It should finally be noted that NFD models have also close relations with very simplified models based on thermodynamical concepts.

Fireball [216, 489] and firestreak [149, 335] models or the row-on-rows [251] models are typical of these approaches. Indeed, in these models, basically, the NFD assumption of local equilibrium is further simplified to a global equilibrium.

3.2 Microscopic one-body descriptions of collision dynamics

As discussed in section 3.1, heavy-ion collisions in the nucleonic regime require fully microscopic descriptions as no simple global variable can account for the whole dynamics, at least in the entrance channel. We are thus bound to explore microscopic approaches to describe such collisional processes, at an even more microscopic level than the NFD models used at high beam energies (section 3.1.4). The kinetic equations developed for describing heavy-ion collisions in the nucleonic regime borrow some concepts and techniques from the two extreme beam energy regimes (Coulomb barrier, $E_{\text{lab}} \sim 10$ MeV/u and high energies, $E_{\text{lab}} \gtrsim 500$ MeV/u) between which they take place. Hence we first briefly recall the dynamical conditions associated with these two energy regimes.

At low energy the mean free path λ of nucleons inside a nucleus is 'large', compared to the size of the system itself, which justifies approaches of the mean-field type (section 2.1.3.1). The de Broglie wavelength λ_{B} of the nucleons is also large and the description of the system hence resorts to quantum mechanics. The theory which is adapted to this energy range is the time-dependent Hartree–Fock (TDHF) approach [72, 341]. On the other hand, at high energy, the mean free path and the de Broglie wavelength both become small as compared to the size of the system, even compared to the distance between the nucleons themselves. The dynamics is then dominated by elementary collision processes between nucleons rather than by the mean field and one can use a *classical* description of the motion of the nucleons. We have seen that at very high beam energies (in the relativistic domain) hydrodynamical models constitute a relevant and efficient approach (section 3.1.4). But the most microscopic model which is finally adapted to this high-energy range and to 'lower' energies ($E_{\text{lab}} \gtrsim 500$ MeV/u) is the intra nuclear cascade INC [137, 422, 501].

Between these two extreme regimes the dynamics is a mixture—mean field and elementary processes act together and one progressively passes, when increasing the beam energy, from a quantal to a partly classical, or rather semi-classical, regime. The term '*semi*-classical' reflects the fact that a minimal quantal property has to be preserved: fermionic statistics (section 2.3.1.4). In this intermediate energy domain where mean-field effects coexist with elementary processes and where the de Broglie wavelength is indeed small but not negligible, one has to devise a new approach. The models developed so far for describing this energy range are Boltzmann-like kinetic equations complemented by a mean field, and the statistics is explicitly taken into account in the collision term [344, 473]. Let us now see how these kinetic equations can be obtained—between mean-field and collisional models.

3.2.1   Two cornerstones: TDHF and INC

3.2.1.1   Nuclear dynamics at low energy: TDHF

In a natural extension of ground-state properties (sections 2.1 and 2.2) low-energy nuclear dynamics is adequately described at the level of mean-field theories. The time-dependent extension (TDHF, [159]) of the Hartree–Fock approach (section 2.2.3) thus represents the basic dynamical description here. The TDHF approximation is again based on a variational principle acting in the *ansatz* space of Slater determinants. Once the A-body wavefunction has been supposed to be of the form (2.39), the TDHF equation can thus be obtained from a variational principle on the action I [426]

$$\delta I = \delta \int dt \, \langle \Psi | \left(\boldsymbol{H} - i \frac{\partial}{\partial t} \right) | \Psi \rangle = 0. \tag{3.16}$$

Note that, without the Slater form hypothesis (2.39), this variational principle would provide exactly the time-dependent Schrödinger equation of the initial A-body problem.

The variational principle (3.16) leads to a set of A-coupled, one-body, Schrödinger equations for the single nucleon kets $|\phi_i\rangle$:

$$i \frac{\partial |\phi_i\rangle}{\partial t} = \boldsymbol{h}[\{|\phi_i\rangle, i = 1, \ldots, A\}] |\phi_i\rangle \tag{3.17}$$

which can be recast into the compact form

$$i \frac{\partial \boldsymbol{\rho}}{\partial t} = [\boldsymbol{h}[\rho], \rho] \tag{3.18}$$

where ρ is the *one*-body density matrix (2.41). In the expressions (3.17) and (3.18) $\boldsymbol{h}$ is the mean-field Hamiltonian as defined in the static Hartree–Fock approximation (2.44). It is a functional of the one-body kets $|\phi_i\rangle$ *via* the density matrix ρ. It is interesting to note that the TDHF approximation replaces the initial *linear* A-body Schrödinger equation by an ensemble of A-coupled, *non-linear*, one-body equations.

It should be noted that the applications of the TDHF approximation were highly simplified by the use of effective forces, which are themselves functionals of the local density, such as Skyrme forces (section 2.2.1.3). In the latter case the potential part of the mean-field Hamiltonian $\boldsymbol{h}$ reduces to a functional of $\varrho(\boldsymbol{r}) = \rho(\boldsymbol{r}, \boldsymbol{r})$. Many results have thus been obtained in this framework, in the case of low-energy heavy-ion reactions since the second half of the 1970s [72, 341]. However, studies of reactions at beam energies of the order of the Fermi energy have shown that the mean field alone does not allow us to understand the fusion phenomena between projectiles and targets which are observed experimentally in central reactions. In such cases, TDHF calculations are unable to account for the stopping of the projectile into the target: 'TDHF' nuclei inadequately cross each other in these central collisions. The missing aspect has long been identified and lies in the absence of a dissipative mechanism in TDHF. The question of extending TDHF to such a dissipative energy domain and thus to complement the mean field by proper 'two-body collisions' effects constituted a central and highly debated question in the early 1980s [209]. Formal and numerical difficulties finally hindered a proper implementation of such 'extended mean-field calculations' [209] apart from some rare test cases [266, 267]. As we shall see later, semi-classical methods, in turn, took the lead in this energy domain.

3.2.1.2 *High-energy nuclear reactions: INC*

In the intra nuclear cascade (INC) model [137, 139, 501] the ket $|\Psi(1, \ldots, A)\rangle$ is replaced by an ensemble of stochastic samplings of A *classical* particles. And the dynamics of the global system is reduced to elementary collisions between these particles *via* a collision cross-section σ. It should be noted that INC was originally formulated in a relativistic framework, unavoidable at the high beam energies, for which it was suited. For the sake of simplicity, and in order to have a more direct link to the non-relativistic Boltzmann equation we are interested in, we shall restrict ourselves here to an 'academic' non-relativistic formulation of INC. We shall also, in the same spirit, omit inelastic channels in nucleon–nucleon collisions. The dynamics of the system then reduces to A coupled classical equations for each sampling of the positions and momenta $\{r_i, p_i\}$:

$$\frac{\mathrm{d}r_i}{\mathrm{d}t} = \frac{p_i}{m} \tag{3.19}$$

$$\frac{\mathrm{d}p_i}{\mathrm{d}t} = \frac{\mathrm{d}p_i}{\mathrm{d}t}\bigg|_{\mathrm{coll}} \tag{3.20}$$

where the index 'coll' refers to a collision algorithm between the particles. At instant t, two particles '1' and '2' of velocities v_1 and v_2 will collide with each other during the forthcoming time step Δt, if two criteria are fulfiled: a kinematical condition and a dynamical one. The kinematical criterion requires that the two particles pass through each other during Δt ('1' travels $|v_1|\Delta t$ along v_1, and '2' $|v_2|\Delta t$ along v_2), namely go through the points corresponding to their distance of closest approach d_{min}. If this first condition is fulfiled, one compares d_{min} to a distance $d_{\mathrm{max}} = \sqrt{\sigma/\pi}$ representing the 'range' of the nuclear interaction, *via* the elementary nucleon–nucleon cross-section σ. If these two conditions are fulfiled, then the two particles do collide during Δt following t. And one samples the new momenta of the two particles 'after' the collision, while preserving total momentum and kinetic energy.

The (relativistic) INC algorithm has been widely used, and with success, for describing nuclear reactions at high beam energy [139]. From a formal point of view the non-relativistic model turns out to provide simulations of the Boltzmann equation for the *classical* phase space distribution $f(\boldsymbol{r}, \boldsymbol{p}, t)$:

$$\frac{\partial f(\boldsymbol{r}, \boldsymbol{p}, t)}{\partial t} + \frac{\boldsymbol{p}}{m} \cdot \frac{\partial}{\partial \boldsymbol{r}} f(\boldsymbol{r}, \boldsymbol{p}) = I_{\text{coll}} \tag{3.21}$$

with

$$I_{\text{coll}}[f] = \int d^3 p_2 \, d^3 p_3 \, d^3 p_4 \, W_{\text{INC}}(12, 34)$$
$$\times [f(\boldsymbol{r}, \boldsymbol{p}_3, t) f(\boldsymbol{r}, \boldsymbol{p}_4, t) - f(\boldsymbol{r}, \boldsymbol{p}_2, t) f(\boldsymbol{r}, \boldsymbol{p}, t)] \tag{3.22}$$

where $W_{\text{INC}}(12, 34)$ is the collision rate

$$W_{\text{INC}}(12, 34) = \frac{1}{m^2} \frac{d\sigma}{d\Omega} \delta(\boldsymbol{p} + \boldsymbol{p}_2 - \boldsymbol{p}_3 - \boldsymbol{p}_4) \delta\left(\frac{p^2}{2m} + \frac{p_2{}^2}{2m} - \frac{p_3{}^2}{2m} - \frac{p_4{}^2}{2m}\right).$$
$$\tag{3.23}$$

Note that the Boltzmann equation (3.21) is indeed the standard equation used for classical dilute gases [249]: there is no mean field and Pauli blocking is not taken into account in the collision term. In practice, however, some INC simulations include a pseudo Pauli blocking, by forbidding collisions between particles with too small relative momentum (typically two-thirds of the Fermi momentum).

By construction, the two-body collisions included in a cascade model provide a dissipative mechanism which was missing in mean-field approaches. Still, in the nucleonic regime, even if INC provides a reasonable stopping of the projectile into the target it is not sufficient to account properly for typical fusion (or incomplete fusion) reactions as observed in the Fermi energy domain. Indeed, the fusion residue is basically unstable as no attractive interaction between nucleons allows the system to be bound in a satisfactory way. Hence one should include the attractive (at low energy) mean field in INC in order to account for such bound exit channels objects. It turns out that it was finally this option (and not the one that consisted of extending TDHF to a dissipative energy domain, section 3.2.1.1) which was favoured from the mid 1980s on, and which has led to the overwhelming successes of the so-called BUU model. The BUU is a semi-classical kinetic equation and, rather than introducing it heuristically, we shall spend some time in section 3.2.2 justifying it and showing how it is related to the vast domain of kinetic theory.

3.2.2   The BBGKY hierarchy for kinetic equations

In this section we introduce an approach to kinetic equations in the framework of the BBGKY hierarchy of reduced density matrices. First we present the quantal formalism before passing, at the last step of the derivation, to the semi-classical approximation.

3.2.2.1   *BBGKY and its truncation*

The idea of the BBGKY (Born, Bogoliubov, Green, Kirkwood, Yvon) hierarchy [25] consists of replacing the original Liouville–von Neumann A-body dynamical equation (2.54) by an ensemble (hierarchy) of $A - 1, A - 2, \ldots, 2$, one-body coupled equations, to figure out schemes of successive approximations to the original global A-body problem. In addition to the A-body density matrix $D = \rho_{1\ldots A}$ (section 2.3.1.1) we introduce, more generally, k-body density matrices ($k < A$) defined as

$$\rho_{1\ldots k} = \frac{A!}{(A - k)!} \, \mathrm{Tr}_{k+1,\ldots,A} \, \rho_{1\ldots A} = \frac{1}{A - k} \, \mathrm{Tr}_{k+1} \, \rho_{1\ldots k+1} \qquad (3.24)$$

where $\mathrm{Tr}_{k+1,\ldots,A}$ denotes the trace on the degrees of freedom associated with $k + 1, \ldots, A$ bodies. In particular, the one-body density matrix may be written as

$$\rho_1 = \frac{A!}{(A - 1)!} \, \mathrm{Tr}_{2,\ldots,A} \, \rho_{1\ldots A} = \frac{1}{A - 1} \, \mathrm{Tr}_2 \, \rho_{12}. \qquad (3.25)$$

These definitions beg some comments. First, the notation $\rho_{1\ldots k}$ (equation (3.24)), although compact, is dangerous. It would probably be more accurate to write $\rho^{(k)}(1 \ldots k)$, in order to identify the order k of the reduced matrix *and* the variables $1 \ldots k$ properly. For the sake of compactness, we admit that the number of indices fixes the order, so that we denote by $\rho_2 = \rho^{(1)}(2)$ the density matrix of order 1 for particle '2'. In addition, it is obvious that the notation 'i' is a compact one for characterizing the quantum state of the particle—without specifying the quantum numbers or the variables under consideration.

As already discussed, one-body effects play a central role in nuclear physics (section 2.1). But it is crucial to note here that the one-body density matrix ρ (equation (2.41)) of the TDHF theory, i.e. the one built from the one-body kets constituting the Slater determinant (2.39), in general differs from the density matrix ρ_1 which is built as a trace on $\rho_{1\ldots A}$. While ρ (equation (2.41)) follows the TDHF equation (3.18), the density matrix ρ_1 (equation (3.25)) follows a different evolution equation (3.28), which is the reduction of the exact equation (2.54) to the one-body level. In this respect, the dynamics of ρ_1 (without any approximation) is *reduced but exact*, while the TDHF one is *approximate* [27].

Once we have defined the reduced density matrices, it is easy to obtain equations for their time evolution while applying partial traces $\mathrm{Tr}_{k+1\ldots A}$ to the Liouville–von Neumann equation (2.54) followed by the A-body density matrix $\rho_{1\ldots A}$. One then obtains a hierarchy of equations of the form

$$\mathrm{i} \frac{\partial \rho_{1\ldots k}}{\partial t} = \sum_{i=1}^{k} \left[K_i + \sum_{j<i}^{k} V_{ij}, \rho_{1\ldots k} \right] + \sum_{i=1}^{k} \mathrm{Tr}_{k+1}([V_{ik+1}, \rho_{1\ldots k+1}]). \qquad (3.26)$$

The kth-order equation for the reduced density matrix $\rho_{1\ldots k}$ is coupled to the $(k+1)$st-order equation *via* $\rho_{1\ldots k+1}$. This coupling between two successive orders results from the two-body nature of the interaction V_{ij} and would become more complex if three, four, $\ldots$-body interactions were taken into account in the total Hamiltonian H (equation (2.40)). The *whole* hierarchy of coupled equations is equivalent to the original Liouville–von Neumann equation (2.54). Its interest, however, lies in the possibility of truncation it offers, namely in the fact that one may keep only part of the hierarchy, provided one makes an approximation of the form

$$\rho_{1...k_0+1} \simeq \rho_{1...k_0+1}[\rho_1, \rho_{12}, \ldots, \rho_{1...k_0}] \tag{3.27}$$

for a well chosen k_0-order. Let us now see more explicitly how this truncation mechanism works and what kind of (kinetic) equations it allows us to obtain.

For illustration, we first consider the simplest truncation of the hierarchy, which focuses on the first-order equation

$$i\frac{\partial \rho_1}{\partial t} = [K_1, \rho_1] + \text{Tr}_2([V_{12}, \rho_{12}]). \tag{3.28}$$

In order to 'close' this equation (and thus truncate the hierarchy) one has to make a hypothesis about the form of the two-body density matrix ρ_{12} as a function of the one-body density matrices ρ_1 and ρ_2. The simplest hypothesis consists in assuming that 1 and 2 are not correlated, namely that $\rho_{12} = \rho_1 \rho_2$ (the Hartree approximation), or, if one takes into account the Pauli principle, that $\rho_{12} = \mathcal{A}_{12}(\rho_1 \rho_2)$ (the Hartree–Fock approximation)[1]. For transparency, we shall restrict ourselves to the Hartree approximation in the formal discussions, and we shall eventually restore the antisymmetrization, at the end of the calculation, when necessary. Truncating the hierarchy with the hypothesis $\rho_{12} = \rho_1 \rho_2$ immediately leads to an equation for the one-body density matrix alone, which reads:

$$i\frac{\partial \rho_1}{\partial t} = [K_1, \rho_1] + \text{Tr}_2[V_{12}, \rho_2 \rho_1] = [K_1 + \text{Tr}_2(V_{12}\rho_2), \rho_1] = [h_1, \rho_1]. \tag{3.29}$$

This is the TDHF equation (equation (3.18)), up to antisymmetrization.

3.2.2.2 A quantum Boltzmann equation

How to go beyond the time-dependent mean field? In the framework of the BBGKY hierarchy, it is natural to look for a truncation scheme at second-order, namely an approximation of the form $\rho_{123} \simeq \rho_{123}[\rho_{12}, \rho_3]$. We shall then solve the second equation of the hierarchy to obtain ρ_{12} as a function of ρ_1 and ρ_2, which will allow us to rewrite the first equation of the hierarchy in terms of the one-body density matrix alone, thus providing what one usually calls a kinetic equation. We thus have to consider the first two equations of the BBGKY hierarchy, equation (3.28) and

$$i\frac{\partial \rho_{12}}{\partial t} = [K_1 + K_2 + V_{12}, \rho_{12}] + \text{Tr}_3([V_{13} + V_{23}, \rho_{123}]). \tag{3.30}$$

Note that keeping a dynamical equation for the two-body density matrix ρ_{12} (equation (3.30)) precisely allows us to treat in a consistent way, at least

---

[1] In the $|r\rangle$ representation one explicitly obtains:
$\langle r_1 r_2 | \mathcal{A}_{12}(\rho_1 \rho_2) | r_1' r_2' \rangle = \langle r_1 | \rho_1 | r_1' \rangle \langle r_2 | \rho_2 | r_2' \rangle - \langle r_1 | \rho_1 | r_2' \rangle \langle r_2 | \rho_2 | r_1' \rangle$.

approximately, correlations *via* ρ_{12}. This possibility did not truly exist when imposing an explicit form of ρ_{12} in the mean-field truncation scheme.

We first discuss the 'simple' case of a dilute system, while overlooking the fermionic statistics, which will lead us to a quantal Boltzmann-like equation [432]. Next we shall adapt the derivation to the case of a dense system of fermions, such as the atomic nucleus [84]. In order to 'derive' the quantum Boltzmann equation we make two hypotheses: (i) we admit that the system is dilute (which amounts to taking $\rho_{123} \simeq 0$) and that the two-body potential V_{12} has a short range; and (ii) we make the hypothesis of *molecular chaos*, which will allow us to break time correlations between particles. Before proceeding, it is, nevertheless, necessary to note the lack of consistency in making the approximation $\rho_{123} = 0$. Rigorously speaking, such a hypothesis leads to $\rho_{12} = \rho_1 = 0$ by successive traces (equation (3.24))—which is not welcome; but the hypothesis $\rho_{123} \simeq 0$ is a standard assumption in order to obtain the Boltzmann equation phenomenologically. A 'clean' truncation of the hierarchy should, in principle, allow us to preserve the relation (3.24) [413], as it is, for example, the case of the scheme leading to TDHF. For the sake of simplicity we shall, nevertheless, keep, as a first step, the 'brutal' truncation $\rho_{123} = 0$.

Let us consider a pair of particles 1, 2 which interact at instant t. As the system is dilute we can neglect the effect of the other particles '3' on this elementary process. However, the interaction itself does correspond to a strong correlation between 1 and 2, which means that $\rho_{12} \neq \rho_1 \rho_2$, at least during a certain time interval. In fact, according to (i) this strong correlation remains localized in time, because of the short range of V_{12}. There is hence an instant t_0 before t, at which the correlation between 1 and 2 is weak so that $\rho_{12}(t_0) \simeq \rho_1(t_0)\rho_2(t_0)$: this is precisely the molecular chaos assumption (ii). This means that the elementary collisions between particles are independent of one another. Before a given collision, memory of the correlations ($\rho_{12} \neq \rho_1 \rho_2$), due to preceding collisions, is thus cancelled, which introduces a distinction between *before* and *after* the given collision: it is the origin of the irreversibility of the Boltzmann equation. Note that although the decorrelation condition $\rho_{12}(t_0) = \rho_1(t_0)\rho_2(t_0)$ bears some similarity with the TDHF truncation scheme (section 3.2.1), its meaning is very different. In TDHF $\rho_{12} = \rho_1 \rho_2$ *at each instant* and, as a consequence, TDHF is microscopically reversible; here, it is *only at instant t_0* that the factorization is possible and the resulting equation becomes irreversible.

Once decorrelation has been assumed at $t_0 \ll t$, one can simply treat the interaction between 1 and 2 in terms of standard scattering theory using a T matrix approach [420]. This allows us to write $\rho_{12}(t)$ in terms of $\rho_1(t)$ and $\rho_2(t)$ as $\rho_{12}(t) = \Omega_{12}\rho_1(t)\rho_2(t)\Omega_{12}^{+}$ where $\Omega_{12} = V_{12}^{-1} \cdot T_{12}$. Thus one finally obtains an equation for the one-body density matrix:

$$i\frac{\partial \rho_1}{\partial t} = [K_1, \rho_1] + \text{Tr}_2([V_{12}, \Omega_{12}\rho_1\rho_2\Omega_{12}^{+}]) = [K_1, \rho_1] + I_{\text{coll}} \quad (3.31)$$

which is a non-developed form of a quantum Boltzmann equation. The first commutator corresponds to the free propagation (with kinetic operator K only) and the second one represents the collision term, which is quadratic in the one-body density matrix. The reduction to the usual form of the Boltzmann equation (3.21) would require some more calculations, which are not crucial for our purpose and can thus be skipped. We would rather like to see to what extent this derivation may be adapted to the nuclear case.

3.2.2.3 *The nuclear case*

In the nuclear case we have to modify the original hypotheses (i) and (ii). The hypothesis of a dilute system is hardly justified and the elementary interaction V_{12} cannot really be considered to be short range, as the relative distance between two nucleons is, in fact, comparable to the range of the nuclear interaction. The hypothesis (i) of the former derivation has hence to be revisited. Furthermore, the Pauli principle has to be properly taken into account! Still, once these difficulties have been identified, it turns out that the strategy of the former derivation may, to a large extent, be adapted, provided a 'good' truncation scheme is used [84]. Too 'simplistic' truncation schemes lead to serious formal difficulties [413].

The key point here is to realize that the interaction between particles 1 and 2 can be treated in terms of Brückner theory (section 2.2.1.2). This amounts to replacing T_{12} by G_{12} in the elementary scattering process and to take care of Pauli blocking in accessing exit channel states. In order to 'recover' the hypothesis of short range of the elementary interaction, one furthermore splits the action of G_{12} into two contributions: the long range part of the interaction shows up as a mean field U, deduced by an average from the real part of G_{12}; and the short range part of the interaction (associated with the imaginary part of G_{12}) leads to the appearance of a Boltzmann-like collision term. One finally obtains a quantal Boltzmann equation, adapted to the nuclear context, which reads [84]:

$$i\frac{\partial \rho_1}{\partial t} - [K_1 + U_1, \rho_1] = I_{\text{coll}} = 2\pi i \, \text{Tr}_2\{[G_{12}\mathcal{A}_{12}(\rho_1\rho_2)G_{12}^+(1-\rho_1)(1-\rho_2)$$
$$- G_{12}^+\mathcal{A}_{12}((1-\rho_1)(1-\rho_2))G_{12}\rho_1\rho_2]\delta(E_{12} - H_{12}^0)\}. \tag{3.32}$$

Comparing this with equation (3.31), one can notice the appearance of (i) a mean field $U \sim \text{Tr}_2 \, \Re(G_{12}\mathcal{A}_{12}\rho_2)$ and (ii) a collision term modified by Pauli blocking factors $(1-\rho)$. The mean-field part is computed here from the real ($\Re$) part of the averaged (ρ_2) effective interaction G_{12}, up to antisymmetrization ($\mathcal{A}_{12}$). The imaginary part of G_{12} enters the collision integral (see also section 2.1.3.2). Note that the bare interaction is now replaced by an effective interaction (entering U) which accounts for medium effects ($T_{12} \rightarrow G_{12}$). As in the case of the elementary Boltzmann equation, it would, nevertheless, be necessary to continue the derivation, in order to obtain a more 'usual' form of this kinetic equation. We shall again skip that part of the derivation and focus on the semi-classical counterpart of equation (3.32).

3.2.3 Semi-classical kinetic equations

3.2.3.1 Semi-classics

The intermediate energy domain we are considering in this book is admittedly a dynamical regime in which semi-classical aproximations are likely to be useful [91, 92, 97]. Fully classical pictures, such as the ones provided, for example, by INC models, are obviously unsuitable as basic quantum effects such as the Pauli principle are still fully active in the nucleonic regime. But the de Broglie wavelength is already small enough so that a fully quantal picture is not necessary (figure 2.3). Semi-classical approximations, in the sense that they retain basic quantum properties such as the Pauli principle, have thus been extensively used over the years in the nucleonic energy domain. They have been particularly used for understanding dynamical features in terms of kinetic equations, which represent the semi-classical counterparts of the equations just derived (section 3.2.2).

  The BBGKY hierachy was originally derived for classical systems. We shall, nevertheless, not repeat the formal derivation of section 3.2.2. It is more satisfying, in many respects, particularly in terms of a proper account of the Pauli principle, to perform the semi-classical approximation onto the quantum kinetic equation, once derived, rather than to try to derive it from scratch, within inserting *ad hoc* Pauli principle effects. Still, we recall a few definitions for completeness.

  The classical BBGKY hierarchy [234] is built starting from the A-body phase space distribution $f(\boldsymbol{r}_1, \boldsymbol{p}_1, \ldots, \boldsymbol{r}_A, \boldsymbol{p}_A) = f_{1\ldots A}$. The latter is a function of the positions and momenta of the A particles, and the corresponding reduced distributions are then simply defined as

$$f_{1\ldots k}(\boldsymbol{r}_1, \boldsymbol{p}_1, \ldots, \boldsymbol{r}_k, \boldsymbol{p}_k) = \frac{A!}{(A-k)!} \int \mathrm{d}\boldsymbol{r}_{k+1}\, \mathrm{d}\boldsymbol{p}_{k+1} \ldots \mathrm{d}\boldsymbol{r}_A\, \mathrm{d}\boldsymbol{p}_A\, f_{1\ldots A}$$

(3.33)

where traces reduce here to simple integrals as particles are represented only by their position and momentum. Integrating over all the particles, but one, leads, as in the quantum case, to the one-body density distribution in phase space, better known as the phase space distribution $f_1(\boldsymbol{r}, \boldsymbol{p}, t) = f(\boldsymbol{r}, \boldsymbol{p}, t)$. The interpretation of $f(\boldsymbol{r}, \boldsymbol{p}, t)$ is simple in the purely classical case: $f(\boldsymbol{r}, \boldsymbol{p}, t)\, \mathrm{d}^3 r\, \mathrm{d}^3 p > 0$ represents the number of particles in a phase space volume $\mathrm{d}^3 r\, \mathrm{d}^3 p$ around point $(\boldsymbol{r}, \boldsymbol{p})$. A proper reduction of the *classical* BBGKY at lowest order, making the simplest assumption $f_{12} = f_1 f_2$ (with the same notational convention as in section 3.2.2), again leads to a mean-field equation, the so-called Vlasov equation (see later), which, if derived that way, is a purely classical object, particularly without reference to any quantum statistics.

3.2.3.2 The Vlasov and BUU equations

The Vlasov equation may be obtained as the (semi)-classical limit of equation (3.29) or of equation (3.18) [289, 398]. The 'simplistic' classical limit is obtained by replacing the one-body density matrix ρ_1 by a one-body phase space distribution f_1 and commutators by Poisson brackets

$$[\cdot,\cdot] \rightarrow \{\cdot,\cdot\} = \cdot_r\partial\partial_p\cdot - \cdot_p\,\partial\partial_r \qquad (3.34)$$

where the location of the index explains in which direction (to the left or to the right) the derivation acts. One thus obtains the Vlasov equation [482]:

$$\frac{\partial f_1(r_1,p_1,t)}{\partial t} = \{h_1, f_1\} = \frac{\partial h_c}{\partial r_1}\frac{\partial f_1}{\partial p_1} - \frac{p_1}{m}\frac{\partial f_1}{\partial r_1} \qquad (3.35)$$

where the classical one-body Hamiltonian reads:

$$h_1(r_1,p_1) = h_c(r_1,p_1) = \frac{p_1^{\,2}}{2m} + \int dr_2\,dp_2\,V_{12}f_1(r_2,p_2). \qquad (3.36)$$

Note that, although derived from the quantum world, nothing at the level of this Vlasov equation (3.35) seems to distinguish a classical from a semi-classical approximation. In fact this point is very subtle and deeply hidden in what the one-body distribution is assumed to represent. When coming from the quatum world, the one-body distribution usually represents the Wigner transform [398] of the one-body density matrix and, as such, may take negative values, which renders a truly classical picture invalid. This difficulty may be overcome by using other representations of the one-body distribution such as the well-known (smooth and positive) Husimi distribution [252, 289, 456]. We shall, nevertheless, not enter this interesting but complex discussion and only remember that the simple semi-classical approximation (3.34), in fact, overlooks subtle, although understood, difficulties. For practical purposes we shall content ourselves with the fact that, once 'solved' formal questions, the interest of the *semi*-classical approximation over a direct classical approach lies in the fact that one can allow the Pauli principle into the semi-classical world, as a remnant of the quantum Pauli principle contained in the one-body density matrix. As an example, a nucleus in its ground state will thus be described by a one-body density matrix $f_{\mathrm{gs}} = \theta(\epsilon_F - h_c)$, the semi-classical counterpart of the ground-state one-body distribution $\rho_{\mathrm{gs}} = \theta(\epsilon_F - h)$ (rather than by a purely classical ground-state distribution $\propto \delta(r)\delta(p)$). With this word of caution at hand, let us now consider the semi-classical kinetic equations used in the nucleonic regime.

In heavy-ion physics, only the semi-classical version of equation (3.32) is used. The nuclear Boltzmann equation then reduces (approximation of weak gradients [84]) to the familar form:

$$\frac{\partial f_1}{\partial t} + \{f_1, h_c\} = I_{\mathrm{coll}}^{\mathrm{UU}} \qquad (3.37)$$

where the collision term, similar to the phenomenological one of Uehling and Uhlenbeck [344, 473], reads

$$I_{\mathrm{coll}}^{\mathrm{UU}} = \frac{\nu}{h^9}\int dp_2 p_3\,dp_4\,W(12,34)[(1-f_1)(1-f_2)f_3 f_4 - (1-f_3)(1-f_4)f_1 f_2]$$

$$(3.38)$$

(ν is the degeneracy) as a function of the collision rate

$$W(12,34) = \frac{1}{m^2}\frac{\mathrm{d}\sigma}{\mathrm{d}\Omega}\delta(\boldsymbol{p}_1 + \boldsymbol{p}_2 - \boldsymbol{p}_3 - \boldsymbol{p}_4)\delta(\epsilon_1 + \epsilon_2 - \epsilon_3 - \epsilon_4). \qquad (3.39)$$

In these latter equations we have used the usual notation: $f_i = f(\boldsymbol{r},\boldsymbol{p}_i,t)$ and $\epsilon_i = p_i^2/2m + U_i = h_\mathrm{c}(\boldsymbol{r}_i,\boldsymbol{p}_i)$ (equation (3.36)). In addition, the differential cross-section $\mathrm{d}\sigma/\mathrm{d}\Omega$ may be evaluated from $\boldsymbol{G}_{12}$ ($\mathrm{d}\sigma/\mathrm{d}\Omega \sim |\boldsymbol{G}_{12}|^2$), which reflects the fact that there is only *one* nucleon–nucleon interaction, the effects of which are arbitrarily split between mean-field and residual interactions. This holds at the quantal (equation (3.32)) as well as at the semi-classical level (equation (3.37)). This implies that *a priori* one should not choose independently the mean field and cross-section entering the collision term. Let us finally note that equation (3.37) does indeed lead to a mean-field-type equation at low beam energy. In this case, most of the collisions are Pauli blocked and $I_\mathrm{coll}^{\mathrm{UU}} \sim 0$. Equation (3.37) then reduces to the Vlasov equation (3.35), which is nothing but the semi-classical analogue of TDHF.

Equation (3.37) together with the collision term (equation (3.38)) is the basis of all BUU, VUU, LV, BNV, ... simulations, which have been developed over the last 15 years. These calculations have been applied to many situations in heavy-ion collisions (see, for example, [45, 396]), and with some successes; some examples will be presented in the following.

3.2.3.3   *On the numerical method of resolution of kinetic equations: Vlasov, BUU*

The basic idea of the numerical methods, called test-particle methods, used to solve Boltzmann-like kinetic equations is to replace (project) the one-body distribution $f(\boldsymbol{r},\boldsymbol{p},t)$ by an ensemble of N *numerical* (hence classical) particles [45, 70, 390, 454, 488]

$$f(\boldsymbol{r},\boldsymbol{p},t) \simeq \frac{A}{N}\sum_{i=1}^{N}\mathcal{G}(\boldsymbol{r} - \boldsymbol{r}_i, \boldsymbol{p} - \boldsymbol{p}_i) \qquad (3.40)$$

where $\mathcal{G}$ is a normalized function representing the numerical particle (Dirac, B-spline, Gaussian, etc). In the case of the Vlasov equation the time evolution of $f(\boldsymbol{r},\boldsymbol{p},t)$ is then found by solving Hamilton equations of motion for the test particles:

$$\frac{\mathrm{d}\boldsymbol{r}_i}{\mathrm{d}t} = \frac{\partial h_\mathrm{c}}{\partial \boldsymbol{p}_i} * \mathcal{G} \qquad (3.41)$$

$$\frac{\mathrm{d}\boldsymbol{p}_i}{\mathrm{d}t} = -\frac{\partial h_\mathrm{c}}{\partial \boldsymbol{r}_i} * \mathcal{G} \qquad (3.42)$$

where h_c is the classical one-body Hamiltonian (equation (3.36)). In the case of BUU one has to complement the second equation (equation (3.42)) by a term of the form $\mathrm{d}p_i/\mathrm{d}t_{|coll}$ representing a collision algorithm. BUU collision algorithms are directly based on the INC algorithm (section 3.2.1.2). The Pauli principle is taken into account here in the exit channel: the collision takes place with a probability equal to the product of the phase space occupation numbers around the points towards which particles are scattered. Note that the Pauli principle also enters (indirectly) into the evaluation of the effective interaction (see section 2.2.1.2) and of the cross-section in the collision term.

Test-particle algorithms call for some comments. First, one has to note that one switches from a *physical one-body* problem for $f(\mathbf{r}, \mathbf{p}, t)$ to a *numerical N-body* problem for the test particles. And the latter classical N-body problem is treated in a molecular dynamics way [390]. Notice that N is usually much larger than A, at least in nuclear physics applications [396]. But the important point here is that these N numerical particles are classical by nature and hence obey Boltzmann, and not Fermi–Dirac, statistics [390]. One thus expects that in the course of the evolution the system relaxes towards a Boltzmann equilibrium for the N particles. This is effectively what calculations show [256, 391].

Systematic calculations of this relaxation effect have been performed [390, 391] and have allowed us to clarify the difficulties linked to the use of test particles to simulate kinetic equations describing fermions. Relaxation towards Boltzmann is the rule, even if this relaxation can be delayed by an appropriate choice of the numerical parameters and/or in the BUU case. In addition, a formal analysis of the test-particle algorithms shows that these methods do not lead to an explicit resolution of the Vlasov equation, in particular, when $\mathcal{G}$ is extended [289, 390]. One has hence to remain cautious in the interpretation of Vlasov or BUU simulations, with respect to the equations they are supposed to solve. It should finally be noted that a possible solution to this problem of fermionic stability in BUU/Vlasov simulations has recently been proposed [392]. The idea is to complement the standard test-particle dynamics by a sort of collision term, which enforces the Pauli condition. The first test-case results are extremely encouraging.

3.2.4   Kinetic equations and beyond

3.2.4.1   Some hidden difficulties

The BBGKY hierarchy provides a coherent theoretical framework, which allows us to obtain, more or less systematically, approximations of the A-body problem in terms of kinetic equations. The nuclear Boltzmann equation (section 3.2.2.3) thus takes into account in a coherent way mean-field and elementary processes. The mean field U and the elementary cross-section $\mathrm{d}\sigma/\mathrm{d}\Omega$ can both be deduced from the same Bruckner matrix $\mathbf{G}_{12}$, which describes interactions between nucleons in a nuclear medium. Hence we have at hand a tool which should allow us to describe an intermediate regime between mean-field and elementary collision dynamics. A word of caution is, nevertheless, necessary here, concerning the content of the G_{12} interaction. Strictly speaking G_{12} is a time-dependent object. Indeed, for a given elementary interaction V_{12} (which is time independent) the construction of G_{12} depends on the structure of phase space (in particular on the accessible 'empty' states). Remember the key role played by Pauli blocking and the Fermi energy in the determination of G_{12} in the simple static case (section 2.2.1.2). In the course of a heavy-ion collision the structure of the phase space, and hence of accessible states, depends on time, evolving from two occupied Fermi spheres at the beginning (separated by the beam relative momentum, figure 6.1), to a diffuse Fermi sphere asymptotically, once relative motion has been fully relaxed. A proper account of this dynamical content of G_{12} represents a huge effort, which has hardly been attacked practically [268, 421]. In the vast majority of cases this question is thus simply overlooked. However, even with such a problem set aside, the obtaining of the Boltzmann equation is not straightforward in the nuclear context.

In order to obtain this nuclear Boltzmann equation we have overlooked some difficulties. The major problem lies in the timescales. In the nucleonic regime, we have essentially to consider three microscopic times: (i) the duration of an elementary collision $\tau_{\mathrm{coll}} \sim$ 3–5 fm/c; (ii) the average time interval between two collisions $\Delta t_{\mathrm{coll}} \lesssim$ 10 fm/c; and (iii) and the characteristic evolution time of mean field $\tau_{\mathrm{mf}} \sim$ 10 fm/c. It turns out that the decoupling of these microscopic timescales is extremely questionable:

$$\tau_{\mathrm{coll}} \not\ll \Delta t_{\mathrm{coll}} \lesssim \tau_{\mathrm{mf}}. \qquad (3.43)$$

This weak decoupling has crucial formal consequences. First, it becomes difficult to 'decouple' the two-body density matrix ρ_{12} at 'some' instant t_0 before the collision. In fact this decoupling ($\rho_{12}(t_0) \sim \rho_1(t_0)\rho_2(t_0)$) has only a meaning in comparison to the duration of an elementary collision τ_{coll}, which is supposed to be infinitely short compared to any other timescale. It then becomes possible to break the correlation between particles 1 and 2 at t_0, and also to do it in terms of the global density matrix ρ_{12}, at this instant. In other words one overlooks the physics at a timescale of order τ_{coll}. The molecular chaos assumption is thus in trouble in the case of a weak decoupling of timescales. In turn, the Markovian nature of the collisions, namely the fact that succcessive collisions are independent from each other, also becomes questionable. One should thus *a priori* treat elementary collisions while accounting for the whole preceding 'history' of the nucleons. These memory effects can indeed be introduced into a Boltzmann-type kinetic equation, but it leads to an equation with a much more complex structure than the ones we have been considering here [219].

Another class of difficulties lies in the semi-classical approximation performed at the end of the derivation, in order to obtain the final form equation (3.37). This last approximation hides subtle difficulties. First, passing to the semi-classical (or classical) regime puts all nucleons on their mass shell. In other words, the classical description allows us to define, at any instant, the energy of each particle, provided one knows its position and momentum. This property is not necessarily welcome in the context of heavy-ion collisions, where off-shell effects may *a priori* show up [146, 304]. In addition, the 'elementary' classical limit (equation (3.34)) corresponding to $\hbar \to 0$ is, in fact, not trivial and should be treated with some caution [289].

Finally, a last class of difficulties concerns the numerical simulations of the nuclear Boltzmann equation. Numerous algorithms have been developed for simulating this equation and we shall not discuss their relevance here, but we have to keep in mind the fact that the numerical resolution of such equations raises many difficulties (section 3.2.3.3). We refer the interested reader to relevant references on this topic [45, 396] and [256, 390, 391, 462, 488].

3.2.4.2 Beyond the Boltzmann equation

We have previously seen how to obtain a Boltzmann-like kinetic equation which is adapted to the nuclear case. Let us now take the nuclear Boltzmann equation for granted and wonder whether this approach does provide a satisfactory theory for heavy-ion collisions.

In order to bring together some pieces of the answer to this question, let us briefly consider the results obtained in the analysis of heavy-ion collisions in the Fermi energy domain. Many phenomena have been observed in detail, and some successes have been obtained in the theoretical interpretation of the measurements, in particular, concerning the formation of hot nuclei [45, 449] (see also chapter 7). But the nuclear Boltzmann equation has also reached its limits, for example, in the study of rare particle production (mesons below threshold, etc) [118] (section 6.3) or for understanding the formation of intermediate mass fragments, which, notably represents an important de-excitation channel of hot nuclei (see chapter 8). In addition, the dynamics of nuclear collisions is strongly dissipative, in particular, in the entrance channel. Standard arguments of out-of-equilibrium statistical physics hence suggest the occurrence of large fluctuations during the early phases of the collisions, which are definitely not accounted for at the level of an 'average' description such as that provided by the Boltzmann equation [28].

Hence it seems crucial to go 'beyond' the nuclear Boltzmann equation, in order to account for the variety of the phenomena observed in the course of the nuclear collisions in which we are interested. In this respect, two types of approach have been developed so far—beyond the nuclear Boltzmann equation. In the first category, one remains in the spirit of kinetic equations while introducing stochastic extensions of these equations. These models are known under the generic name of Boltzmann–Langevin. They have mainly been formulated as Langevin extensions of BUU [1, 17, 18], and, to a lesser extent, at Fokker–Planck level [382]. In the second class of approach one tries to return directly to the original A-body problem, but while accounting for the quantum nature of the nucleons in an approximate way. These methods are molecular dynamics methods and nowadays constitute an active direction of research, as an alternative to one of the kinetic equations. In the following section we focus on Boltzmann–Langevin approaches; molecular dynamics calculations will be discussed in section 3.3.

3.2.5 Stochastic extensions of kinetic equations

3.2.5.1 An elementary derivation

The possible application of a Langevin approach to kinetic equations relies on the following idea. We consider a system, the global evolution of which (through the Boltzmann collision term) is slow compared to the timescale associated with elementary processes (which is *a priori* a typical case of application of the Boltzmann equation). One may then consider the collision term as the mean effect of collisions on the one-body distribution $f(r, p, t)$, which hence can (has to) be complemented by a fluctuation term, analogous to a Langevin force. One thus switches from a description of the system in terms of *one* distribution function $f(r, p, t)$ to an *ensemble*:

$$f(r, p, t) \longrightarrow \{f_\lambda(r, p, t), \lambda = 1, \dots, N_{\text{ens}}\} \tag{3.44}$$

such that

$$f(r, p, t) = \frac{1}{N_{\text{ens}}} \sum_{\lambda=1}^{N_{\text{ens}}} f_\lambda(r, p, t) \tag{3.45}$$

and one thus considers a type of diffusion process in the abstract space of one-body distributions.

The simplest case of a linearized Boltzmann equation provides a tutorial derivation [52], in the spirit of simple Brownian motion (sections 3.1.2.1 and 3.1.2.2) and shows how to complement a Boltzmann-like collision term with a fluctuation term. We consider a simple Boltzmann equation for classical particles, in the relaxation time approximation (τ)

$$\frac{\partial f}{\partial t} + \frac{p}{m} \cdot \nabla_r f = -\frac{f - f_\infty}{\tau} \tag{3.46}$$

where f_∞ represents the Maxwell–Boltzmann equilibrium distribution. By analogy with Brownian motion, we write a stochastic extension of equation (3.46) as

$$\frac{\partial f}{\partial t} + \frac{p}{m} \cdot \nabla_r f = -\frac{f - f_\infty}{\tau} + \delta I_{\text{coll}}(r, p, t) \tag{3.47}$$

where $\delta I_{\text{coll}}(r, p, t)$ represents a 'fluctuating' collision term of zero mean $\langle \delta I_{\text{coll}} \rangle = 0$ and correlation function

$$\langle \delta I_{\text{coll}}(r, p, t) \delta I_{\text{coll}}(r', p', t') \rangle \propto 2\delta(t - t')\delta(r - r')\frac{1}{\tau f_\infty}. \tag{3.48}$$

The correlation function (3.48) is local in r space $(\delta(r - r'))$ and the process is Markovian $(\delta(t - t'))$, as the elementary collisions enter in a Boltzmann collision term. Note finally that the correlation function equation (3.48) is proportional to the 'intensity' of the collision term $(1/\tau)$, which reflects the strong links (fluctuation-dissipation theorem) which exist between the average dissipation (Boltzmann collision term) and fluctuations (fluctuating collision term). The resulting Boltzmann–Langevin equation (3.47) hence constitutes a stochastic extension, 'à la Langevin', of the Boltzmann equation.

3.2.5.2   *The Boltzmann–Langevin equation in the nuclear context*

Is it possible to establish a connection between the previous heuristic arguments and the general BBGKY scheme? In other words, is it possible to identify, in the framework of the truncation scheme of the BBGKY hierarchy, a term, which could be interpreted as a stochastic term? The extension of these arguments (section 3.2.5.1) to the nuclear case in fact requires some complementary derivations. First, one should notice that equation (3.46) is linear, as is the case of the elementary Brownian motion (equation (3.1)). The first difficulty hence lies in the extension of the simple Langevin picture to non-linear situations. We are interested in strongly out-of-equilibrium processes, for which relaxation time approximations (equation (3.46)) hardly make sense. We shall hence gratify ourselves with a further approximation, to which we shall come back later. We assume that there exists a timescale τ_{BL} during which fluctuations are small, so that a linear approach, around a (not necessarily linear) Boltzmann equation remains conceivable. This hypothesis amounts to assuming that

$$\tau_{\mathrm{coll}} \ll \tau_{\mathrm{BL}} < \tau_I, \tau_{\mathrm{mf}} \tag{3.49}$$

where τ_I represents the characteristic time associated with the (global) effect of the Boltzmann collision term I_{coll} (equation (3.21) or equation (3.37)), and τ_{mf} a characteristic mean-field time.

We now come back to our truncation scheme for the BBGKY hierarchy (section 3.2.2) starting with the simple case of section 3.2.2.2 and we consider the propagation of density matrices during τ_{BL}, but *without* the hypothesis of molecular chaos (which would lead to the Boltzmann equation), namely while writing [17, 18] that at time t_0

$$\boldsymbol{\rho}_{12}(t_0) = \boldsymbol{\rho}_1(t_0)\boldsymbol{\rho}_2(t_0) + \delta\boldsymbol{\rho}_{12}(t_0) \tag{3.50}$$

together with the condition $\langle\delta\boldsymbol{\rho}_{12}(t_0)\rangle_{\tau_{\mathrm{BL}}} = 0$. The molecular chaos hypothesis $\delta\boldsymbol{\rho}_{12}(t_0) = 0$ is thus replaced by a weaker condition on the average $\langle\cdot\rangle_{\tau_{\mathrm{BL}}}$. This restores some correlations in the two-body density matrices, while discarding them on average, which allows us to recover, on the average, the Boltzmann evolution. The nature of this average, nevertheless, deserves some comment. It is not an average over the ensemble of all the one-body density matrices. One restricts oneself here to an average over the time interval τ_{BL}, which amounts to considering only a sub-ensemble of the ensemble of one-body density matrices, in the vicinity of an average density matrix which follows the Boltzmann equation. This subtle distinction is, in fact, crucial because it allows us to adapt the previous derivation (section 3.2.5.1) to the non-linear regime [1, 17, 387]. For the rest of the derivation, however, one only needs to keep in mind the importance of τ_{BL}, which allows us to transpose the linear formalism to more general cases.

The time propagation of the two-body density matrix (equation (3.50)) leads to a new equation for the one-body density matrix, which can be written in a schematic form as

$$i\frac{\partial \boldsymbol{\rho}_1}{\partial t} \simeq [\boldsymbol{K}_1, \boldsymbol{\rho}_1] + \mathrm{Tr}_2([\boldsymbol{V}_{12}, \boldsymbol{\Omega}_{12}\boldsymbol{\rho}_1\boldsymbol{\rho}_2\boldsymbol{\Omega}_{12}{}^+])$$
$$+ \mathrm{Tr}_2([\boldsymbol{V}_{12}, \text{propagation of } \delta\boldsymbol{\rho}_{12} \text{ from } t_0 \text{ to } t]) \qquad (3.51)$$

in the simple case of the quantum Boltzmann equation (3.31). The last step of the derivation consists in making the further approximation of treating the last term of equation (3.51) as a stochastic collision term δI_{coll}, which we characterize by its zero average value $\langle \delta I_{\mathrm{coll}} \rangle_{\tau_{BL}} = 0$, and its correlation function during τ_{BL}, which may be expressed in terms of the gain and loss terms of the Boltzmann collision term: $\langle \delta I_{\mathrm{coll}}(\boldsymbol{r}, \boldsymbol{p}, t) \delta I_{\mathrm{coll}}(\boldsymbol{r}', \boldsymbol{p}', t') \rangle_{\tau_{BL}} \sim$ (gain + loss) terms. The (crucial) hypothesis of treating the last term of equation (3.51) as a stochastic term may look somewhat artificial. However, if the condition (3.49) is indeed fulfiled it is likely that this term does represent a fluctuation compared to the collision term $\mathrm{Tr}_2([\boldsymbol{V}_{12}, \boldsymbol{\Omega}_{12}\boldsymbol{\rho}_1\boldsymbol{\rho}_2\boldsymbol{\Omega}_{12}{}^+])$. In the nuclear case (section 3.2.2.3), as for the derivation of the simple Boltzmann equation (section 3.2.2.2), the bare interaction is replaced by a Brückner matrix ($\boldsymbol{\Omega}_{12} \to \boldsymbol{\Omega}'_{12}$) and one has to properly account for the effects of the Pauli principle, but the strategy of the derivation remains similar.

The last step consists of finally taking the semi-classical limit, and one obtains a stochastic extension of the nuclear Boltzmann equation:

$$\frac{\partial f_\lambda}{\partial t} + \{f_\lambda, h\} = I_{\mathrm{coll}}^{\mathrm{UU}}[f_\lambda] + \delta I_{\mathrm{coll}}^{\mathrm{UU}}(\boldsymbol{r}, \boldsymbol{p}, t) \qquad (3.52)$$

where one recovers a Uehling–Uhlenbeck type collision term $I_{\mathrm{coll}}^{\mathrm{UU}}$ (equation (3.38), acting on fluctuating single-particle distributions f_λ) and a fluctuating collision term $\delta I_{\mathrm{coll}}^{\mathrm{UU}}$ which is characterized by a zero average and a correlation function $C(\boldsymbol{r}, \boldsymbol{p}, \boldsymbol{p}', t)$ during τ_{BL}. The latter is local in real space and Markovian, as the collision term itself.

This formulation of the Boltzmann–Langevin equation is directly copied from the Langevin description of Brownian motion. A formulation in terms of a Fokker–Planck picture is also possible and was proposed in [382] and further tested, in particular, in [104] (see also [124] for a general presentation). Each fluctuating single-particle distribution f_λ then becomes a 'coordinate' of a generalized distribution function $\tilde{W}[\{f_\lambda\}]$ for which one writes a diffusion (Fokker–Planck type) equation, the associated transport coefficients being themselves expressed in terms of the f_λ. The resulting diffusion equation is quite involved and has only been solved in schematic (low dimensional) systems [104, 124]. Approximate calculations in realistic cases of nuclear collisions were, in turn, performed in the more flexible Langevin approach [451–453]. Altogether, the Boltzmann–Langevin equation (3.52) has focused a lot of principle studies for several years. The interested reader may refer to [1] and to the references therein for some examples of application.

3.2.5.3 *Some remarks on the Boltzmann–Langevin equation*

The Boltzmann–Langevin equation *a priori* represents a noticeable improvement over the Boltzmann equation (section 3.2.2), as it allows us to estimate widths around average values of observables. Let us note, on this occasion, that Boltzmann–Langevin average values do not necessarily reduce to Boltzmann ones, because of the non-linearity of the equations. The Boltzmann–Langevin equation does reduce to the Boltzmann equation only locally in time (over τ_{BL}). But large fluctuations and/or differences in the global average values may appear over long times. In turn, Boltzmann–Langevin events may explore large domains of the abstract space of one-body distributions $f(\boldsymbol{r}, \boldsymbol{p}, t)$. This is precisely the interest of this approach, which *a priori* allows us to explore physical situations in which fluctuations may play an important role [1, 40, 502].

One should, however, note that the Boltzmann–Langevin equation does not solve all the problems raised by the Boltzmann equation; it actually even raises new ones. For example, in the nuclear case, the difficulties linked to the justification of the Markovian approximation are obviously transferred from the Boltzmann to the Boltzmann–Langevin equation. In fact the problem is even more crucial in the latter case. We have indeed seen the crucial role played by the characteristic time τ_{BL}. As in nuclear physics $\tau_{coll} \ll \tau_{mf}$, the existence of a timescale τ_{BL}, which would take place in between the duration of a collision τ_{coll} and the mean-field evolution τ_{mf}, is only marginally justified. Furthermore, τ_{BL} has to be long enough to allow a large number of elementary collisions to take place, although this difficulty is not specific to the nuclear case.

Another major difficulty lies in the finite nature of the system (the nucleus) to which one would like to apply a stochastic description. We have only characterized the fluctuating collision term δI_{coll}^{UU} by its average value and its correlation function. In the general case for which the 'bath' in contact with the Brownian degree of freedom is not thermal, the functional form of the Langevin force is not *a priori* uniquely determined by its first two moments only (section 3.1.2.2). In the case we are interested in, it is clear that the bath cannot be thermal because the system is finite and isolated, hence microcanonical. We raise here a fundamental difficulty related to preservation of the conservation laws in a finite system in which one wishes to apply a stochastic model. To our knowledge this problem remains practically open.

A last difficulty concerns the numerical simulations of the Boltzmann–

Langevin equation. We have already noted the problems encountered in the simulations of the Boltzmann equation (section 3.2.3.3). To these difficulties are added those specific to the Boltzmann–Langevin equation, due to the complexity of the fluctuating collision term [1]. As a consequence the calculations performed up to now have been restricted either to schematic cases [104, 105, 123, 124] or to approximate methods in realistic cases [451, 452].

3.2.6   An alternative approach: Stochastic TDHF

An alternative approach, both to kinetic equations and to the molecular dynamics methods (section 3.3), but still in the realm of one-body theories, has been proposed recently [1, 388]. The goal is to provide a theoretical framework for stochastic extensions of TDHF, without referring to the kinetic equation stage. Two ideas underlie this approach: (i) a perturbative treatment of the residual interaction V_{12} on a well chosen time interval τ_{STDHF}, around a TDHF trajectory; and (ii) a projection of the correlated states built in (i) onto an ensemble of Slater determinants. To some extent this amounts to reconsidering the derivation of the Boltzmann–Langevin equation but, this time, starting from the mean-field evolution and not from a Boltzmann-like kinetic equation. This also amounts to treating all the elementary collisions on an equal footing, without trying to share their effects in a collision term and/or a fluctuating collision term.

The hypothesis of a perturbative treatment on a time interval τ_{STDHF} imposes, as for the time interval τ_{BL} of Boltzmann–Langevin (equation (3.49)), a condition of loss of coherence of the TDHF trajectories:

$$\tau_{\text{coll}} \ll \tau_{\text{STDHF}} \ll \tau_{\text{mf}}. \tag{3.53}$$

Indeed, the dynamics has to remain dominated by the mean field but, simultaneously, the residual interaction should be sufficient to allow transitions between the states of the system, and even a sufficient number of transitions, in order to justify a statistical hypothesis. The use of the notation τ_{coll} can be misleading here. In the framework of STDHF, the duration of a collision does not exist directly as such (section 3.2.2). But the condition of development of transitions, that one has to impose to τ_{STDHF}, in fact reduces to comparing τ_{STDHF} to the τ_{coll} introduced earlier (section 3.2.4.1).

Under the assumption (3.53) the propagation of an initially uncorrelated state of density matrix D_N (Slater determinant) between an instant 0 and τ_{STDHF}, while accounting for the residual interaction, leads to expressing the density matrix of this state $\mathcal{D}_N$ (correlated after τ_{STDHF}) as a statistical superposition of uncorrelated states D_M (Slater determinants):

$$\underbrace{\mathcal{D}_N(\tau_{\text{STDHF}})}_{\text{correlated}} = \sum_M \underbrace{W_{MN}}_{\text{transition rate}} \underbrace{D_M(\tau_{\text{STDHF}})}_{\text{uncorrelated}} \tag{3.54}$$

where the weights W_M can be expressed by the Fermi golden rule [388].

Once the elementary propagation (3.54) has been defined, one can build an ensemble of stochastic trajectories according to the scheme:

$$
\left\{
\begin{array}{l}
D_N \overset{\tau_{\text{STDHF}}}{\rightarrow} \{D_M, W_M\} \\
\qquad D_{M_0} \overset{\tau_{\text{STDHF}}}{\rightarrow} \{D_L, W_{M_0 L}\} \\
\qquad\qquad \cdots
\end{array}
\right\}_{N=1,\ldots}
\qquad . \qquad (3.55)
$$

Starting from a Slater state D_N one chooses, after the first STDHF time step (by sampling according to the W_M weights) a new Slater determinant D_{M_0}. And one iterates the process starting with D_{M_0}. One thus obtains one trajectory. An ensemble of trajectories built this way constitutes the STDHF ensemble representation of the dynamical process under study.

What does STDHF bring that is new compared to the approaches we have introduced earlier? It may seem, at least superficially, that STDHF does not look similar to the previously discussed stochastic processes (section 3.2.5). In particular, the timescale τ_{STDHF} plays a crucial role in the derivation and leads to a coarse grained description of the dynamics. One should, however, remember that the Boltzmann–Langevin equation exhibits the same difficulty through τ_{BL}, even if τ_{BL} does not explicitly appear in the time derivatives. One could actually rewrite STDHF as a differential equation—provided the restriction that the time derivatives would be considered *modulo* τ_{STDHF}. The apparent difference between STDHF and Boltzmann–Langevin hence only appears in the presentation and not to the nature of the theories (recall, for example, that in the Boltzmann equation τ_{coll} is 'set to 0' (section 3.2.4.1)). But STDHF does have an advantage over Boltzmann–Langevin: first, STDHF relies on a transparent hypothesis and, second, it does not require an explicit separation of the effect of the residual interaction between a collision term and a fluctuating collision term. This has two consequences: (i) mean-field fluctuations appear directly in the dynamics, while in Boltzmann–Langevin they are 'filtered' by the collision term; and (ii) one can hope to remove some of the numerical difficulties linked to a proper treatment of the collision term.

Very few practical applications of the STDHF formalism have been performed so far. The first attempts to direct simulations were reported in [455]. One should also mention the molecular dynamics-like calculations of [350] which, although not explicitly connected to STDHF, have been heavily inspired by this theory.

3.3   Molecular dynamics approaches

Kinetic theory is a 'traditional' approach to the many-body problem. It allows us, as we have seen, to write down the equations for the one-body distribution $f(\mathbf{r}, \mathbf{p}, t)$ in phase space. The quantal nature of the nucleus leads to the reformulation of kinetic theory on a quantal basis but the idea of the dominance of one-body effects does continue and seems well justified for heavy-ion collisions in the nucleonic regime. In the classical case the capability of today's computers allows us to consider on a serious basis an alternative approach: molecular dynamics (MD) [248]. In MD we study the motion of all the particles of a system under their mutual interactions. In principle, at least in the classical case, an MD calculation leads to a complete solution of the problem for an isolated system (microcanonical). Extensions to canonical [247, 345] or grand canonical [107] ensembles are actually also available. The applicability of MD methods to systems of quantal interacting particles, nevertheless, remains an open problem.

In a quantal context the idea of an MD method amounts to replacing the A-body ket representing the state of the system under study by a set of A classical particles characterized only by their positions and momenta

$$|\Psi_{1...A}\rangle \longrightarrow \{r_i, p_i, i = 1, \ldots, A\} \qquad (3.56)$$

the evolution of which are followed in time. This time evolution is performed according to Hamilton's equations of motion:

$$\frac{\mathrm{d}r_i}{\mathrm{d}t} = \frac{\partial H_\mathrm{c}}{\partial p_i} \qquad (3.57)$$

$$\frac{\mathrm{d}p_i}{\mathrm{d}t} = -\frac{\partial H_\mathrm{c}}{\partial r_i} \qquad (3.58)$$

where H_c denotes the classical A-body Hamiltonian corresponding to H (equation (2.40)). The model is hence fully specified as soon as one knows H_c and these Hamiltonian equations do correspond to an 'exact' solution of the classical problem. In contrast to kinetic equations such an MD approach hence contains *all* the correlations between particles. But these correlations are strictly *classical*. It is hence *a priori* unnatural to apply an MD method to describe a strongly correlated system such as the nucleus. Still, we have seen that the dynamics may be a bit forgiving in this respect, but only to some extent. Let us see what types of MD methods have been proposed for describing heavy-ion collisions.

3.3.1 Classical molecular dynamics

The model proposed by Pandharipande during the second half of the 1980s is based on a Hamiltonian of the form

$$H_\mathrm{c} = \sum_{i=1}^{A} \frac{p_i^2}{2m} + \sum_{i<j} V_{ij} + \text{Coulomb} \qquad (3.59)$$

where the two-body potential $V_{ij} = V_{ij}(r = |r_i - r_j|)$ is built from a Lennard-Jones potential [411]

$$V_{ij} = \begin{cases} V^{\mathrm{LJ}}[(r^{-12} - r^{-6}) - ((r_\mathrm{c}^{\mathrm{LJ}})^{-12} - (r_\mathrm{c}^{\mathrm{LJ}})^{-6})], & r < r_\mathrm{c}^{\mathrm{LJ}} \\ 0, & r > r_\mathrm{c}^{\mathrm{LJ}} \end{cases} \qquad (3.60)$$

or from two truncated Yukawa potentials [288]

$$V_{ij} = \begin{cases} V_1(e^{-\mu_1 r} - e^{-\mu_1 r_c^Y}) - V_2(e^{-\mu_2 r} - e^{-\mu_2 r_c^Y}), & r < r_c^Y \\ 0, & r > r_c^Y. \end{cases} \qquad (3.61)$$

This model was proposed to study dynamical processes in the course of collisions, initially in argon gas, without necessarily pretending to be realistic from the nuclear point of view. As the authors themselves say:

> We stress that the classical argon balls used in this study are not intended to be mock nuclei, but instead to provide simple systems whose time evolution can be studied exactly (...) Direct comparisons with nuclear data are difficult (...) It is possible to define a classical system such that its density, binding energy and compressibility are similar to nuclear matter. [411]

The methodology is hence clear in these calculations. The systems under study exhibit some energetic similarities with nuclei but the goal is mainly to study the dynamics of these systems, without aiming at a realistic nuclear description. Systematic comparisons have, nevertheless, been performed between these calculations and a Boltzmann-like description, for collisions of pseudo 'nuclei' $(A = 50) + (A = 50)$ at beam energies of order 450 MeV/u. Ingredients entering in the Boltzmann equation are then carefully deduced from the original Hamiltonian. These comparisons exhibit a good agreement between Boltzmann and MD calculations [288]. This is, however, not surprising, in view of the algorithm used for simulating the Boltzmann equation (section 3.2.3.3). Note also that the questions related to the Pauli principle are explicitly overlooked in these calculations, as only strictly classical particles are considered here, even at the Boltzmann level.

3.3.2 Molecular dynamics with the Pauli potential

The obvious defect of MD methods in nuclear physics lies in the fact that they provide a completely *classical* description of nucleons. This fundamental question was attacked at the end of the 1970s in a series of papers by Wilets [108, 493, 494], who was aiming at developing classical but realistic models of heavy-ion collisions. This method was also studied later by the Berkeley group [163, 164, 166]. The idea of these works is to restore the Pauli principle in a minimal way by means of a repulsive potential in phase space. One thus hopes to be able to enforce a maximal occupation (2 or 4) of elementary phase space cells of size $(\Delta r \Delta p)^3 \sim h^3$ through a specific interaction between the particles. And this interaction has, of course, to act globally in phase space, namely in both r and p spaces.

The Hamiltonian of the system can then be written as

$$H_c = \sum_{i=1}^{A} \frac{p_i^2}{2m} + \sum_{i<j}[V_N(\boldsymbol{r}_i - \boldsymbol{r}_j) + V_P(\boldsymbol{r}_i - \boldsymbol{r}_j, \boldsymbol{p}_i - \boldsymbol{p}_j)] + \text{Coulomb} \quad (3.62)$$

where one can identify a nuclear V_N and a Pauli V_P interaction term. In the original calculations [493] V_N is built from three Yukawa terms:

$$V_N(\boldsymbol{r}_i - \boldsymbol{r}_j) = V_N(|\boldsymbol{r}_i - \boldsymbol{r}_j|) = V_N(r_{ij}) = \sum_{n=1}^{3} V_n \frac{e^{-\mu_n r_{ij}}}{r_{ij}} \quad (3.63)$$

and the Pauli potential takes the form

$$V_P(\boldsymbol{r}_i - \boldsymbol{r}_j, \boldsymbol{p}_i - \boldsymbol{p}_j) = V_P^0 e^{-\alpha((\boldsymbol{r}_i - \boldsymbol{r}_j)\cdot(\boldsymbol{p}_i - \boldsymbol{p}_j))^4}. \quad (3.64)$$

In the more recent model of the Berkeley group [163] the nuclear part has the form of a modified Lennard-Jones potential:

$$V_N(r_{ij} = |\boldsymbol{r}_i - \boldsymbol{r}_j|) = V_N^0 \left(\left(\frac{r_1}{r_{ij}}\right)^{p_1} - \left(\frac{r_2}{r_{ij}}\right)^{p_2}\right) \frac{1}{1 + \exp(\alpha(r_{ij} - d))} \quad (3.65)$$

and the Pauli potential is Gaussian:

$$V_P(\boldsymbol{r}_i - \boldsymbol{r}_j, \boldsymbol{p}_i - \boldsymbol{p}_j) = V_P^0 \exp\left(-\frac{(\boldsymbol{r}_i - \boldsymbol{r}_j)^2}{2\sigma_r^2}\right) \exp\left(-\frac{(\boldsymbol{p}_i - \boldsymbol{p}_j)^2}{2\sigma_p^2}\right). \quad (3.66)$$

The parameters of V_N are adjusted on the binding energy per nucleon and the saturation density of nuclear matter, as well as on the incompressibility modulus of nuclear matter and the nucleon–nucleon cross-section. The parameters entering V_P are adjusted so that one recovers a reasonable Fermi energy. Note finally that some attempts to include the Heisenberg uncertainty principle have also been proposed on the basis of a phase space potential, by Wilets's group [262].

Numerous calculations have been performed with these formalisms, for nuclear collisions [108,493,494] and for the thermodynamical properties of nuclei [163–166]. These calculations have also served as a basis for a method of early recognition of intermediate mass fragments formed during heavy-ion collisions leading to multifragmentation [166]. However, these MD methods with a Pauli potential raise many difficulties. One can first wonder whether it makes sense, at all, to introduce a Pauli potential? Would that suggest the existence of a fifth interaction? In addition, one knows that, quantally, the Pauli principle acts on all particles together (remember the antisymmetrization operator $\mathcal{A}_{1...A}$), and not on a pair of nucleons, and with a 'finite range'. The latter criticism might, however, be attenuated in view of recent results obtained in plasma physics [263]. Another difficulty in this method lies in the structure of the ground states it produces. Nuclei appear as 'crystals'.

3.3.3 'Quantum' molecular dynamics

The quantum molecular dynamics (QMD) model was introduced in the second half of the 1980s as a substitute to BUU and aims (as the acronyms indicates) at providing a realistic description of heavy-ion collisions while accounting for some quantal effects [3–5]. Two major aspects make it different from standard MD methods: (i) particles have a size; the usual Dirac distribution $\delta(r - r_i)\delta(p - p_i)$ is replaced by a classical Gaussian 'wavepacket'; and (ii) some properties of quantum scattering are restored through a stochastic component of BUU type (section 3.2.3.3).

The original Hamiltonian used in these calculations has the form [5]

$$H_c = \sum_{i=1}^{A} \frac{p_i^2}{2m} + \sum_{i<j} [V_{\text{local}}(r_i - r_j) + V_{\text{Yuk}}(r_i - r_j) + V_{\text{vel}}] + \text{Coulomb.} \quad (3.67)$$

It contains a 'local' term of Skyrme type (section 2.2.1.3), expressed as a function of the local density $\rho(r)$

$$V_{\text{local}} = t_1 \delta(r_i - r_j) + t_2 \delta(r_i - r_j)\rho^\gamma((r_1 + r_2)/2) \quad (3.68)$$

a Yukawa term V_{Yuk} (see equation (3.61)) providing a description of surface and a velocity-dependent term

$$V_{\text{vel}}(r_i - r_j, p_i - p_j) = t_4 \ln^2(t_5(p_i - p_j)^2 + 1)\delta(r_i - r_j). \quad (3.69)$$

The local potential is adjusted on saturation, the Yukawa one on surface properties and the velocity-dependent potential on the real part of the optical potential (section 2.1.3.2).

A collision algorithm, inspired from BUU (section 3.2.3.3), is furthermore added to the effect of these potential terms. One should note here the somewhat paradoxical character of this latter ingredient. As MD in principle represents a complete description of A-body classical dynamics, it is hard to justify the existence of 'residual' interactions, as in the case of kinetic descriptions. One can at best invoke the restoration of quantal properties [5]. One would thus obtain an MD in which a stochastic component is expected to restore some quantal effects.

The QMD model has been used in numerous calculations; these calculations have led to some results. Several studies have been devoted to multifragmentation [4, 5]. Relativistic extensions (RQMD, [436]) and versions incorporating isospin (IQMD, [235]) are also available. For a recent review of these various (numerous) versions of QMD, we refer the reader to [236] and references therein. Some questions, nevertheless, remain open. First, one has to admit that QMD does contain a lot of phenomenology. More annoying, however, is the lack of a link to a clear theoretical framework, connected to well known methods of the quantal A-body problem. In particular, the reader familiar with BUU will surely have recognized the strong relationship between QMD and BUU: the Skyrme mean field, two-body collisions, etc. This link becomes even clearer when one considers more precisely the BUU algorithm (section 3.2.3.3). At last, it turns out that in QMD (as in BUU, actually) the Pauli principle is taken into account only at the level of two-body collisions, which might be questionable to fully justify the 'quantum' label of QMD.

3.3.4 Fermionic molecular dynamics

3.3.4.1 Limitations of standard MD methods

In nuclear physics, the key problem in MD methods lies in the treatment of the Pauli principle. Nucleons are fermions and it is difficult to speak of nuclei without the Pauli principle. This question is simply overlooked in strictly classical MD calculations (section 3.3.1). MD methods with a Pauli potential directly address this question (section 3.3.2), but with disputable success. The QMD model (section 3.3.3) takes into account some aspects of the Pauli principle, but through two-body collisions, which is somewhat contradictory with an MD picture.

In view of the difficulties encountered with taking the Pauli principle in classical MD methods properly into account, a new class of approaches was developed at the beginning of the 1990s, the so-called fermionic molecular dynamics: FMD for fermionic molecular dynamics [181] and AMD [351, 352] for antisymmetrized molecular dynamics.

3.3.4.2 Original FMD and AMD models

The originator of the idea of the FMD method was H Feldmeier [181–183], who suggested representing the A-body ket of the system by a Slater determinant built out of peculiar one-body kets, namely Gaussian wavepackets:

$$|\Psi(1,\ldots,A)\rangle = \mathcal{A}_{1\ldots A}\left(\prod_{i=1}^{A}|\mathcal{G}_i\rangle\right) \qquad (3.70)$$

where the kets $|\mathcal{G}_i\rangle$ are localized wavepackets of $|r\rangle$ representation

$$\langle r|\mathcal{G}_i\rangle = \mathcal{N}_i \mathrm{e}^{\mathrm{i}\varphi_i} \exp\left(-\frac{(r-\langle r_i(t)\rangle)^2}{2\sigma_i(t)} + \mathrm{i}\langle p_i(t)\rangle\right) \qquad (3.71)$$

and which are characterized by the phase space parameters $(\langle r_i\rangle, \langle p_i\rangle)$ and by the width σ_i ($\mathcal{N}_i$ is the normalization and φ_i the phase). Each nucleon is thus represented by one Gaussian wavefunction. Spins may also be taken into account. The equation of motion is obtained from the same variational principle as TDHF (section 3.2.1.1), namely

$$\delta\int \mathrm{d}t\,\langle\Psi|\left(H - \mathrm{i}\frac{\partial}{\partial t}\right)|\Psi\rangle = 0.$$

This variational principle leads to equations of motion for the $\langle r_i \rangle$, $\langle p_i \rangle$, σ_i, which are the dynamical variables of the model. Note that because of the hypothesis of antisymmetrization (equation (3.70)) the $\langle r_i \rangle$, $\langle p_i \rangle$ cannot be directly interpreted as the classical coordinates of nucleons.

Two versions of this model have been developed. In the original FMD model [181] the potential part of the Hamiltonian H has a simple form built out of two Gaussians:

$$H_{\text{FMD}} = \sum_{i=1}^{A} \frac{p_i^2}{2m} + \sum_{i<j} \left(V_1 \exp \left(\frac{(r_i - r_j)^2}{2\sigma_1^2} \right) - V_2 \exp \left(\frac{(r_i - r_j)^2}{2\sigma_2^2} \right) \right).$$

(3.72)

In the AMD version [351], the FMD Hamiltonian is complemented by a three-body zero-range term V_{123}, as in QMD (equation (3.68)). In addition, a standard two-body collision algorithm of BUU type (section 3.2.3.3) is added to the dynamics. This *classical* algorithm raises some practical problems in this case. Indeed, as already mentioned, 'classical' positions and momenta of the nucleons are not directly accessible from the AMD or FMD wavefunctions. The authors of [351] are thus led to reconstruct classical coordinates R_i, P_i from the $\langle r_i \rangle$, $\langle p_i \rangle$. The BUU algorithm is then used on the R_i, P_i and its effect is, in a last step, tranferred back to the $\langle r_i \rangle$, $\langle p_i \rangle$. Extensions of AMD including a stochastic component to simulate wavepacket diffusion effects have also been proposed [355].

Over the years the FMD and AMD methods have been developed further and have served as a basis either for direct applications to nuclear collisions [356] or the development of extensions of the theory (see section 3.3.4.4). They have also been investigated in terms of their statistical properties [349, 353, 354, 414]. We refer the reader to [356] and [183] for further references on these works.

3.3.4.3 Discussion

What do FMD models bring? By construction the Pauli principle is preserved, which is clearly a positive point. However, if one examines the hypotheses used for building these models more closely, one soon realizes that the two constitutive hypotheses (form of the wavefunction and variational principle) are nothing but a degraded version of TDHF (section 3.2.1.1). In TDHF the form of the A-body ket is also a Slater determinant of one-body kets, but without any restriction on the form of the one-body kets. The FMD hypothesis is thus a TDHF hypothesis but in a sub-space of kets which is restricted compared to the totally accessible space. Correlatively, the introduction of the BUU algorithm into AMD makes it a type of extended TDHF. Finally, one should note that FMD, as AMD, are extremely complex models at the numerical level. One can thus wonder what is the gain offered by these models compared to, say, TDHF or an extended TDHF.

To close this section, it is interesting to make a few final remarks concerning the links between these FMD methods and other 'similar' approaches. First, one can note that if one releases the antisymmetrization condition (3.70) the FMD dynamics of the Gaussians simply reduces to the dynamics of the test particles used for simulating kinetic equations (section 3.2.3.3). FMD methods should also be compared to another simulation method, used in molecular physics, the so-called Car–Parinello method, in which antisymmetrization is enforced on model electronic wavefunctions (plane waves) [115]. Finally, it is instructive to see how FMD methods might be related to STDHF. We have seen that these MD methods, in fact, represent degraded versions of TDHF. One can thus imagine extending them with STDHF—in other words, to produce degraded STDHF models. This idea was recently explored in the AMD framework [350]. But here again one may wonder whether it is necessary to resort to AMD, while TDHF is a well established theory, on the formal as well as on the numerical levels [341]. STDHF hence seems to constitute an interesting alternative to the previously introduced methods in the energy range we consider here. In this respect, it also complements earlier (non-stochastic) extensions of TDHF [19, 26].

3.3.4.4 *The correlator operator*

As previously discussed most FMD calculations have been done up to now at the pure 'mean field' [183] or 'extended mean field' [356] level. But a recent extension of the theory has been proposed which should allow the development of a true, beyond the mean field, approach. As for any mean-field theory, the further step in FMD consists in accounting for correlations not included in the mean field. The idea here is to focus on short range correlations and to enforce them by modifying the relative wavefunction at short distance by means of a unitary correlator operator $C = \exp(-iG)$, involving a two-body operator G which is expressed in terms of the relative distance between the two interacting nucleons. Alternatively, the effect of the correlator operator is to renormalize the interaction at short distance, which allows us to use a standard FMD Slater determinant as a many-body wavefunction, but with a modified Hamiltonian. The first results obtained with this approach seem encouraging. The basic ground-state properties of nuclei are properly recovered and dynamical calculations can now go truly beyond the mean field. It would be an interesting and useful task to link this new approach with more standard theories of the nuclear many-body problem, in particular, the methods used in dynamical problems, in connection with the BBGKY hierachy.

3.4   Conclusion of the chapter

We have devoted this chapter to basic theoretical tools for describing the dynamics of heavy-ion collisions in the nucleonic regime. These approaches can be viewed, to some extent, in the general framework provided by out-of-equilibrium statistical physics, in terms of the reduction of a complex dynamics to a few relevant degrees of freedom. In the present case, one-body descriptions typically constitute such a set of relevant variables, as discussed at length in this chapter. In particular, we have seen how to derive kinetic equations, and to what extent such approaches can be justified in a nuclear context. Kinetic equations, in terms of the celebrated BUU model, have thus constituted a key working tool in this field of physics since the mid 1980s.

Due to their inadequacy to depict all the various situations encountered in heavy-ion collisions, BUU-like approaches have been progressively complemented by more involved theories, or replaced by MD-like approaches. Still, BUU remains a basic tool of investigation in the field. Stochastic extensions of BUU, although formally appealing, have not yet reached a 'flexibility' and 'robustness' threshold which would allow them to be routinely used in realistic cases. In turn, the various MD methods, in particular, QMD-based models, have found a wide range of applications because of their simplicity and in spite of their defects. Altogether, there is no theory at hand which fulfils both the formal and practical requirements to become a basic tool, beyond BUU, in spite of appealing attempts, as provided by stochastic TDHF and the correlator extension of FMD.

Chapter 4

Basic experimental and analysis tools

This chapter aims to describe the basic tools used in the experimental investigations of nuclear reactions in the Fermi energy range. In the first part we consider the experimental tools (namely the beam facilities and the detectors) that have been designed to achieve a good detection of all the various reaction products. A major breakthrough in the investigation of nuclear collisions has been the large development of 4π devices which are nowadays used almost exclusively. These detectors, which cover almost the whole space around a reaction, allow the recording of kinematically quasi-complete events. They are described in section 4.1.2 after a brief description of the accelerators used in the energy range discussed in this book (section 4.1.1).

As a consequence of the use of complex devices, a large body of information is obtained on an event-by-event basis. Therefore, specific 'reduction' techniques have been developed for the metrology of nuclear reactions. First, collisions must be 'visualized' (section 4.2.1) in order to have a first idea of the topology of the reaction, as discussed in section 4.2.3.4. Then, they must be sorted according to the impact parameter (section 4.2.3), which is a key parameter in disentangling reaction mechanisms at a given beam energy (chapter 5). The characterization of excited sources produced in the course of the collision is discussed in section 4.3. Basic techniques to estimate excitation energies (nuclear calorimetry), rotational energies, temperatures (nuclear thermometry) and timescales (nuclear chronometry) are described. Finally, the simulation tools which are needed to disentangle detection biases as well as autocorrelations in the data analysis are briefly described (section 4.4) before concluding the chapter (section 4.5).

4.1 Experimental tools

Basically, heavy-ion-induced reactions are processes during which an initial relative kinetic energy is shared among various degrees of freedom leading to outgoing products reminiscent of the underlying energy-sharing mechanisms.

The initial kinetic energy is obtained from accelerator facilities which are briefly described in section 4.1.1. The outgoing products are analysed in dedicated devices which are presented in section 4.1.2

4.1.1 Beam facilities

Heavy-ion facilities can be sorted into four classes depending on the velocity of the projectiles (or the energy, expressed in MeV/u) they deliver:

- low-energy domain: [5–15 MeV/u];
- intermediate or Fermi energy domain: [15–200 MeV/u];
- relativistic energy domain: [200 MeV/u–10 GeV/u];
- ultra relativistic energy domain: [>10 GeV/u].

Each energy range provides access to a specific class of physical properties. The low-energy domain mainly addresses collisions dominated by mean-field effects. The Fermi energy domain is the realm of nucleonic dynamics. The collisions described in this book correspond mainly to this second domain. It is a transition region between the low and relativistic energy regions as will be detailed in chapter 5. It thus involves several competing mechanisms. Relativistic collisions essentially access sub-nucleonic degrees of freedom, and lead to consideration of, for instance, the physics of π's and Δ's in the nuclear medium. Finally, the ultra-relativistic energy domain is devoted to the search for the QGP (section 1.3.2).

Sorting according to beam energy not only corresponds to the physical properties but also to the technical limitations. Technical limitations lie in the increasing size of the facilities needed to attain large energies and/or in the increasing relativistic effects which have to be taken into account above 100 MeV/u. Most of the existing machines are based on cyclotrons [114] even if linear accelerators (LINACS) have been built for limited energy ranges (10–20 MeV/u). Synchrotrons have to be used in the GeV/u range or above. For energies exceeding a few tens of MeV/u, several successive accelerators are often used with a stripping foil between two machines in order to improve the acceleration efficiency by increasing the charge state of the accelerated ions. The first machine can be a tandem (Catania) or a cyclotron (GANIL). The second machine is generally a cyclotron (or a synchrotron). Such a complication can partly be avoided by using ECR (electron cyclotron radiation) sources delivering high-charge-state ions [13] but large intensity beams are better achieved by using the stripping method. Figure 4.1 is a non-exhaustive compilation of the existing facilities in the intermediate energy domain of interest in this book. In figure 4.1 the various thresholds are indicated. The broken horizontal arrow corresponds to the pion production threshold in a nucleon–nucleon collision. The physics of interest in this book lies mainly below this line since it concerns non-excited hadronic matter. The full horizontal arrow in figure 4.1 corresponds to an available centre-of-mass energy equal to the total binding energy for symmetrical

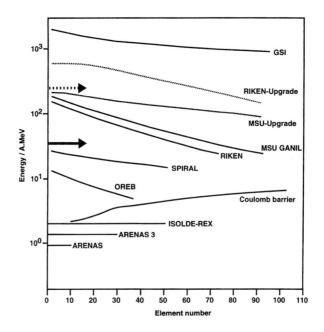

Figure 4.1. Compilation of the existing facilities in the nucleonic regime. The maximum acceleration energy (ordinate) is indicated as a function of the atomic element number (abscissa) for each facility. Also indicated are the Coulomb barrier (full line) for each element, the so-called pion threshold (broken horizontal arrow) (see section 2.2.1.1) and the incident energy per nucleon corresponding, in symmetric collisions, to available centre-of-mass energies equal to the mean binding energy in nuclei (full horizontal arrow).

projectile–target systems. In this book, we shall mainly focus on the energy range below 200 MeV/u.

As an example of the possibilities offered by a typical facility in the nucleonic domain, we show in figure 4.2 a list of the available beams at GANIL after the first cyclotron (SME, Sortie Moyenne Energie beam) and for the whole machine. The accelerated charge states and typical intensities are also indicated in electrical intensity (nA or μA) or in particle per second. Note that GANIL beams are delivered by a set of two major cyclotrons with primary injection from an ECR source into a third primary small cyclotron. For a given facility larger energies are reached for lighter projectiles. The reason for this lies in the maximum energy obtained from a cyclotron:

$$E_{\text{max}} = KZ^2/A \qquad (4.1)$$

where Z is the charge state of a projectile of mass number A and K is a characteristic parameter of the cyclotron. The decreasing tendency of the curves

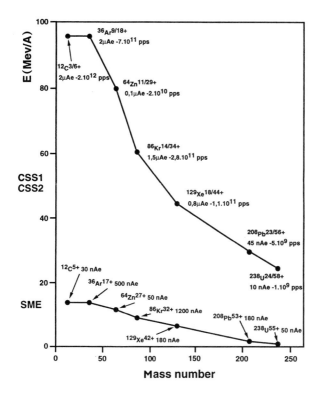

Figure 4.2. Compilation of the available beams at GANIL after the first cyclotron (bottom points) and after the whole machine (upper points). Beam energy (in MeV/u) is plotted as a function of mass number. Also indicated are the corresponding currents (in A or pps (particles per second)).

in figure 4.1 reflects the increasing difficulty in strongly stripping heavy atoms. Typical values of K are 115 (Louvain La Neuve), 380 (GANIL), 450 (Lanzhou), 520 (College Station), 800 (Catania) and 1200 (East Lansing). For about ten years, cryogenic cyclotrons have also been built, which ensure a smaller machine size and reduced power consumption.

4.1.2   Detectors

4.1.2.1   Introduction

In an experiment, there is a need to characterize the outgoing products in order to identify the relevant physical mechanisms governing the response of the system to the collision. Two quite different situations are observed in heavy-ion-induced reactions depending on the number of outgoing massive products. If one or two massive outgoing products are produced, the corresponding kinematics is very simple and may be analysed by using small solid angle detectors. For instance, if only two massive fragments are emitted as in fission, one fragment is emitted in the plane defined by the beam and the partner fragment. Both massive products can thus be detected and identified by using limited area detectors located at selected angles. The situation is far more complicated if more than two heavy products are produced, because the corresponding angular correlations are not simply defined. In this case, it is necessary to cover a sizeable fraction of the whole solid angle around the reaction chamber, in order to detect with a large efficiency the emitted products. Heavy-ion-induced reactions in the Fermi energy domain correspond to this second situation because many nuclear species are often produced during the reaction. This is the main reason why it has been necessary to build dedicated multidetectors covering most of the solid angle surrounding the target. These detectors are known as 4π devices.

In addition to the problem due to the number of produced species, two further difficulties have to be considered. The first one is that many kinds of particles are emitted: heavy fragments or residues (typical charge number Z exceeding 10), intermediate mass fragments (IMF: $3 \leq Z < 10$), light-charged particles (LCP: $Z = 1, 2$), neutrons, gamma-rays and mesons such as pions or even kaons. The second difficulty lies in the wide energy range to be covered for each kind of particle. For instance, a proton evaporated by the target may be rather slow (typically a few MeV) whereas it can acquire a very large kinetic energy when emitted by the projectile. It is thus not possible to build an ideal detector covering a 4π solid angle in which *any* fast or slow particle would be identified in nature (charge, mass), geometrically (emission angle) and energetically (kinetic energy). Instead, it turns out that the large solid angle detectors which have been built are dedicated to some specific kinds of particles: charged particles or gamma-rays or neutrons.

4.1.2.2   Detection of charged products

The criteria used to define the technical choices are essentially the total geometrical efficiency, the charge and/or mass identification, the energy measurement, the granularity and the energy range.

The first generation of 4π detectors was built in the 1970s and in the 1980s. Let us cite as examples the Plastic Ball [22], the Streamer Chamber [441] installed at Berkeley, the NAUTILUS ensemble [53,87,360,406] at GANIL, DIOGENE [8] at SATURNE, AMPHORA [168] at Grenoble, the DWARF BALL and WALL from Washington University [445], and the array called MEDEA in Catania [315], which is, however, mostly devoted to gamma-rays. Various identification methods were used in these devices: $\Delta E - E$ for the plastic ball, or $\Delta E - ToF$ (time of flight) in NAUTILUS, trajectory reconstruction with momentum measurement in a magnetic field (Streamer Chamber or DIOGENE), etc. The quantity ΔE is here the energy left in the detector by a particle during its passage, while E is the total energy deposited when the particle is stopped in the detector. By virtue of the Bethe formula [264], the correlation between E and ΔE allows us to measure the charge of the detected product. The ToF is the time interval between the emission of the particle and its detection, which allows a determination of the velocity of the particle. The $\Delta E - E$ method requires a full stopping of the detected products in solid state detectors constituting detection telescopes. Such instruments consist of a stack of at least two detectors. The study of the signals (and their correlations) delivered in each module of the telescope allows us to discriminate and identify products according to their atomic number and, to some extent, to their mass number. In contrast to the $\Delta E - ToF$ method for which identification is properly achieved, this method is useful only above a sizeable threshold corresponding to the thickness of the detector. In both cases ($\Delta E - E$, $\Delta E - ToF$), only charge (Z) identification is achieved on a large Z range; mass identification is also possible but only for light species ($Z < 5$). This drawback can, nevertheless, be overcome if magnetic methods are used, for which both a ΔE and a curvature radius, providing a charge-to-mass ratio, can be measured.

The second generation of detectors, which are now in operation at GANIL, SIS and MSU, rely on the same detection techniques ($\Delta E - E$, $\Delta E - ToF$) but with large improvements in geometrical efficiency, thresholds, identification and energy resolutions.

In FOPI [208], both NAUTILUS and DIOGENE techniques are used: a $\Delta E - ToF$ measurement is performed in the forward gas–plastic wall. Its angular aperture ($1.2°–30°$) is sufficient to allow a large coverage of the centre-of-mass forward hemisphere for symmetrical collisions, and its granularity (764 moduli) reduces the pile-up to a 10% probability in a given cell. The term 'pile-up' refers here to a situation in which two products emitted in the same event are detected in the same detector, thus leading to a pile-up of the corresponding electronic signals and to a bad characterization (energy and nature) of the products. In FOPI, charge identification is achieved up to $Z = 20$ and thresholds are 15 MeV and 50 MeV/u for protons and $Z = 15$, respectively. At larger angles particle trajectories in a magnetic field are analysed by using drift chambers. A combined measurement of their curvature radius, energy loss in the gas counter and ToF in a scintillator barrel allows us to identify detected products both in mass and charge.

Reconstruction of the trajectories is also the key point of the ALADIN set-up [250]. This small angular acceptance system ($\pm4.7°$ and $\pm4.5°$ in horizontal and vertical planes respectively) covers most of the kinematical region associated with the decay of projectile-like nuclei in peripheral relativistic inverse kinematics reactions (section 5.2). Any $Z \geq 2$ product is identified with a 90% efficiency from a ΔE (in the time projection chamber (TPC) + plastic wall) $- ToF$ (in the wall) measurement. The corresponding energy is obtained from trajectory reconstruction. ALADIN is hence well suited for projectile-like multifragmentation (see chapter 8), but one has to keep in mind that protons are undetected and that the information on the remaining parts of the system (target-

Table 4.1. Comparisons of performances of various multidetectors involved in experiments below 100 MeV/u bombarding energy.

Detector	Indra	MINIBALL	CHIMERA	Amphora	Isis
Number of cells	336	215 (188)	1192	140	162
Nature of detectors	Gas Si	Phoswich	Si	CsI	Gas Si
	CsI	(plast-CsI)	CsI	Phoswich ($<15°$)	CsI
Angular range	2–176°	9(14)–160°	1–176°	2–164° 14–86.5°	93.5–166°
Geometrical efficiency	90%	89%	94%	82%	80%
Charge identification range	1–50	1–18	Yes	1–9	1–15
Mass identification	H–He–Li–Be	H–He	Yes	H–He	H–He–Li
Energy thresholds	1 MeV/u	$Z = 3(10)$	Very low	$Z = 2(10)$	Very low
	all Z	$>1.5(2.5)$ MeV/u		$>4.0(8.0)$ MeV/u	

like and participant zone in the intermediate rapidity region) is thus poor, even if an ensemble of plastic scintillators has been added close to the target. FOPI and ALADIN are well suited to SIS experiments for which the kinetic energies of the detected products are often high. Similarly, the TPC of the EOS [206] collaboration is well suited for this energy range because it has a large efficiency and a correct charge resolution for light nuclei. Below $E_{lab} = 200$ MeV/u, charged-particle kinetic energy ranges are lower and $\Delta E - E$ methods give very good results. It is the main reason why this latter method is widely used in many devices suited to this energy range: INDRA [375] (GANIL), MINIBALL [157] (MSU), CHIMERA [6] (Catania), MULTICS [254], or ISIS [276] (Indiana). These various equipments have large geometrical efficiencies (from 80 to 94%, see table 4.1). They differ by the choice of ΔE or E detectors (hence, by the corresponding thresholds), by the associated electronics (in particular the energy range covered) and by the methods used to calibrate detectors. A ΔE gas counter is, for example, a good choice to achieve low detection thresholds; however, it is necessary to take care of the corresponding dead areas. The choice of silicon detectors ensures optimal Z identification (above $Z = 50$ for INDRA) and energy resolution. A granularity of several hundreds of cells allows small pile-up effects (5% double hits for 320 detectors and 40 particles in INDRA [375]). Table 4.1 gives a summary of the main characteristics of charged-particle multidetectors used below 100 MeV/u beam energy. Figure 4.3 shows the geometry and the detectors constituting the INDRA multidetector which has been used in various campaigns of measurement at GANIL and GSI. To conclude this part about the detection of charged particles, let us mention the possibility of using nuclear emulsions to 'visualize' (by means of tracks) complex events [417]. However, this technique does not allow us to accumulate large statistics.

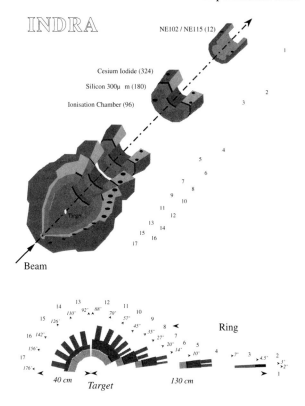

Figure 4.3. Schematic views of the multidetector INDRA. In the lower part of the figure, the geometry of the detector is shown. It consists of 17 rings labelled from 1 to 17 covering 2–176°. The solid angles of each ring have been optimized in order to limit the multi-hits and to obtain a good granularity without exhorbitant cost (i.e. of the associated detectors and dedicated electronics). A small dead zone around 90° allows the target to be placed in the centre of the detector. The device is very compact except in the very forward direction. The upper part of the figure displays the combination of detectors (NE102/NE115, CsI scintillators, silicon detectors and ionization chambers) which constitute each individual cell. The total energy and the identification of each detected particle is achieved by measuring the energy left in the various stages of a given cell. From [375].

4.1.2.3   Detection of neutral products

Besides charged-particle detectors, other dedicated large solid angle devices have been constructed to detect neutrons or gamma-rays. Neutrons are difficult to detect and characterize because they can be scattered from one detector to another. Hence, it has been necessary to consider two extreme situations. In neutron balls consisting of a large vessel filled with a liquid scintillator doped with Gd [200], the total efficiency is quite large (the geometrical efficiency can be very close to unity) but it is only possible to count the total number of emitted neutrons. In discrete neutron detectors, it is possible to measure the neutron kinetic energy by using the ToF technique but it is difficult to cover a sizeable fraction of space because of the cross-talk problem between neighbouring neutron cells. Indeed, this process corresponds to the situation in which a particle leaves a signal in more than one detection cell because it has been scattered from one detector to another. For instance, in DEMON, the geometrical efficiency is only about 1% [54, 460]. An intermediate situation between discrete counters and a single vessel can be achieved by using paddles of scintillators as developed by the LAND collaboration [56]. However, it is difficult to attain a general description of a collision only from neutron detection. Nevertheless, neutron balls may be used to sort the collisions according to the measured neutron multiplicity, which provides a measure of the dissipated energy in the reaction (see section 4.3.1.2). On the other hand, discrete neutron detectors may be used to obtain precise information on neutrons emitted in collisions which have been sorted by using information obtained from other (charged-particle) detectors.

The detection of gamma rays has also been investigated in large solid angle devices [315]. Of special interest are large energy gamma rays, because they retain a memory of the early stage of a nucleus–nucleus collision (section 6.4). The TAPS multidetector [346], which has a large efficiency for such high-energy (>30 MeV) gamma-rays, has been used to obtain interesting results in this field, both at GANIL and GSI. For such high-energy gamma-rays, the interaction in a given cell of the multidetector produces many particles (a shower) which can fire not only this cell but also neighbouring cells. Such a shower may be correctly analysed only by adding the signals obtained in neighbouring counters. This is the reason why such devices are organized in groups of large-size and high atomic number scintillators (BaF_2 in TAPS). Again, as for discrete neutron counters, a proper sorting of the events may be achieved only by using additional (charged-particle) counters.

It turns out from this brief detector survey that the experimental situation for nucleus–nucleus collisions is not ideal. Proper 4π charged-particle detection is more or less achieved today but this is not the case for all other particles. In fact, it is indeed nearly impossible to reach this goal in a satisfying way for *both* charged and neutral particles.

4.2   Analysis tools

We now come to a discussion of the analysis tools that have been designed to handle the large body of information obtained from experiments.

4.2.1 'Visualizing' nuclear collisions

Various reaction mechanisms are observed in nucleus–nucleus collisions (see chapter 5). To obtain an idea of the general 'topology' of the collisions, a good solution consists of examining bi-dimensional velocity plots of the emitted products. As a simple example, let us consider the case of a collision leading to one excited product with no angular momentum. Since the subsequent decay is isotropic, the centre-of-mass velocity spectrum of the corresponding emitted particles is similar in any direction. This property may be evidenced in a bi-dimensional plot where the absissa and ordinate are, respectively, the 'parallel' to the beam ($v_\parallel$ or v_{par}) and 'perpendicular' to the beam ($v_\perp$ or v_{perp}) velocity components. The variable in the third axis is the corresponding cross-section. Figure 4.4 is a schematic picture of the expected invariant cross-section isocontours in the case of three well-defined (and well-separated) isotropic sources. Circles are observed because the cross-sections have been normalized according to the elementary differential volume $2\pi v_\perp \, dv_\parallel \, dv_\perp$. An interesting feature is that such a cross-section $\frac{1}{v_\perp} \frac{d^2\sigma}{dv_\parallel \, dv_\perp}$ is invariant in a Galilean transformation since the elementary differential volume is conservative (see the appendix for more details). This means that, if the emitting source is not at rest in the reference frame, the isocontours will appear as circles centred at the source velocity. Such invariant cross-section plots may thus be used to characterize sources (isotropic emission in the present case), but also to measure their recoil velocities in the reference frame.

Note finally that when considering relativistic kinematics, one has to replace $v_\parallel$ and v_{perp} by the quantities y and β_{perp}: y is called the rapidity and is defined by

$$y = \frac{1}{2} \ln \left[\frac{E + p_\parallel}{E - p_\parallel} \right] \tag{4.2}$$

where E and $p_\parallel$ are the total particle energy (including mass energy) and the linear momentum parallel to the beam. The rapidity y may also be written as

$$y = \frac{1}{2} \ln \left[\frac{1 + \beta_\parallel}{1 - \beta_\parallel} \right] \tag{4.3}$$

where $\beta_\parallel$ is the velocity parallel to the beam, normalized to the velocity of light, and $\beta_\perp$ (β_{perp}) is the corresponding quantity perpendicularly to the beam. It is easy to check that y reduces to $\beta_\parallel$ for small β values as is the case in the non-relativistic regime.

4.2.2 Some remarks on the concept of 'sources'

The discussion of the previous section is based on the fact that an equilibrated, zero spin, excited nucleus is a particle source decaying isotropically. The underlying physical idea here is that the system has forgotten the way it was

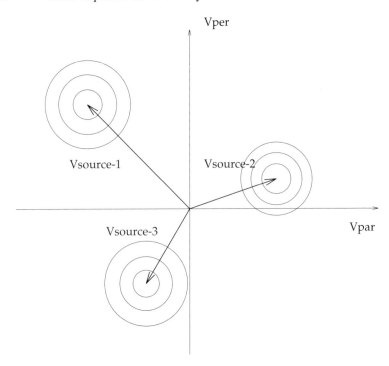

Figure 4.4. A schematic view of an isocontour plot for the invariant cross-section $\frac{1}{v_\perp} \frac{d^2\sigma}{dv_\parallel\, dv_\perp}$ in the case of three well separated sources with velocities $V_{\text{source-}i}$.

formed, in particular, the direction of the beam. This is the case if the lifetime of the source is long enough compared to the time needed to form it. However, the situation may be more complicated in some cases. For instance, if some angular momentum is brought into the system in the entrance channel, the source decay can exhibit some anisotropy reflecting its finite angular momentum. However, even in this case, the decay will be isotropic in the plane perpendicular to the spin vector. Similarly, we will see that nucleus–nucleus collisions can induce some compression of the released source(s), which can then expand. In this case, the isotropy will also be preserved, provided that the compression is purely radial. This means that a source behaviour (i.e. some isotropy in the decay step) holds when the stored energy has three origins: thermal (complete equilibrium or uniform phase-space occupation), radial compression–expansion and rotation. Note also that the source pattern will be distorted if some particles are emitted early, i.e. before a complete memory loss of the beam direction. These particles are called pre-equilibrium particles. Their behaviour will be discussed in chapter 6.

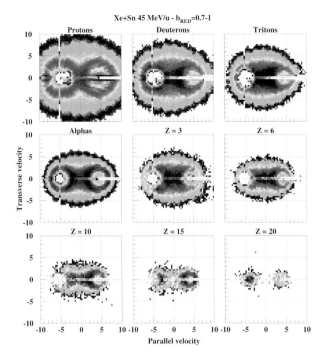

Figure 4.5. Isocontour plots of $v_{\mathrm{par}}-v_{\mathrm{perp}}$ (in cm ns^{-1}) for selected events observed in Xe+Sn collisions at 45 MeV/u for various emitted species from protons up to $Z = 20$. Events have been selected with the techniques described in section 4.2.3: they correspond to mid-central collisions. Two sources of emission associated with circular contours are seen, especially for protons, alphas and heavy fragments ($Z = 20$). However, deviations (too large abundance) from this simple scenario are also observed at $v_{\parallel}$ close to zero (see section 5.3.2.2). From [63].

Figure 4.5 is a typical example of invariant cross-section plots. It corresponds to semi-peripheral events selected according to the impact parameter of the reaction. This latter quantity has been estimated with the help of the method described in section 4.2.3. The system is Xe+Sn at 45 MeV/u. Relativistic effects may be neglected in this case. Two sources are clearly recognized, whatever the considered particle, even if a closer look at some plots reveals an enhancement of emitted particles at mid-velocity (i.e. close to $v_{\parallel} = 0$). This effect will be discussed in section 5.3.2.2. When the kinematics is favourable, as in the present case, the concept of an emitting source is very useful to analyse the data. Still, it is impossible to build such bi-dimensional spectra (as shown in figure 4.5) on an event-by-event basis, i.e. by using the particles of a single collision. Indeed, too few particles are emitted in a single event and, thus, no clear pattern would emerge in such a case. Hence it is necessary to add the contributions of as many, and as similar as possible, events which necessitates an efficient and reliable sorting procedure. As the amount of information obtained in a given event is quite large, the reduction of the information is thus a prerequisite for a suitable study of complex experimental events. This dedicated procedure can be achieved by using global variables built on an event-by-event basis, and by considering the characteristics of all detected particles.

4.2.3   Event sorting and reduction of the information

4.2.3.1   Impact parameter estimation

An important issue is to achieve event sorting according to the impact parameter b, which is a key parameter to understanding the reaction mechanism at a given beam energy (chapter 5). As a matter of fact, and in contrast to other physical systems, the initial conditions of heavy-ion collisions cannot be prepared and/or controlled in that respect. This is due to the obvious impossibility of directly measuring or controlling b (there is no instrument to measure distance or focus particles at the femtometre level!). Therefore, one has to rely on indirect methods, based on quantities which are expected to vary with the impact parameter. One has, nevertheless, to keep in mind that this will never be perfect, for two reasons: (i) because it will always be model dependent; and (ii) because the number of detected particles is a finite number so that fluctuation effects will play a role.

Basic impact parameter estimates are based on a guess about a possible monotonic relationship between a physical observable X and b. One can then define the reduced impact parameter by the relation [365]:

$$\frac{b(X)}{b_{\max}} = \sqrt{\int_X^\infty \frac{\mathrm{d}P(Y)}{\mathrm{d}Y}\,\mathrm{d}Y} \qquad (4.4)$$

where P is an observable proportional to the cross-section and $b_{\max}$ is the maximum impact parameter of the considered reaction. Examples of such a procedure for four different observables are shown in figure 4.6. The chosen variables involve either particle multiplicities (all particles (N_c) or limited to $Z = 1$ (N_1) in figure 4.6) or kinematical variables such as the transverse energy E_t (see equation (4.5)). In the first case, it is assumed that the most violent collisions correspond to the largest energy deposit and thus to the largest particle multiplicities. In the second case, the basic idea resides in the fact that energy dissipation induces a transfer of kinetic energy from the initial beam direction to the other directions, in particular, the transverse direction, perpendicular to beam direction.

The total transverse energy (E_t), taking into account the N-detected

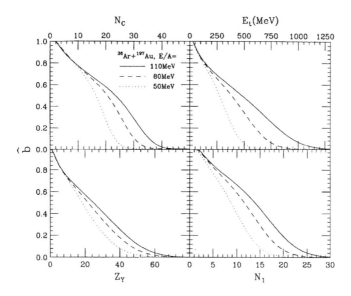

Figure 4.6. Reduced impact parameter (equation (4.4)) extracted from four quantities: the detected particle multiplicity N_c, the total transverse kinetic energy E_t, the mid-rapidity charge Z_y (see [365] for details) and the hydrogen multiplicity N_1. The different curves correspond to the three incident energies indicated in the first panel. From [365].

particles, is then defined as

$$E_t = \sum_{i=1}^{N} E_i \sin^2(\theta_i) \tag{4.5}$$

in which E_i and θ_i are, respectively, the kinetic energy and the laboratory detection angle of particle i. Of course, such a variable is meaningful only if the inefficiency of the experimental set-up does not introduce strong biases. Such defects can be reduced by restricting the summation to those particles for which threshold effects are less disturbing, such as light-charged particles (LCPs, $Z = 1, 2$). The corresponding transverse energy $E_{\text{trans}12}$ has been used in figure 4.5 as a sorting parameter. The energy transferred to the transverse direction E_t is sometimes normalized to its parallel component E_{par}. One may then use E_{rat} defined as

$$E_{\text{rat}} = \frac{E_t}{E_{\text{par}}} = \frac{\sum_{i=1}^{N} E_i \sin^2 \theta_i}{\sum_{i=1}^{N} E_i \cos^2 \theta_i} \tag{4.6}$$

which is here calculated in the non-relativistic approximation, and evolves in the range [0–2]. This quantity is close to zero for gentle (peripheral) collisions and reaches its maximum value for central head-on collisions.

Another impact-parameter sorting has been used recently in FOPI experiments [9, 10]. It is estimated from the azimuthal angular distribution of emitted particles through the so-called directivity

$$D = \frac{|\sum \boldsymbol{p}_t|}{\sum |\boldsymbol{p}_t|} \qquad (4.7)$$

where the sum runs over all detected particles. This quantity is close to zero for an isotropic azimuthal emission.

An important remark at this point concerns finite number effects which can induce strong fluctuations. They will be larger if the multiplicity of involved particles is smaller. For instance, fluctuation effects are larger on the transverse energy of LCP's ($E_{\text{trans}12}$) than on the total transverse energy E_t. Generally speaking, it is better to insert as much information as possible to define a sorting variable. This can be achieved by using several crossed sorting parameters. As an example, figure 4.7 shows correlations between various sorting variables for the reaction Ar+Au at 50 MeV/u. But the width of the correlation between two parameters gives a good indication of their ability to achieve an accurate sorting only if the variables of interest are not obviously correlated. This condition is not easily fulfiled; for instance, in figure 4.7, the total multiplicity N_c is obviously correlated with the hydrogen isotope multiplicity N_1 (including protons, deuterons and tritons) since the second one is included in the first one, and the measured correlations are then optimistic indications of the 'sorting power' of the selected variables.

4.2.3.2 Tensor analysis

Powerful methods consist of defining a single variable preserving the major part of the information collected by the experimental set-up. Variables measuring the shape of an event are good candidates here. They are obtained from a tensor constructed on the momenta [291]:

$$F_{ij} = \sum_{\nu=1}^{N} \gamma p_i^{\nu} p_j^{\nu} \qquad (4.8)$$

where F_{ij} are the tensor components. The sumation is usually performed by considering only fragments but can also be done with all detected particles. The $p_{i,j}$ are the momentum coordinates of each particle in the centre-of-mass (the indices i and j refer to Cartesian coordinates); γ is a weighting factor characterizing the tensor: momentum tensor if $\gamma = \frac{1}{p}$; energy tensor if $\gamma = \frac{1}{2m}$ (m being the particle mass); velocity tensor if $\gamma = \frac{1}{mp}$.

The energy tensor is most often used to analyse the experimental data. It can be diagonalized to reduce the event shape to an ellipsoid, the main axis of which

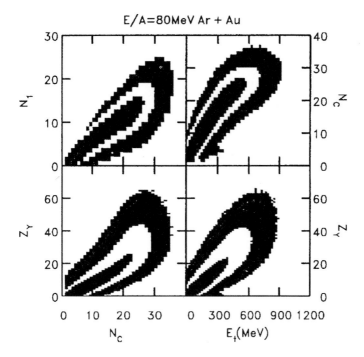

Figure 4.7. Experimental correlations between various sorting parameters for the Ar+Au system at 50 MeV/u: charged-particle multiplicity N_c, total transverse energy E_t, hydrogen isotope multiplicity N_1, mid-rapidity charge Z_y (see [365] for a definition of this variable). From [365].

is the eigenvectors of the tensor. The corresponding eigenvalues $\lambda_1, \lambda_2, \lambda_3$ are then used to define the event shape. The λ's are generally normalized to unity:

$$\lambda_1 + \lambda_2 + \lambda_3 = 1 \qquad (4.9)$$

and ordered according to

$$\lambda_1 \leq \lambda_2 \leq \lambda_3. \qquad (4.10)$$

Several quantities may then be defined from this tensor analysis:

- the sphericity
$$S = \tfrac{3}{2}(1 - \lambda_3) \qquad (4.11)$$

- the coplanarity
$$C = \frac{\sqrt{3}}{2}(\lambda_2 - \lambda_1) \qquad (4.12)$$

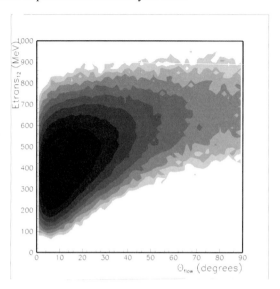

Figure 4.8. Correlation between $E_{\text{trans}12}$ and θ_{flow} for Xe+Sn collisions at 50 MeV/u. For small values of θ_{flow}, $E_{\text{trans}12}$ evolves rapidly until a soft saturation around 600–700 MeV. Then, in this range of energy dissipation, the flow angle populates a large angular domain and can be used as a sorting variable as discussed in section 5.3.3. From [342].

- the flow angle θ_{flow} between the beam axis and the λ_3 axis, namely the largest eigenvalue of the reconstructed ellipsoid.

An illustration of the use of such variables is discussed in section 5.3.3. Let us, nevertheless, take an example of the correlation between $E_{\text{trans}12}$ and θ_{flow}, as shown in figure 4.8 for the Xe+Sn system at 50 MeV/u. In this case, the energy tensor has been calculated using only 'massive' fragments ($Z \geq 3$). It is worthwhile to note that there are no other correlations than physical ones between $E_{\text{trans}12}$ and θ_{flow}, since the former is built with the characteristics of light particles while the latter is calculated with those of the fragments. For moderate values of $E_{\text{trans}12}$ corresponding to the less dissipative collisions, the flow angle θ_{flow} remains close to the grazing angle. In this region, the impact parameter is directly related to the value of $E_{\text{trans}12}$ and may be sorted according to different $E_{\text{trans}12}$ cuts [296]. For the most central collisions, θ_{flow} turns out to be a more discriminating variable. Hence, central collisions are better sorted according to θ_{flow} cuts (see section 5.3.3). We see on this example that a given indicator can be better suited for a given impact parameter range.

Other global variables have been proposed which are also correlated with the global shape of the event. For instance, the Fox momenta [190] are defined by the relation:

$$H_e = \frac{\sum_{ij} |\boldsymbol{p}_i||\boldsymbol{p}_j| P_e(\cos\theta_{ij})}{N} \qquad (4.13)$$

where P_e is the Legendre polynominal of order e, θ_{ij} is the relative angle between particles i and j, $\boldsymbol{p}_{i,j}$ their centre-of-mass linear momenta. In equation (4.13) N is a normalization factor such that $H_0 = 1$. Such a sorting method has been used in [306]. A unique source decaying isotropically (spherical shape) leads, for example, to $H_0 = 1$ and $H_\ell = 0$ for $\ell \neq 0$.

4.2.3.3 *Multi-dimensional analysis*

Up to now, we have only considered the correlation between a few (generally two) global variables. Recently, a very general procedure has been proposed to sort the events as a function of as many variables as possible [156]. Such a technique, based on multi-dimensional reduction techniques, allows us to search systematically for the best combination of variables that help to classify the events according to some prescription. A typical problem is to find the best combination of variables that can isolate, on an event-by-event basis and unambigously, those collisions which lead to a single isolated source in nuclear collisions [308]. Such methods are widely used in other fields of science and constitute a general framework for the statistical analyis of complex data. This very powerful technique, however, needs to be systematically calibrated and checked with the help of simulations. These are briefly discussed in section 4.4.

To summarize this rapid description of sorting variables, one may recall that none of them is perfect because of finite particle number effects. Some of them are better suited for peripheral reactions, others for more violent collisions. In any case, the impact parameter cannot be extracted from data with a precision better than about 10–20%. Some variables are better suited for a given experimental set-up if they are relatively insensitive to the experimental deficiencies, uncertainties or thresholds. However, generally speaking, the most powerful variables are those using the full information extracted from the events, namely multiplicities and kinematical quantities.

4.2.3.4 *Separation and reconstruction of sources*

The result of event sorting may be that invariant cross-section plots reveal circular contours which are signatures of sources. The next step then consists of a separation of these sources on an event-by-event basis. This means that for each event one aims at attributing each detected product to a given source. Several methods have been developed for this purpose. They all use the fact that particles or fragments emitted from a given moving source have velocities focused along the source velocity. In that sense, particles or fragments emitted from a recoiling source look like a 'jet' in the direction of the velocity of the source. This description is fully correct if the source velocity is large compared with the velocities of particles and fragments in the source frame.

In the thrust method, for example, the attribution of each product of an event to a given source is obtained by maximizing the quantity:

$$T = \frac{|\Sigma_i \boldsymbol{p}_i| + |\Sigma_j \boldsymbol{p}_j| + \cdots}{\Sigma_k |\boldsymbol{p}_k|} \tag{4.14}$$

in which $p_{i,j,...,k}$, are the linear momenta of the various products in the centre-of-mass frame. Each sum of the numerator corresponds to a given source and the algorithm consists of searching the 'best' repartition between various sources, i.e. the repartition which maximizes the numerator. The denominator is simply a normalization factor such that T evolves in the range $[0, 1]$.

In the MST minimum spanning tree (MST) method, one seeks for the 'best' way of connecting N products in momentum space, i.e. the way which minimizes the total path. This path includes $N - 1$ bonds. It is possible to distinguish two different sources by cutting the longest bond. The application of this method needs to define the 'length' of a bond. In our case, it may be defined in the velocity space and the square of the relative velocity between two products is generally used.

As already stressed, all these methods are relevant only if the involved sources are clearly disconnected in velocity space. We will see that, unfortunately, the event topology is often not so simple and that products may be emitted at mid-rapidity between two main sources. The attribution of such products to one given source can therefore be misleading.

4.3   Relevant variables and source characterization

We have seen in figure 4.5 how $v_\parallel - v_\perp$ plots may be useful in identifying emission from underlying sources. These sources correspond to excited nuclei. An important aspect of the study of nuclear collisions is the proper characterization of such excited species and the measurement of their excitation energy, temperature and lifetime. Indeed, excitation energy and temperature are the key quantities in studying nuclear thermodynamics, which is one of the main goals of the study of hot nuclear species. Finally, timescales are obvious crucial quantities for detailed discussions of nuclear dynamics and the time development of nuclear collisions towards thermalization.

In this section we describe the experimental methods used to obtain these physical quantities (excitation energy, temperature, etc) putting more emphasis on the techniques than on the results. The latter will be extensively discussed in chapters 5 and 8. We therefore assume in the following discussions that the sources to be characterized have been properly separated and reconstructed with the methods described in the previous section. The question of the formation of such sources in nuclear collisions will be addressed in chapter 5.

4.3.1   Nuclear calorimetry

The methods of nuclear calorimetry aim to estimate the total excitation energy or the excitation energy per nucleon ϵ^* deposited in sources in the course of the reaction. Let us recall (see section 4.2.2) that ϵ^* is the sum of the thermal, compression and rotational energies. It is the part of the incident energy which has been dissipated during the collision, the remaining part being observed as pre-equilibrium particles or as the kinetic energies of the released sources. As we shall see later ϵ^* can be measured in two ways, either by considering the energy of all emitted particles from the sources or by a subtraction method. This latter consists of measuring the part of the energy that has not been dissipated and subtracting it from the total available energy in the collision. Before discussing these two methods, it is, nevertheless, necessary to estimate, on an experimental basis, the degree of equilibrium attained by the sources, the energy of which one aims at evaluating.

4.3.1.1   Experimental signatures of equilibrium

As previously explained, we focus here on sources which have decayed after a formation step during which they have lost the memory of the incoming beam direction. This means that the corresponding fraction of the incident kinetic energy has been shared among various degrees of freedom: thermal, compressional and rotational degrees of freedom. The experimental criterion which may be used to test such a memory loss of the entrance channel beam direction is the isotropy of the angular distributions of emitted particles, in the reference frame of the studied object. Examples of such (partially) equilibrated systems are displayed in figures 4.9 and 4.10. In figure 4.9, the charge distributions of the IMFs emitted in central Xe+Sn collisions at 50 MeV/u are shown for various centre-of-mass angular ranges. All distributions can be superimposed except for very forward–backward angles for which a (small) non-equilibrated component is apparent. Such an anisotropic component can be associated with remnants of the projectile and target which did not fuse in the course of the reaction. A very similar behaviour is observed for the same system when focusing on α-particles, as shown in figure 4.10, in which the kinetic energy distributions have been compared at various centre-of-mass angles. Here again, all distributions are superimposed (without any arbitrary normalization) except for very forward and backward angles for which the spectra exhibit a high-energy part associated with fast pre-equilibrium emission. Once an experimental signature of equilibrium or an experimental selection of particles reflecting the excitation energy stored in the source has been attained, the next step consists of measuring the excitation energy deposited in the system.

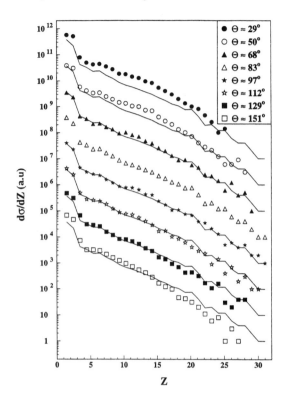

Figure 4.9. Charge distributions of the IMFs ($Z > 2$) emitted in central Xe+Sn at 50 MeV/u collisions for various centre-of-mass angular bins. These collisions correspond mainly to the formation of a single source decaying isotropically. The lines correspond to the distribution at a mean angle of 83° (same as open triangles). For clarity, each distribution has been shifted by an order of magnitude but no normalization has been applied from one angle to another. From [305].

4.3.1.2   *Calorimetry with the balance equation and/or the neutron multiplicity*

The balance equation method takes into account the kinematical characteristics of the particles emitted by a source by considering an energy balance. The deposited excitation energy per nucleon ϵ^* is thus defined as

$$\epsilon^\star = \frac{\sum_{i=1}^{M} E_{\text{kin}}^i + Q^i}{A_{\text{meas}}} \qquad (4.15)$$

where E_{kin}^i is the kinetic energy of product i in the source frame, M the multiplicity of particles emitted from the source, Q^i the binding energy of product i and A_{meas} the reconstructed mass of the source. This latter quantity

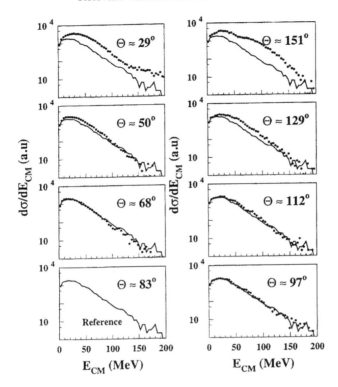

Figure 4.10. Centre-of-mass kinetic energy distributions of the α particles emitted in central Xe+Sn 50 MeV/u collisions for various centre-of-mass mean emission angles. The full line in each panel is the distribution for a mean angle of $83°$. From [305].

is not easily accessed since, most of the time, neither the neutrons nor the masses of the emitted products are measured. When neutrons are not measured, their multiplicity is estimated from mass balance or from proton multiplicity by assuming a given value for the isospin ratio $R = N/Z$ of the source. The same procedure is used for the mass numbers which are estimated by assuming the most probable isotope associated with each measured atomic number. The average kinetic energy of the neutrons can be obtained from the proton spectra by subtracting an effective Coulomb barrier usually of the order of a couple of MeV.

A very similar method can be used with the help of the neutron multiplicity measured on an event-by-event basis by neutron detectors such as ORION [200]. The main drawback of this method is that it relies, to a large extent, on the capability of the experimental apparatus to detect correctly *all* particles emitted by the source. Therefore, it is very sensitive to the thresholds and dead zones of the detector and the estimated quantity must very often be corrected or at least checked with help of simulations.

4.3.1.3 Calorimetry with the 'subtraction' method

This method is somehow complementary to the previous one. It consists of obtaining the excitation energy per nucleon ϵ^* by subtracting from the total available energy non-thermal components of the motion of both light particles and massive fragments. This is achieved by considering the kinematics of the reaction.

Collective non-thermal motion is easy to identify in some well-defined situations. In fusion-like reactions in which only one massive excited nuclear object (the so called 'compound') is produced, the excitation energy is obtained directly by subtracting the 'recoil' energy[1] of the compound from the available energy. However, there may be some problems with the fast emission of light particles (the so-called pre-equilibrium stage of the reaction (see chapter 6)). In this case, the relation between the linear momentum transferred to the recoiling compound nucleus and the corresponding energy deposit is no more straightforward.

As will be discussed in chapter 5, collisions may end up with two massive outgoing products reminiscent of the entrance channel characteristics of the reaction. In this case, simple two-body kinematics allows us to obtain the total kinetic energy E_{TKE} (also denoted TKE in the literature) by the following expression:

$$E_{\text{TKE}} = \tfrac{1}{2}\mu V_{\text{rel}}^2. \tag{4.16}$$

In this equation, V_{rel} is the relative velocity between the two massive outgoing products and μ is the reduced mass of the system. The excitation energy per nucleon then reads:

$$\epsilon^* = \epsilon_{\text{CM}} - \frac{E_{\text{TKE}}}{A_p + A_t} \tag{4.17}$$

in which ϵ_{CM} is the available centre-of-mass energy per nucleon and A_p and A_t are, respectively, the mass number of the projectile and target. As before, pre-equilibrium processes may affect this estimation of ϵ^*.

Measurements of the excitation energy from the relative velocity of the reconstructed partners of the reaction have been used mainly for heavy systems in which the Wilczynski plots (see section 5.3.2.1) have been built [88,280,313,314]. The main drawback of this method is that it only gives an average value of the total energy dissipated in both partners. Thus, no information is available on the sharing of the dissipated energy among the two partners of the reactions and one has to rely on some specific assumptions (equal temperature or equal excitation-energy sharing between the two partners) to assign a given value of the excitation energy to each source.

---

[1] This is the term used to denote the kinetic energy of the excited nucleus in the laboratory frame.

4.3.2 Nuclear thermometry

We have seen in section 2.3.2.3 that the description of an excited nucleus in terms of temperature is somewhat ambiguous because the correlation between temperature and excitation energy is not clear for a small system. In other words, microcanonical and canonical approaches are not fully equivalent for small systems.

From an experimental point of view, this conceptual difficulty is, however, not so serious because the observations are not performed on a single nucleus but on a collection of nuclei which have been sorted as explained in section 4.2. Because of this cumulative effect, it is meaningful to describe an excited nucleus in terms of a mean temperature: the successive observations of N similar nuclei is equivalent to the observation of a unique larger system for which the small size limitations are softened.

Another conceptual but more serious problem is related to the fact that the system is open. When a particle is emitted (evaporated) as described, for example, by the statistical model (see section 2.4), it is taken away from the system, in contrast to the situation of an equilibrated system in which the evaporated particles are kept in the vapour phase surrounding the liquid phase. In the nuclear case, the system can thus exchange matter and energy with the outside: in other words it is metastable. This question of the theoretical description of a hot and open system is discussed in section 7.1.2. Here, we handle the problem from the experimental point of view of extracting a temperature from the characteristics of the detected light particles and fragments.

In section 2.4.1 the statistical model was used to calculate the emission probability of a given particle with a defined kinetic energy. In this framework the temperature may be introduced and is correlated to the density of states of the nuclei which remain after particle emission (equation (2.83)). The measurement of temperatures in nuclear collisions is thus basically a matter of measuring the density of state $\rho(E^*)$. In the microcanonical description, this is equivalent to a measurement of the probability of occurrence of the selected channel. This general method may be applied in three specific cases which are described here [41]:

- by studying the slopes of maxwellian kinetic energy spectra which leads to the so-called 'kinetic' temperatures (upper part of figure 4.11);
- by studying discrete state population ratios of selected clusters which leads to a determination of the 'internal excitation' temperature (middle part of figure 4.11);
- by studying double isotopic yield ratios: the so-called 'double ratio' or 'chemical' temperatures (lower part of figure 4.11).

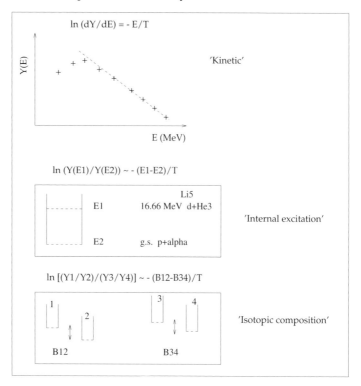

Figure 4.11. A schematic description of the three methods discussed in the text that are used to measure temperature T in nuclear collisions. All the methods are based on an experimental determination of the density of states ρ of the decaying hot nucleus as a function of the excitation energy (see equation (2.83)). However, they differ by using the properties of different decay products. The first method (upper part, developed in section 4.3.2.1) is based on the light-particle-evaporation process and thus assumes a direct connection between the kinetic energy 'density of states' of the evaporated particles and ρ. The second method (middle part, discussed in section 4.3.2.2) is based on cluster emission and assumes that the relative probability to excite different states of a given cluster is directly linked to the density of states ρ of the parent nucleus by means of a Boltzmann factor. The last one (bottom part, section 4.3.2.3) is similar to the previous one but uses the yields of four different emitted particles whose difference in their binding energies is denoted by Bij.

4.3.2.1 Determination of 'kinetic' temperatures

In section 2.4.1, the relative probability of emitting a particle i with a kinetic energy ϵ has been calculated as:

$$P_i(\epsilon) = \frac{\epsilon - B_i^{\text{Coul}}}{T^2} \exp\left[-\frac{\epsilon - B_i^{\text{Coul}}}{T}\right] \qquad \epsilon \geq B_i^{\text{Coul}} \qquad (4.18)$$

where B_i^{Coul} is the Coulomb barrier for emitting particle i from the excited nucleus. This relation means that the kinetic energy spectrum $dN_i/d\epsilon$ exhibits a maxwellian shape with a high-energy exponential fall-off reflecting the nucleus temperature. This method of measurement of temperatures has been widely and successfully used in the literature.

One should, however, note that this method only gives an average estimate of the temperature along the decay chain. Indeed, since it is impossible to attribute an emission time to each observed particle, the quantity $dN_i/d\epsilon$ is time-averaged and can only provide a so-called *apparent* temperature. Furthermore, since the emission of the various species is governed by phase space constraints, there is a hierarchy in the various emission probabilities so that different particles are not emitted with the same timescales and thus reflect different temperatures. The averaging effect along the decay chain may in turn be unfolded by comparing measurements for various excitation energies [231].

A common procedure to obtain the initial temperature is to confront the data with the predictions of statistical models (see section 2.4). In such models, the sequential effects as well as the competition between the emission of various particles are taken into account. The comparison with experimental data of the spectra generated by such models using different input parameters allows us to 'trace back' the initial temperature [142, 324].

A general drawback of this method is related to kinematical effects. Kinematical problems arise from the fact that the observed kinetic energy of the emitted particle is a convolution of the velocity of the emitted particle, relative to the emitting source, with the source velocity itself. This latter quantity can be poorly known for several reasons. In particular, successive recoils of the decaying nucleus after each emission induces uncertainties which become increasingly important with the mass number of the emitted particle.

One has also to be cautious when the excitation is sufficient to lead to the multifragmentation of the system (a process which is extensively discussed in chapter 8). In this case, equation (4.18) has to be revisited to take into account the fact that all the particles are emitted in a single step from the bulk of the nuclear system. Their kinetic energy distribution is then the internal kinetic energy distribution of particles belonging to a gas admixture. In the Boltzmann approximation (which may be valid when the density of the nuclear system becomes low), one then obtains

$$P_i(\epsilon) \sim \frac{\sqrt{\epsilon - B_i^{\text{Coul}}}}{T^2} \exp\left[-\frac{\epsilon - B_i^{\text{Coul}}}{T}\right] \qquad \epsilon \geq B_i^{\text{Coul}} \qquad (4.19)$$

which looks like equation (4.18) except for the square root of $\epsilon - B_i^{\text{Coul}}$ in the forefactor. One speaks of surface emission when equation (4.18) is valid, i.e. when particles are sequentially emitted as was considered in section 2.4. This is the case at limited excitation energies. Conversely, volume emission (equation (4.19)) corresponds to a one-step emission from a disintegrating system

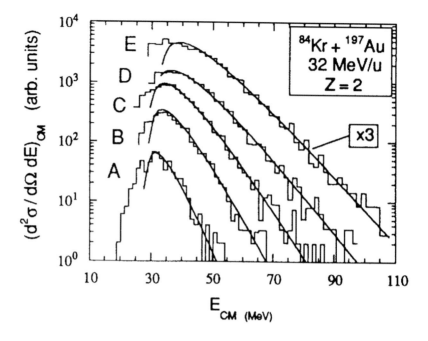

Figure 4.12. Kinetic energy distribution for $Z = 2$ particles emitted in Kr+Au collisions at 32 MeV/u. Particles have been detected at backward angles in which evaporation is the dominant process. Each curve corresponds to a given neutron multiplicity M_n bin (increasing from curve labelled A to curve labelled E). M_n is a measure of the total excitation energy deposited in the system. Data have been fitted with maxwellian-like distributions expected in a thermal process. From [135].

and applies at larger excitations. A further modification has to be considered if the measured kinetic energy does not reflect only a thermal decay but also a collective expansion energy. This point is quantitatively discussed in section 8.4.2. Finally, one has to pay attention to pre-equilibrium emissions (see section 6.1.1) which can strongly affect the high-energy tails of spectra. This drawback may, however, be eliminated by analysing energy spectra in directions where pre-equilibrium emission is minimized (see section 6.1.1). This is usually the case at backward angles for asymmetrical systems involving a light projectile and a heavier target (see figure 4.12). In this angular range, particle kinetic energy spectra show monotonic maxwellian-like behaviours. An analysis of the slopes of the distributions provides a measure of the temperature of the emitters.

4.3.2.2 Ratios of populations of excited states

In order to avoid the drawbacks of the kinetic energy spectra method, several years ago the use of a method based on the study of excited states of composite particles was proposed. In this method, instead of measuring a continuum of states by the kinetic energy spectra as in the previous section, the population of a few excited states of a given emitted product (cluster) is analysed.

Let us consider a composite particle with two discrete excited states labelled, respectively, 1 and 2. The excitation energy of the states, measured from the ground state, are, respectively, E_1 and E_2 with spin s_1 and s_2. It is possible to use the formalism of the statistical model (section 2.4.1) to calculate the ratio of the probabilities of observing the two selected states. The probability of observing a given product with a kinetic energy ϵ is given by equation (2.81). This equation, for the emission of the considered composite particle in a given state $i = 1, 2$, reads:

$$P_i(\epsilon)\, d\epsilon = \frac{\rho(E_f^*(i))}{\rho(E_0^*)} (2s_i + 1) \frac{4\pi p^2}{h^3} \sigma_c(\epsilon)\, d\epsilon \tag{4.20}$$

in which E_0^* and $E_f^*(i)$ are the energy of the emitter before and after emitting the considered cluster in a state i. The value of $E_f^*(i)$ depends on the respective energies of the two excited states according to the following balance equation:

$$E_f^*(i) = E_0^* - E_i - Q - \epsilon \tag{4.21}$$

where $-Q$ is the Q value of the process when considering particles in their ground state and ϵ is the kinetic energy of the cluster in the emitter frame. Over an ensemble of measurement, the mean value $\bar{\epsilon}$ is independent of the excited state of the emitted cluster. Thus, the ratio of the two probabilities reads:

$$\frac{P_1}{P_2} = \frac{2s_1 + 1}{2s_2 + 1} \frac{\sigma_{c1}(\bar{\epsilon})\rho(E_f^*(1))}{\sigma_{c2}(\bar{\epsilon})\rho(E_f^*(2))}. \tag{4.22}$$

A usual approximation consists of assuming that the cross-section for the inverse process (i.e. capture) does not depend on the internal state of the cluster. Then, from

$$E_f^*(1) - E_f^*(2) = E_2 - E_1 \tag{4.23}$$

and from equation (2.92) for $\rho(E_f^*(i))$, one ends up with the following relation:

$$\frac{P_1}{P_2} = \frac{2s_1 + 1}{2s_2 + 1} \exp\left[\frac{-(E_1 - E_2)}{T}\right]. \tag{4.24}$$

The same result can be obtained by assuming that the emitted product is in thermal equilibrium with a heat bath at temperature T. The excited states are then populated according to a Boltzmann law. The ratio of state populations of the emitted clusters thus provide a measure of the ratio of the state population of the emitter itself.

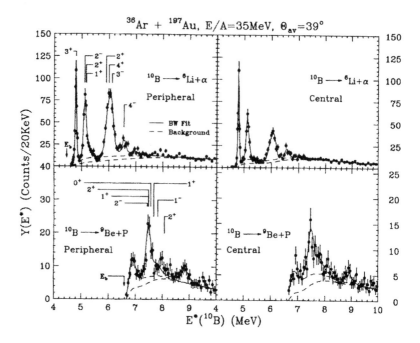

Figure 4.13. Yields for various decay channels of ^{10}B emitted in Ar+Au reactions at 35 MeV/u. The abcissa is the excitation energy of excited ^{10}B reconstructed by considering the kinematics of the two associated decay products (^{6}Li and α in the upper case, ^{9}Be and proton in the lower case). Each peak in the distribution is associated with one or several discrete excited state(s) whose spin and parity is (are) indicated. The broken line is the estimated background contribution. The left panels are for peripheral collisions and the right panels for central collisions. From [503].

Before coming to the experimental aspects of the technique, let us mention a few inherent conceptual problems. First, it is implicitly assumed that the emission of the composite particle in a given state is solely dictated by phase space constraints: this means that there is no 'hidden' dynamical effect in the emission probabilities and (as already mentioned) that the inverse capture cross-section $\sigma_c(\epsilon)$ is the same for both excited states. This also means that neither the question of the pre-formation of such states in the emitter nor the question of possible final-state interactions that could affect the measured emission probabilities are considered. From an experimental point of view, the observation of discrete states of a given composite particle is achieved by measuring the relative energy spectra of the associated coincident decay products (mainly charged particles [503, 504], but also neutrons [150]). Therefore, each discrete excited state is associated with a peak in the excitation energy spectrum (figure 4.13). The number of counts in each peak gives (after subtraction of the background) the population associated with each discrete state. In this case, the background is due to uncorrelated particles, i.e. to particles which do not result from a single composite particle decay.

Apart from the problem of background subtraction, the main drawback of this method certainly lies in side-feeding effects. This effect follows from the fact that all observed resonances are not emitted by the source of interest but may result from the decay process of other more massive clusters which were previously emitted. These 'side-fed' resonances thus do not reflect the actual temperature of the emitter and lead to an underestimation of the temperature. Correcting for such effects relies, to a large extent, on actual knowledge of all discrete states up to massive clusters and is thus a delicate procedure, which requires the use of quantum statistical models [226, 232, 269, 471]. But it turns out that side-feeding effects are minimized when considering discrete states with a large energy difference ΔE. Furthermore, the method has several attractive advantages. It may be applied on various fragments leading to several independent temperature measurements and it is frame-independent since only relative energies are needed.

4.3.2.3 Double ratio of isotopic yields

In the framework of the statistical approach, the probability of fragment or particle emission is mainly governed by phase space constraints, in particular, by the separation energy associated with the emission process. This property is readily seen in equation (2.81). It has been used in section 2.4.3 to express the probability for neutron emission as a function of the neutron binding energy (equation (2.95)). The separation energy is nothing but the opposite of the chemical potential. In decay processes involving evaporation from the surface of a hot nucleus at a density close to the saturation density, the chemical potentials are easily estimated from ground-state properties. However, there are situations in which decay processes occur at low density (see chapter 8). The probabilities of formation of a given cluster then depend on the chemical potential (see equation (8.11)). This latter is related to the chemical potentials for free protons μ_Z and free neutrons μ_N as given by equation (8.12). These two quantities are not easily accessible experimentally. Albergo *et al* [11] have thus proposed a method to measure the temperature in which the chemical potentials are not needed. The idea is to study ratios of populations of clusters differing either by a single neutron or by a single proton so that the chemical potentials cancel in the final expression. Let us define R as the following double ratio of particle yields $Y_{ij} = Y(A_i, Z_j)$

$$R = \frac{Y(A_i, Z_i)}{Y(A_{i+1}, Z_i)} \frac{Y(A_{j+1}, Z_j)}{Y(A_j, Z_j)} = R_1 R_2. \qquad (4.25)$$

According to equation (8.11), the first ratio R_1 in equation (4.25) reads (R_2 is obtained in a similar way):

$$R_1 \sim \frac{(2s_{i,i}+1)\exp(\mu_{i,i}/T)\exp(B_{i,i}/T)}{(2s_{i+1,i}+1)\exp(\mu_{i+1,i}/T)\exp(B_{i+1,i}/T)} \qquad (4.26)$$

in which the B's are the binding energies of each considered cluster (i,j) with mass number A_i and charge Z_j and with spin $s_{i,j}$. According to equation (8.12), the exponential terms containing the μ's give altogether a factor 1. Combining the spin factors into a single variable α_s and the binding energies according to

$$\Delta B = B(A_i, Z_i) - B(A_{i+1}, Z_i) - B(A_j, Z_j) + B(A_{j+1}, Z_j) \qquad (4.27)$$

one finally obtains

$$R = \alpha_s \exp(\Delta B/T). \qquad (4.28)$$

By measuring R experimentally, the temperature is readily obtained as

$$T = \frac{\Delta B}{\ln(R/\alpha_s)}. \qquad (4.29)$$

This method has been applied first to ^{3}He, ^{4}He, ^{6}Li and ^{7}Li species by the ALADIN group [372] but other isotopes have been used by the INDRA [299] and MSU [471] collaborations.

This technique suffers from the same defect of side-feeding effects as found in the technique using discrete state population ratios. In [471] it is suggested that the side-feeding corrections would be system-independent for a given double pair of particles. However, this result is not established for significantly large temperatures. As a matter of fact, there is a question concerning the sensitivity of the method for temperatures exceeding 5 MeV. In [303], it is concluded that the double pair ^{3}He, ^{4}He, ^{6}Li, ^{7}Li may still be used in this temperature range, but reverse conclusions are obtained in [299, 471].

A cross-comparaison of these three thermometric methods and their consequences regarding nuclear thermodynamics will be extensively discussed in section 8.4.3.4.

4.3.3   Nuclear rotation

As mentioned in the introduction to this section, nuclear calorimetry and thermometry require the estimation of ordered collective motion. Rotation and expansion of the sources are typical examples of such motions. Expansion will be discussed and estimated in section 8.4.2. We focus here on nuclear rotation.

At non-zero impact parameters, large angular momenta (spins) can be transferred to the partner(s) of a reaction, resulting in an intrinsic rotational motion of the nucleus(i) and thus to anisotropic emission of particles or fragments. Let us take the example of Ar+Au collisions at 30 MeV/u for an impact parameter

$b = 2$ fm. A classical estimate of the available angular momentum reads (with the notation of section 4.3.1.3)

$$L = \frac{\mu V_{\text{rel}} b}{\hbar} \tag{4.30}$$

which gives in this case $L \sim 78\hbar$. Now, if the decay of such 'rotating' nuclei is governed by statistical features the probability for emitting a particle removing a rotational energy E_{rot} is

$$W(E_{\text{rot}}) \sim \exp\left(-\frac{E_{\text{rot}}}{T}\right) \tag{4.31}$$

in which T is the temperature of the source. The rotational energy E_{rot} can be written as

$$E_{\text{rot}} = \frac{\hbar^2 J^2 \cos^2(\theta)}{2 I_{\text{eff}}} \tag{4.32}$$

in which J is the value of the 'aligned' spin. The 'aligned' spin is that part of the angular momentum which is perpendicular to the reaction plane defined by the beam axis and the recoil velocity of the rotating source. The variable θ is the angle between the direction of the angular momentum and the angle of emission of the considered particle while I_{eff} is the moment of inertia of the di-nuclear system made of the final nucleus and the emitted particle at contact which is usually estimated with the rigid body hypothesis.

According to equation (4.32), the particle angular distributions are Gaussian-like. Experimentally, such distributions are fitted in order to extract the angular momentum once the temperature has been deduced by the methods described in section 4.3.2. There again, various problems arise because of the uncertainty about the measure of the temperature and also the moment of inertia of the source which enters in the estimate of I_{eff}. Owing to the fact that particles may be emitted at various steps of the reaction, only an average value of the angular momentum is accessible from the data. Therefore, statistical model calculations are often necessary to correct for such effects and to estimate the initial angular momentum that has been transferred to the rotating source. From a general point of view, the angular momentum effects are an increasing function of the mass of the emitted particles. They are of special importance when the excited nuclei decay by fission.

4.3.4 Nuclear chronometry

We have briefly mentioned in the introduction the role of time in the description of hot sources and we shall discuss this issue again in sections 7.2.3 and 8.4.1.1. In the spirit of the present discussion, which is essentially devoted to the basic characterization of excited and well-defined nuclear species, we now come to the discussion of reaction time and nuclear lifetime measurements. Various techniques can be used depending on the considered timescales. Here we focus on the methods used in the lifetime range below 10^{-17} s. For such very small values, the recoil ranges of objects with a typical velocity of $0.1c$ is only 3×10^{-10} m, which makes irrelevant any method based on a direct measurement of the flight path before decay. Various methods can be classified according to their applicability. Long times can be measured by 'smooth' methods such as Ericson fluctuations or crystal blocking techniques. The competition between slow and fast processes, such as fission versus evaporation (see section 7.2.3), can be handled using the so-called neutron clock. In violent collisions other methods must be used taking advantage of the correlations between fragments and particles induced by the long-range Coulomb force.

4.3.4.1 *Ericson fluctuations and crystal blocking techniques*

About 40 years ago, Ericson proposed using energy autocorrelation functions to measure the decay width Γ of a given state [176, 177]. The lifetime is then obtained from the usual relation

$$\tau = \frac{\hbar}{\Gamma}. \tag{4.33}$$

At high excitation energy Γ is much larger than the mean energy spacing Δ of the states. Thus, the cross-section results from the various interferences between all nearby overlapping states. By considering a given final state or a set of final states in a small energy interval, the decay width is measured by building the energy correlation function. Recently, this method has been revisited [7, 116] to be applied at much higher energies corresponding to decay widths as large as 250 keV (2.6×10^{-21} s) while the initial technique essentially covered values from $\simeq 5 \times 10^{-21}$ to $\simeq 10^{-19}$ s [39].

A direct measurement of a collective process such as nuclear fission (see section 7.2.3) can be achieved by using the crystal blocking technique. The idea is to induce the fission of a heavy projectile interacting with one of the nuclei located at the site of a crystal [212, 325]. Fission fragments are detected along a direction corresponding to an axis of the crystal. An effect based on the propagation of the fission fragments in the crystal is used to estimate the timescale of the process. Indeed, in the case of a fast process in the vicinity of a crystal site, the strong electromagnetic interaction of the fission fragments with the crystal will deflect the latter so that very few fragments will escape the crystal. In contrast, for a long process, the fission fragments are produced far from the crystal sites and can thus escape the system. A quantitative estimation of the fission timescale is thus possible: typical values obtained that way (to be discussed in section 7.2.3) range from 10^{-16} to 10^{-19} s.

4.3.4.2 *Intensity interferometry between light particles*

The second-order interferometry method (the so-called intensity interferometry) was originally proposed by Hanbury-Brown and Twiss to measure the angular size of stars [233]. It can also be used in the nuclear context to measure source sizes and lifetimes in nuclear collisions. Lifetimes are estimated by measuring the average time between two successive particle emissions. The method is based on the study of the correlation of pairs of particles (either fermions or bosons) emitted at low relative momentum (for reviews see [15, 35, 307]).

A correlation function $R(q)$ can be built by considering the number $N_{12}(q)$ of measured coincidences between two particles of momenta p_1 and p_2 at relative momentum $q = p_1 - p_2$ and the normalized number of 'singles' $N(p_1)$ and $N(p_2)$. One defines:

$$1 + R(q) = \frac{N_{12}(q)}{N(p_1)N(p_2)} \qquad (4.34)$$

so that $R(q) = 0$ when the two particles are uncorrelated. The interpretation of the experimental correlation functions takes into account the spacetime distribution of the emitted particles as well as their mutual nuclear and electromagnetic interactions. Another important point resides in the quantum statistical nature of the two-particle wavefunction. The single-particle distribution function g inside the source as a function of momentum p, position r and time t is usually parametrized according to

$$g(r, t, p) \sim \exp\left(-\frac{r^2}{R_S^2} - \frac{t}{\tau}\right) Y(p) \qquad (4.35)$$

where R_S is the radius of the source, τ its lifetime and $Y(p)$ its momentum distribution. Starting from such a theoretical spacetime distribution of the matter inside the source, it is possible to build a theoretical correlation function as defined in equation (4.34). This latter is compared with the experimental results as shown in figure 4.14 thus giving access to a determination of the size and lifetime of the source. It appears that the quantities R_S and τ are correlated through relation (4.35). This means that it is possible to measure τ only if one knows the value of R_S from another observable. From this point of view, the measurement only gives access to the 'spacetime size' of the source. However, one has to stress the fact that it is possible to obtain two independent measurements of τ and R_S if one uses the fact that p is a vector and if two independent measurements are performed for q parallel (respectively perpendicular) to $p_{1,2}$ (longitudinal— respectively transverse—correlation function in the caption to figure 4.14). The difficulty usually lies in the statistical accuracy of the measurement which requires huge statistics.

4.3.4.3 Intensity interferometry between large fragments: Coulomb correlations

For violent collisions leading to a large explosion of the matter into many different species (see chapter 5), the intensity interferometry method with light particles exhibits limitations due to strong final-state interactions between the many emitted particles and fragments. Other techniques such as Ericson fluctuations

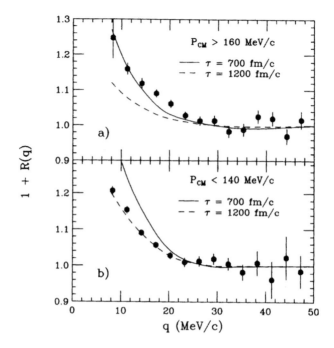

Figure 4.14. Experimental two-neutron correlation function obtained in O+Mg collisions at 130 MeV total incident energy (black points). Calculations for two different values of the lifetime τ have been compared with the data (full and broken lines). The source size R_S (see equation (4.35)) in the calculation has been fixed at 4.4 fm. It has been obtained from a detailed study of the longitudinal and transverse correlation functions (see [131] for more details). The increase of the correlation function at low momentum below 20 MeV/c is a consequence of the attractive S-wave interaction between the two neutrons (the scattering length is equal to 16 fm). Data have been sampled according to the total momentum P_{CM} carried out by the two neutrons (upper and lower panels). Neutron emissions with a large P_{CM} are associated with shorter emission times while low values of P_{CM} correspond to long emission times. This is interpreted as a consequence of the cooling of the compound nucleus: the larger P_{CM} is, the 'hotter' the system is, and the shorter the lifetime is. From [131].

or crystal blocking techniques are obviously not useful for such 'extreme' experimental situations. There is thus a need for a specific method. About 10 years ago, Tröckel *et al* [468] proposed to take advantage of the spacetime correlations between fragments induced by their mutual Coulomb interaction.

For the sake of illustration, this method can be used to estimate the average time τ between two successive fragment emissions from an excited source. To this end, correlation functions [468] are built either from the reduced relative velocity V_{red} or the relative angle θ_{FF} between fragments taken two by two. The idea is that when the emission time becomes short, Coulomb repulsion between fragments due to their proximity in spacetime will produce, in the correlation functions, the so-called Coulomb hole at low V_{red} or θ_{FF}. It is possible to quantify this effect by performing simulations calculating the propagation of fragments which are successively emitted with an adjustable lifetime τ.

In figure 4.15, τ has been estimated for the Ar+Au system at two different bombarding energies by considering the evolution of the quantity

$$\eta(\theta_{FF}) = \frac{N^{corr}(\theta_{FF}) - N^{uncorr}(\theta_{FF})}{N^{corr}(\theta_{FF}) + N^{uncorr}(\theta_{FF})} \tag{4.36}$$

in which $N^{corr}(\theta_{FF})$ and $N^{uncorr}(\theta_{FF})$ are, respectively, the normalized numbers of correlated and uncorrelated pairs of fragments emitted at a relative angle θ_{FF}. Events for which the system decays into three massive fragments ($Z \geq 10$) have been selected. Results shown in figure 4.15 indicate that the decay process is sequential at 30 MeV/u beam energy while it is much faster (within 100 fm/c) at 60 MeV/u. For these two bombarding energies, the excitation energies have been estimated, respectively, at 3 and 5 MeV/u for 30 and 60 MeV/u bombarding energies with the methods previously described (see section 4.3.1). These results will be discussed more extensively in section 8.4.1.1.

4.4 Event generators and simulations

In the preceding sections, we have explored the various methods that have been developed to sort the events, to extract the information and to estimate physical quantities in order to characterize excited sources properly. Such procedures are rather complex and not free from possible biases. The latter can be induced either by the limitations of the experimental apparatus (the so-called experimental filter) or by the analysis methods themselves which may not be free from autocorrelations. Therefore, it is often necessary to qualify (or calibrate) the sorting procedures by performing simulations.

The most powerful descriptions of the reaction are based on the nuclear Boltzmann equation or on quantum molecular dynamics (see chapter 3). But the main difficulty of such simulations lies in the fact that such codes cannot really be used as event generators because they are very computer-time consuming and usually are unable to follow nuclear processes on very long timescales. Hence, highly phenomenological approaches are generally used (see, for instance, [170]). Among the various event generators that have been designed, the FREESCO generator [178, 270] has certainly played an important role and most of the more recent generators have been, to some extent, inspired by this precursor. Such generators generally consider the reaction as a two-step process (see chapter 5):

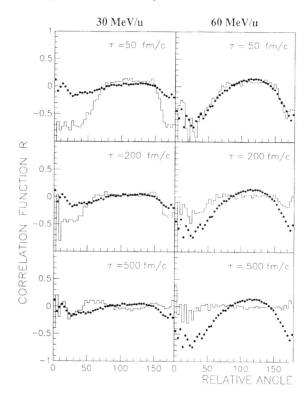

Figure 4.15. Correlation function for the relative angle Θ_{FF} between three massive fragments, taken two by two, detected in Ar+Au collisions at 30 and 60 MeV/u. The bold points are the data; the histograms the results of a computer calculation simulating the Coulomb trajectories of the fragments. The main parameter of the model is the time τ between two successive emissions. For calculations with low values of τ (upper case), a strong depletion is observed at small Θ_{FF} due to the strong Coulomb repulsion between the fragments. Correspondingly, due to the conservation of the total momentum of the three fragments, a similar depletion is observed at large Θ_{FF}. These features are in agreement with the data at 60 MeV/u. For large values of τ (lower case), the correlation function is almost flat since there is no longer a sizeable interaction between the fragments. This is in agreement with the data measured at 30 MeV/u. From [295].

- an entrance channel based on simplified modelling of the encounter of the two incoming nuclei using, for example, interaction potentials such as the ones described in section 5.1;
- an exit channel phase in which the decay of the excited source(s) produced in the entrance channel phase is described by means of statistical models

(section 2.4) This phase is sometimes called the after-burner phase and allows us to follow the propagation of emitted particles on very long timescales.

The major advantage of the phenomenological event generators is their flexibility. Several assumptions regarding the scenario of the collision can be easily simulated and tested by a comparison with the data. To this end, an efficient computerized experimental filter must be built in which the response of the apparatus is carefully taken into account. This allows the simulated events to be filtered and thus a direct comparison with the experimental data to be made.

4.5   Conclusion of the chapter

In this chapter we have successively described the facilities and the detectors that are in use for the experimental study of nuclear collisions in the intermediate energy range. As stressed at the beginning of the chapter, a major improvment in the last years has been the use of sophisticated experimental devices (the so-called 4π detectors of second generation) to obtain high quality data in terms of completeness and accuracy. But such complex physical events require dedicated analysis techniques. The most commonly used tools have been described. The study of global variables allows us to reduce the large amount of available information from the data, in order to extract physical quantities that can be easily compared with theoretical predictions. In particular, the sorting of the events according to the impact parameter of the reaction is an important part of the work. Another important topic concerns the identification and characterization of well-defined sources in terms of mass, charge, shape, excitation energy, temperature, angular momentum, etc. To this end, the techniques of nuclear calorimetry, thermometry and chronometry have been sucessively described and applied to some specific cases. Such techniques will be used throughout the following chapters when discussing the various processes occurring in the course of nuclear collisions.

Owing to the complexity of 4π data analysis, dedicated simulation tools are necessary to check the analysis procedure and to estimate the biases induced by the experimental filter. Such tools are based on event generators which aim at describing nuclear collisions with the help of phenomenological models. Most of the time, they are based on statistical assumptions coupled with a simplified description of the entrance channel of the reaction. Their main goal is thus to provide a very useful link between the data and the microscopic models which can hardly describe the collision from its very beginning to its very end.

Chapter 5

Reaction mechanisms

The study of reaction mechanisms aims mainly to classify collisions according to some global 'topological' features which, following the discussion in section 4.2.2, means the identification and the characterization of source(s). The topology of each collision is, to a large extent, related both to the impact parameter of the collision and to a partial or total transformation of the available centre-of-mass kinetic energy of the relative motion between the two partners of the collisions into disordered motion (heat). This dissipation process is governed by several important ingredients. One of them is the relative velocity between the initial partners of the reaction v_{AA}. The corresponding reduced relative wavelength associated with a nucleon–nucleon collision then reads

$$\frac{\lambda}{2\pi} = \frac{h}{2\pi m v_{AA}} \tag{5.1}$$

where m is the nucleon mass. According to equation (5.1), the following values (in the case of symmetrical systems) of $\lambda/2\pi = 6.5, 2.1, 0.67, 0.24$ fm are obtained for 1, 10, 100 and 1000 MeV/u beam energies, respectively. These values have to be compared with the mean nucleon–nucleon distance in a nucleus (typically 2 fm). If $\lambda/2\pi$ exceeds this distance, a collective behaviour of nucleons during the collision is expected. In other words, mean-field (one-body) effects overcome nucleon–nucleon collisions (two-body) effects. The situation will be reversed if $\lambda/2\pi$ is smaller than the mean nucleon–nucleon distance. According to this criterion, it turns out that mean-field effects are expected to be dominant in the low-energy region (below 15 MeV/u).

The same conclusions may be drawn for the role of the Pauli principle in relation to the evolution of the mean free path as shown in figure 2.3. At low bombarding energy, i.e. when the nucleus–nucleus relative velocity v_{AA} is less than the Fermi velocity v_F ($\sim 0.3c$), nucleon–nucleon collisions are strongly inhibited by the fact that few final states are available in the exit channel. In this case, energy dissipation occurs mainly because projectile nucleons are retained inside the target (and vice versa) by the corresponding attractive potential thus

136

giving rise to one-body dissipation. An increase in the incident energy induces an opening of the available phase space for outgoing nucleons, which leads to some decrease in Pauli blocking effects. In turn, in the limit of relativistic incident energies, dissipation occurs mainly through nucleon–nucleon collisions (two-body dissipation).

Another important ingredient is the available energy per nucleon, i.e. the maximum excitation energy which can be brought into the system. An incident energy of 30 MeV/u, for example, corresponds to an available energy close to the total binding energy for symmetrical nucleus–nucleus collisions. In other words, the low incident energy domain (below 15 MeV/u) will correspond to moderate excitation of final products but much higher excitation can be reached at intermediate or large incident energy. Finally, as already stated, the values of v_{AA} are smaller than the velocity associated with pion production threshold in nucleon–nucleon collisions, which corresponds in vacuum to an energy of 290 MeV/u. This confirms again that, in the intermediate energy domain, nucleon excited states will have a negligible influence.

Reaction mechanisms at 'intermediate' beam energies are better understood when viewed with the help of concepts developed for energy regimes (high or low beam energies) in which reaction mechanisms are well known. We shall thus first recall some basic properties of reaction mechanisms at low (section 5.1) and high (section 5.2) incident energy before directly addressing the Fermi energy domain (section 5.3). In the high-energy regime, a section will furthermore be devoted to a brief description of collisions induced by light projectiles in the multi-GeV range since such reactions provide physical conditions, in particular, in terms of excitation energies, which are comparable to those encounterd in the Fermi energy range.

5.1   Nuclear reactions close to the Coulomb barrier

Reaction mechanisms in the Coulomb barrier region and below 15 MeV/u may be easily classified according to the impact parameter b or the orbital angular momentum l (figure 5.1). Elastic scattering is observed for impact parameters exceeding $b_{max} = R_t + R_p$ values (or the corresponding angular momentum l_{max}) which corresponds to grazing collisions. Slightly below b_{max}, quasi-elastic and transfer reactions are observed in which the projectile and target kinematical properties are only slightly perturbed. They mainly reflect the external orbit properties of the interacting nuclei. Dissipative collisions, also called deep inelastic collisions (DIC), and possibly fusion, are observed for more central collisions (figure 5.1). They are clear signatures of mean-field effects leading to a collective behaviour of the involved nuclei. In the DIC case, projectile and target nuclei are strongly slowed down due to nuclear matter friction (section 3.1.2). For a short time they form a 'quasi-molecular' state before reseparation. During this step nuclei may exchange nucleons. The corresponding lifetime may be estimated

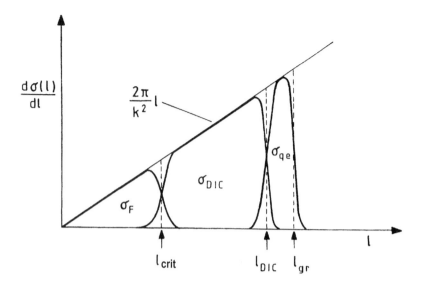

Figure 5.1. A schematic diagram of the partial wave decomposition of the reaction cross-section in low incident energy heavy-ion reactions: the abscissa refers to orbital angular momentum or to the impact parameter; quasi-elastic collisions correspond to $l_{\mathrm{DIC}} < l < l_{\mathrm{gr}}$ ($l_{\max} = l_{\mathrm{gr}}$) while DICs (these are collisions in which the two partners re-separate after a contact phase during which matter and a significant amount of energy are exchanged) are for $l_{\mathrm{crit}} < l < l_{\mathrm{DIC}}$, and fusion associated with a single source in the final state corresponds to $l < l_{\mathrm{crit}}$.

from the rotation angle of the di-nuclear system before decay. As shown in the so called Wilczynski plot (figure 5.2), the energy relaxation turns out to be an increasing function of the rotation angle.

Fusion corresponds to the most central collisions (figure 5.1). The angular momentum that separates the fusion and the DIC regions is called the critical angular momentum (l_{crit}). Its value is governed by the interaction potential between the interacting nuclei which consists of several terms:

- The Coulomb repulsion term evolves as the inverse of the relative distance;
- The nuclear contribution does not play a significant role at large distance, but is attractive for relative distances exceeding slightly the sum of the nuclear radii. This contribution becomes repulsive when the densities of the two nuclei significantly overlap. This is due to the incompressibility of nuclear matter (section 2.2.2.1) and to the fact that nuclear shapes cannot evolve rapidly enough during the first stages of the collision;
- Finally, there is the rotational contribution at finite impact parameter. This

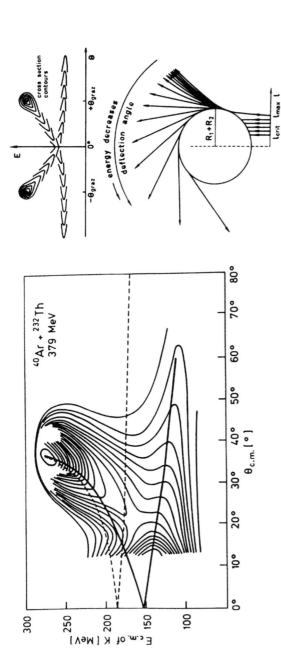

Figure 5.2. Left-hand side: The 'Wilczynski' plot obtained for the system Ar+Th at 379 MeV total beam energy. In such a plot, the centre-of-mass energy of a given product of the reaction (here the K nuclei, the atomic number of which is close to that of the Ar projectile) is plotted as a function of its centre-of-mass emission angle θ_{CM}. A large bump is observed at $E \sim 280$ MeV and $\theta_{CM} \sim 40°$. This corresponds to quasi-elastic collisions ($l \sim l_{max}$ or l_{gr}, see lower right-hand part of the figure). Kinetic energy relaxation is associated with a rotation of the di-nuclear system towards lower values of θ_{CM}. These are collisions which correspond to $l_{crit} < l < l_{max}$ in which a large decrease in the energy is associated with a large deflection angle (see right upper part of the figure). In principle, both negative and positive angles are populated but experimentally only angles of a given sign (positive here by convention) are measured. The broken line (and the similar full line) correspond to model calculations based on trajectory calculations with two different interaction potentials such as the one depicted in figure 5.3. From [495] and [415].

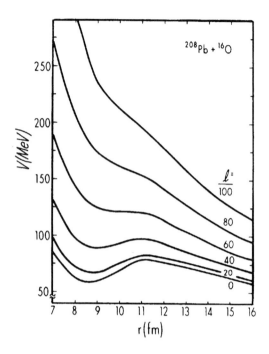

Figure 5.3. Evolution of the interaction potential between two interacting nuclei as a function of the relative distance r. The total potential including the nuclear, centrifugal and Coulomb contributions is shown for various values of the angular momentum l. The critical angular momentum l_{crit} corresponds to the l value for which the pocket of the potential curve disappears (here around $l \sim 60$). From [415].

last term is determined by the combined effects of the angular momentum and the moment of inertia of the system.

An example of such an interaction potential is plotted in figure 5.3 for various angular momenta (or impact parameters). The interaction potential may or may not exhibit a pocket depending on the angular momentum value. Generally speaking, the potential pocket does not exist for angular momentum exceeding l_{crit}. Below l_{crit}, the system is trapped inside the pocket due to the slowing down of the relative motion induced by nuclear friction. The (time-dependent) potential may then evolve slowly towards the adiabatic limit corresponding to a spherical shape of the system (fusion nucleus). In contrast, for angular momenta exceeding l_{crit}, the system is never trapped in a potential pocket and the two partners separate again after a contact phase during which energy dissipation and nucleon exchange occur.

A very important aspect of dissipative collisions at such moderate incident energies is that the reaction process, particularly in the case of fusion, may be divided into two steps which are well separated in time. The first step is the collision itself which leads to a fully equilibrated excited nucleus on a timescale shorter than the decay time. In other words, the formation time is much shorter than the lifetime of the system. This is a strong justification for the use of statistical models to describe the decay stage of the reaction (see section 2.4). The 'purest' illustration of such a scenario is the fusion reaction leading to a compound nucleus, i.e. a fusion nucleus which has forgotten all entrance channel features apart from the excitation energy and angular momentum. In the case of DICs, the situation is slightly more complicated because full equilibrium may not be achieved during the first reaction step, depending on the contact time which is directly linked to the observed rotation angle of the two partners (see figure 5.2). For this reason, the energy relaxation or the excitation energy balance between outgoing partners or the angular momentum transfer may vary from one event to the other.

Generally speaking, a time hierarchy is, nevertheless, observed between the various degrees of freedom. The intrinsic degrees of freedom which are related to thermal excitations are rapidly equilibrated (within a few 10^{-22} s, section 5.3.1). On the other hand, the collective degrees of freedom (shape, mass asymmetry or relative angular momentum) relax much more slowly. Diffusion equations of Fokker–Planck type (section 3.1.2.3) may then be used to describe the time evolution of the shape of the system. In any case, the subsequent decay of released fragments occurs a long time after the initial contact phase. This is due to the fact that a moderately excited nucleus has a decay time larger than the DIC reaction time (typically 10^{-18} s for a nucleus temperature of 1 MeV to be compared with a collision time of 10^{-21} s, see section 2.4.3). In the following we will see that such a situation is no longer valid at higher energies.

5.2   Nuclear collisions in the relativistic energy range

In this section, we consider both heavy-ion collisions at relativistic energies (say between 200 MeV/u and 1 GeV/u incident energies) and collisions induced by light projectiles such as pions, protons, antiprotons up to alphas in the multi-GeV range.

5.2.1   Heavy-ion collisions at a few hundreds of MeV/u

The dissipation process in relativistic heavy-ion collisions (0.2–1 GeV/u range) is dominated by hadronic cascades because the wavelength associated with nucleon–nucleon collisions is shorter than the nucleon size. The corresponding relative velocity between projectile and target nucleons is also much larger than the Fermi velocity v_F. For these two reasons, collisions can be safely described by geometrical concepts leading to the so-called participant–spectator picture:

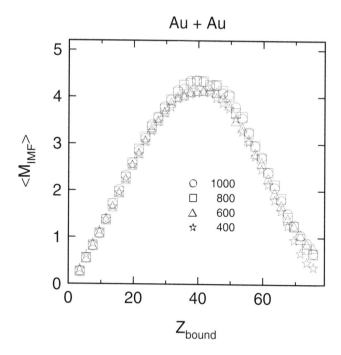

Figure 5.4. Mean intermediate mass fragment (IMF) multiplicity as a function of Z_{bound} (defined as the sum of all detected charges with $Z > 1$) for the reaction Au+Au at $E_{lab}/A = 400, 600, 800$ and 1000 MeV/u. From [418].

nucleons which do not belong to the overlapping zone of the two incoming nuclei do not suffer hard nucleon–nucleon collisions and constitute the spectators while the other ones are the participants.

The physics of the participants has been studied for a long time with the first generation of 4π detectors: the Plastic Ball [442], DIOGENE [8], nowadays with FOPI [394, 395] and by the EoS collaboration [239]. The corresponding deposited energies per nucleon are far beyond the nuclear binding energies. They are outside the scope of this book although some results about these collisions will be mentioned briefly in chapter 8. The physics of the spectators comprises the programme of the ALADIN collaboration (see, for instance, [298] and references therein) and of the EoS collaboration ([239] and references therein). Using mostly inverse kinematics (a heavy projectile bombarding a lighter target) or symmetric systems, advantage is taken of the strong focalization of the particles in the very forward direction to detect properly the decay products of the excited quasi-projectile. It was thought previously that the spectators were released almost cold after separation from the participant zone, their excitation energy originating from their strong 'clean-cut' deformation. This statement was excessive since it turns out that sizeable excitations can, in fact, be reached. Indeed, the ALADIN group has shown that hot spectator nuclei reaching excitation energies comparable to their binding energies can be observed. Since these systems are produced in semi-peripheral collisions, one might expect that the corresponding nuclear matter would not be compressed which is in contrast with nuclear matter involved in hot nuclei produced in central collisions at lower bombarding energies.

Peripheral relativistic collisions may then appear as relevant tools to study hot but uncompressed nuclear matter. They thus bring valuable information on the physics of hot nuclei, without disturbance by compression effects. As explained in section 4.2 there is a need for an impact parameter selection of the collisions. This is achieved here by using a suitable global variable Z_{bound} defined as the total detected charge from fragments of charge ≥ 2. With such a definition and owing to the specificity of the ALADIN detection system, the dominance of the geometry for such collisions leads to the following relation: the larger Z_{bound} is, the larger the impact parameter b is. This means that for peripheral reactions, the quasi-projectile residue has characteristics close to the projectile and is moderately excited. As b decreases, the fraction of the participants with respect to the total system increases and consequently the number of spectators reflected by the value of Z_{bound} decreases. However, since the collision is more violent, the spectators are more excited.

There is thus the interesting possibility of studying, with the same system, a variety of excited sources over a large mass number and excitation energy domain, a situation comparable with the one encountered in the Fermi energy range (section 5.3.3.2). Some universal properties of peripheral reactions in the relativistic energy domain have hence been observed by the ALADIN collaboration. In figure 5.4, the invariance of the IMF multiplicity distribution as a function of Z_{bound} for various incident energies is nicely demonstrated. If one assumes that Z_{bound} is correlated with the deposited energy, this result means that multifragment production mainly reflects the deposited energy rather than the initial projectile velocity. One should also stress that the same is true when varying the target for a given beam energy E_{lab} emphasizing the key role of the geometry in such collisions. The correlation between Z_{bound} and b furthermore allows us to reconstruct the size A_0 of the sources as shown in figure 5.5. It is worth noting that such a correlation between Z_{bound} and A_0 is in very good agreement with the predictions of a geometrical participant–spectator model.

Finally, by using the standard methods of nuclear calorimetry (section 4.3.1), it is possible to reconstruct the excitation energy per nucleon and to follow its evolution as a function of Z_{bound}, as displayed in the bottom part of figure 5.5. Very large values of the excitation energy, exceeding the binding energy per nucleon, can be reached for source sizes from 50 up to 150 mass units. As already stressed, these values are comparable with those obtained in dissipative collisions in the Fermi energy range making the two approaches complementary.

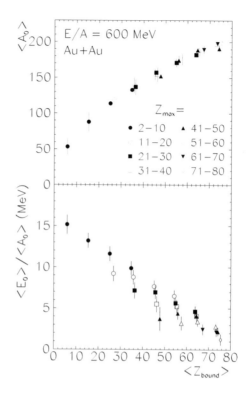

Figure 5.5. Reconstructed average mass (upper part) and excitation energy (lower part) of the decaying spectators as a function of Z_{bound} (abscissa) and as a function of Z_{max} (symbols) defined as the atomic number of the heaviest detected fragment, event by event. From [250].

5.2.2   Collisions with light projectiles in the multi-GeV range

Another experimental approach has proven to be very instructive for the study of nuclear matter far from stability: it consists in studying nuclear reactions induced by light projectiles (for instance p, $\bar{p}$, He isotopes) in the multi-GeV range. The main advantage of these reactions compared to collisions in the Fermi energy range is their apparent simplicity: hadronic cascades minimize collective effects such as rotational or compressional effects. Typical values for the transferred angular momentum are less than $20\hbar$ and nuclear matter stays close to the normal density. Therefore, it is often advocated that such collisions are ideal tools with which to isolate thermal effects in the decay of hot nuclei. However, the situation is not as clean as expected. This is due to the strong relative velocity between the light projectile and the nucleons of the target. Non-equilibrium effects (see

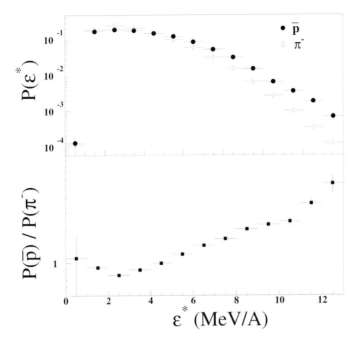

Figure 5.6. Distributions of the excitation energy per nucleon obtained by the method of nuclear calorimetry (section 4.3.1) for π^- (open points) and $\bar{p}$ (black points) reactions on Au at 8 GeV/c. The lower part of the figure is the ratio of the two distributions of the upper panel. It shows that $\bar{p}$ provide a better heating of the target than π^-. It is also worth noting that very large values of ϵ^* (comparable to the binding energy per nucleon) can be reached but with only very small probabilities. From [284].

chapter 6) in the early instances of the process play an important role. The very first dissipation stage is dominated by hard hadronic collisions in which pions are essentially produced (several pions may be produced in a single nucleon–nucleon or hadron–nucleon interactions at such high energy). As our aim is not to go into too much detail, we refer the reader to the works of the various groups involved in such studies (see for instance [211, 271, 327, 348, 370, 457, 481]).

Let us, nevertheless, discuss the relative merits of the different projectiles that have been used up to now in such studies. As an example, a comparison between $\bar{p}$ and π^- projectiles at the same incident linear momentum is shown in figure 5.6. It appears that antiprotons are better suited to heating the target than pions. Indeed, $\bar{p}$'s mainly interact through their annihilation at the surface of the target producing many pions. Although, on the average, about half the pions escape the system for geometrical reasons, the energy deposited remains, nevertheless, larger than for pions due to the huge energy gain in the annihilation process which compensates the geometrical loss described earlier. This is testified by the results shown in the lower part of figure 5.6. The main difficulty of such studies lies in the role of pre-equilibrium particles which are not easily distinguished from particles escaping from an equilibrated system. Nevertheless, it can be established that excitation energies exceeding 5 MeV/u can be reached in such collisions so that they are useful tools for nuclear matter investigations.

5.3   Reaction mechanisms in the Fermi energy range

5.3.1   General remarks

5.3.1.1   *Orders of magnitude*

We consider now the Fermi energy domain, typically between 15 and 200 MeV/u, i.e. the domain in between low (section 5.1) and relativistic (section 5.2) energies. It is a transition region in which both one-body and two-body behaviours are observed and strongly compete. It is interesting to evaluate the corresponding timescales because they play an essential role, in particular, with respect to the degree of thermalization achieved in the course of the collision.

- Energy may be dissipated through elastic nucleon–nucleon collisions (inelastic channels are marginally excited). This is the two-body excitation process. The associated timescale is of the order of the mean time between two successive nucleon–nucleon collisions which can be estimated from simple kinetic theory as:

$$\tau_{nn} = \frac{1}{\sigma_{nn}\rho_0 v} \tag{5.2}$$

  where σ_{nn} is the nucleon–nucleon cross-section in the medium and v the mean velocity (of the order of magnitude of the Fermi velocity, v_F). It is generally admitted that thermalization occurs after a few elementary collisions thus leading to $\tau_{2\text{-body}} = 50$ fm/c. This time is comparable to the traversal time τ_{tr} as already discussed in section 3.1.1.2.
- Energy may be dissipated through the interaction of individual nucleons with the nuclear mean field: this is one-body dissipation with a timescale of the order of $\tau_{1\text{-body}} \simeq R/v_F = 20\text{–}30$ fm/c in which R is of the order of the sum of the radii of the two interacting nuclei.

Whatever the dissipation mechanisms, the thermalization time is comparable to the interaction time since this latter is of the order of $\tau_{\text{inter}} = R/v_{AA}$ in which v_{AA} is the relative velocity between the two partners in the entrance channel of the reaction. A typical numerical value is $\tau_{\text{inter}} \simeq 30$ fm/c for a medium mass system at 50 MeV/u beam energy. From the fact that the thermalization time is comparable with the reaction time, one may assert two important conclusions:

- A sizeable fraction of the available energy may be thermalized during the reaction process itself: it is then possible to create very hot nuclei in nucleus–nucleus collisions in the Fermi energy domain.
- However, a sizeable fraction of the available energy may *not* be thermalized during the collision, leading to a fast emission prior to thermalization. This emission is called pre-equilibrium emission and will be discussed in chapter 6. Of course, from one event to the other, fluctuations will lead to various final situations, i.e. to a varying proportion of pre-equilibrium emission with respect to thermalization. Fluctuations will hence lead to a large panel of outgoing products even for a definite geometry, i.e. for a given impact parameter. This remark has to be related to those of section 4.2: it is highly necessary to perform detailed sorting in order to select a given event topology. Only with such a sorting is it possible to select those hot nuclei for which most of the available excitation energy has been dissipated and thermalized.

As previously mentioned, in the Fermi energy range, the energy dissipated in a nucleus cannot be used to excite intrinsic states of the nucleon. Thus, the whole energy is available to heat the matter. However, part of the energy is also used to excite collective degrees of freedom associated with deformation, rotation and/or compression. The proportion of energy stored in a given mode depends on the typical timescales for the excitation of this mode and also on the initial conditions, i.e. the entrance channel characteristics (see next section). For instance, the amount of rotational motion reflects both the entrance orbital angular momentum and the shapes of the interacting nuclei. Indeed, the deformation of outgoing partners in binary dissipative reactions is related both to the geometry (impact parameter) and to nuclear matter viscosity as is already the case at lower energy (the outgoing fragments in a DIC are released as deformed objects).

It is interesting to evaluate the order of magnitude of such collective energies, and especially the one associated with a radial compressional mode which can be excited in central collisions. As an example, let us consider a 50 MeV/u symmetrical collision: then, the available centre-of-mass energy per nucleon is $\epsilon_{CM} = 12.5$ MeV/u. The amount of energy ΔU needed to put the system at density ρ starting from normal density ρ_0 can be calculated according to equation (2.33). Taking a standard value for K_∞ (~220 MeV), one obtains $\Delta U = 10$ MeV/u to reach $\rho = 2\rho_0$. It thus turns out that ΔU is comparable with ϵ_{CM}. Therefore, sizeable compression may be obtained in the most violent collisions as well as much heating of the system. It is also worth noting that ϵ_{CM} is larger than the binding energy per nucleon B in nuclei, which is typically of the order of 8 MeV/u. Thus, complete dissolution of the system such as nuclear vaporization may be (and has indeed been) observed in these highly dissipative collisions (see section 7.3.2).

From these simple considerations, the Fermi energy range appears to be a very interesting energy region in which to heat and compress pieces of nuclear matter in a broad range of temperatures and densities. Consequently, a large variety of physical processes are expected to play a role as we will see in chapter 7. They are connected to the initial relative repartition of the dissipated energy between various degrees of freedom, but also to the characteristic times associated with the coupling of various collective motions to the 'heat bath' (intrinsic (nucleonic) degrees of freedom, see also section 3.1).

5.3.1.2   *On the modes of storage of energy in nuclear collisions*

In this section we discuss the various energy storage modes mainly with the help of the theoretical results presented in figure 5.7. The upper reduced scale shows the different storage modes of the available energy as a function of time in a central Ar+Au collision at 60 MeV/u beam energy as simulated by a microscopic transport model, namely a Landau–Vlasov calculation. The first instances, up to 40 fm/c, are characterized by a rapid transfer of the available centre-of-mass energy (here denoted E_{CM}) initially in the form of kinetic energy into 'thermal' excitation energy E^*. In the mean time, the collective energy denoted E_{coll} (provided by the incident energy) is strongly reduced indicating the strong stopping of the incoming nuclei. As time goes on, a large amount of E^* goes into a pre-equilibrium plus evaporation component thus cooling the system, while the subsequent expansion towards a lower density of the heated matter induces an increase in the potential energy (denoted here $E(T = 0, \rho)$). The decrease of the potential energy around 90–100 fm/c is associated with a new compression phase (the so-called breathing mode), again with a slight increase in the excitation energy. The three contributions of $E(T = 0, \rho)$ are detailed in the lower part of the figure displayed with a larger energy scale. The contribution called 'cold kinetic energy' is the Fermi motion of the nucleons inside the medium: its decrease above $t = 40$ fm/c corresponds to a decrease of the density of the system while simultaneously the 'interaction' energy $V(\rho)$ (the non-kinetic part of the energy, equation (2.30)) increases, as it costs more and more energy to drive the system to lower density. Finally, the Coulomb energy does not significantly change in the course of the reaction mainly because this last quantity varies slowly with the density of the system. On a longer timescale, a fused system is observed that ultimately decays into many fragments. Simulations of collisions at larger impact parameters would predict another separation of the two incoming partners after a phase of pre-equilibrium emission followed by a subsequent cooling of the two heated nuclei.

Such a theoretical work suggests that the spacetime evolution of the matter involved in a nuclear reaction strongly depends on the transport properties inside the medium in non-equilibrium situations and thus depends on the microscopic ingredients (here the effective force and the nucleon–nucleon in-medium cross-section) used in the calculation. Thus, the understanding of the various energy storage modes constitutes severe tests of the models when the latter are compared with experimental data.

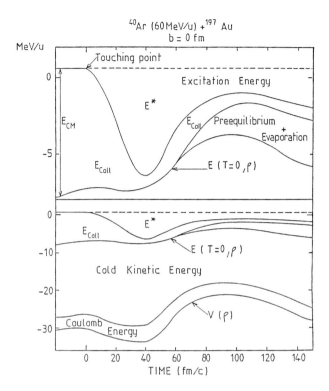

Figure 5.7. Energy transfers in a head-on collision Ar+Au at 60 MeV/*u* as predicted by a Landau–Vlasov calculation. See text for a detailed description of the evolution as a function of time of the various components of the total energy. From [450].

After these two introductory sections, we now come to the core of this chapter by discussing the experimental results about reaction mechanisms in the Fermi energy range. To this end, collisions have been arbitrarily divided into, on the one hand, peripheral and mid-central collisions and, on the other hand, central collisions.

5.3.2 Peripheral and mid-central collisions

5.3.2.1 *The dominance of binary collisions*

Our discussion starts with a description of peripheral and mid-central collisions. Such collisions exhaust a large amount of the total reaction cross-section and are thus easily recognized and characterized. The search for the most central collisions, associated with small cross-sections because of geometry, will be discussed in section 5.3.3.

A general overview of the collisions is accessible, for instance, through a display of isocontours of invariant cross-sections (see section 4.2.1) of $Z = 1, 2$ particles as shown in figure 5.8. Events have been sorted according to the violence of the collision by using the total transverse energy (see section 4.2.3) of light-charged particles ($Z \leq 2$). Due to experimental selections, the elastic and quasi-elastic events have been rejected and the data mainly concern mid-peripheral and central events. For most of them, a binary behaviour is observed. Indeed, Coulomb circles associated with the emission of particles by the two partners of the reaction are clearly seen. This is a general result: in peripheral and mid-central collisions, a quasi-projectile (QP) and a quasi-target (QT) are produced in the collision. If the decay chain is complicated or long, the final products are quite different from the original QP and QT nuclei. Conversely, if it is simple or short, the QP and QT residues still ressemble the initial projectile and target nuclei. In this case, a closer look at the v_{par} projections (lower part of the figure) shows a non-symmetric distribution of the particles with respect to the centre-of-mass velocity of the sources, mostly visible in the first two columns of the figure. In other words, there is an extra particle emission in between the two sources: this is a first hint for the formation of intermediate structures which will discussed in detail in section 5.3.2.2. The two large final products (as observed, for example, in figure 5.8) will be called, respectively, projectile-like and target-like fragments (PLF and TLF, respectively) in the following. The question of how the dissipated energy is shared among the two partners of the reaction has been reviewed in [463]. The PLF and TLF are easily recognized in the detectors only if the QP and QT excitation energies are moderate.

Figure 5.9 is another illustration of the binary character of most collisions but here it concerns all emitted products whatever their atomic number. The two studied systems (Ar+KCl at 52 MeV/u and Xe+Sn at 50 MeV/u) show very similar trends as a function of the impact parameter. Peripheral collisions are associated with the decay of two moderately excited sources (the QP and the QT) whose main products have atomic numbers close to the projectile and target charges. Their velocities are also close to the centre-of-mass velocities corresponding to the incident energy. They are accompanied by the emission of light particles and IMFs over a broad range of parallel velocities. In particular, it is worth noting that a sizeable contribution is observed at mid-rapidity (i.e. close to zero velocity for such symmetric or nearly symmetric systems). As the experimentally determined impact parameter decreases, the relative velocity between the PLF and the TLF reduces as well as their respective atomic numbers. Finally, for the most dissipative collisions, the very notion of PLF and TLF becomes questionable because most of the emitted matter is in the form of light particles and IMF's. Then the question arises as to what extent it is possible to 'group' particles according to some definite decaying sources in such collisions.

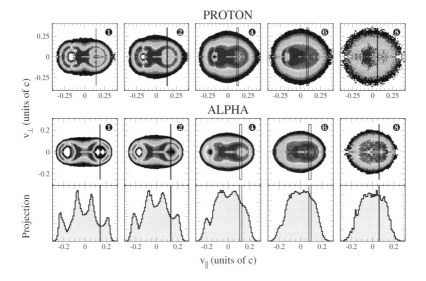

Figure 5.8. Invariant cross-section contour plots for protons and α-particles in a $v_{\parallel}-v_{\perp}$ (in units of c) plane for Xe+Sn collisions at 50 MeV/u. Events have been sorted on an event-by-event basis using the method described in section 4.2.3 (equation (4.4)) with the sorting variable E_{trans12}. This latter is defined as the sum of the transverse energies of all light-charged particles. The centrality increases from left to right (regions labelled 1 to 8). Accordingly, the velocity (indicated by vertical bars) of one of the two sources (the projectile-like source) located in the forward direction decreases as a function of centrality showing the gradual damping of the relative motion of the two partners of the reaction. The lower part of the figure shows the projection along the parallel velocity in the case of α-particles. From [296].

In section 4.2.3.4 several methods were discussed for 'reconstructing' sources using the kinematical characteristics of all detected products. One of them (the 'thrust method') has been used to reconstruct the chemical and kinematical characteristics of the QP and QT in Xe+Sn collisions at various beam energies, and is illustrated in figure 5.10. In this figure, the total relative kinetic energy of the reconstructed quasi-projectile (QP) and quasi-target (QT) nuclei has been correlated with the rotation angle of the system with respect to the beam. These plots are thus quite similar to the Wilczynski plots observed at low bombarding energy (figure 5.2). Again, it turns out that the dissipation is an increasing function of the rotation angle, i.e. the collision time. Very large dissipation (up to the maximum possible values) is observed. This result remains valid when the bombarding energy increases. This means that there is no saturation of the deposited energy. However, the cross-section significantly decreases with increasing dissipation. Further, one should note that the reconstruction method

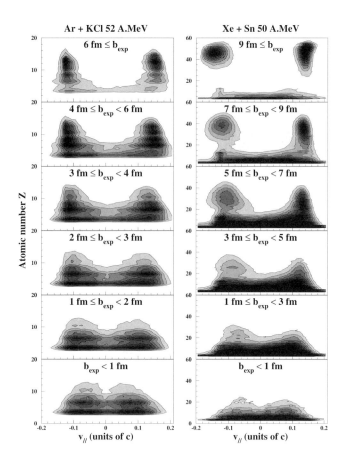

Figure 5.9. Bi-dimensional plots of $v_\parallel$ (in units of c) as a function of the atomic number for the two systems Ar+KCl at 52 MeV/u and Xe+Sn at 50 MeV/u. Determination of the experimental impact parameter is based on the same method as that used for the data displayed in figure 5.8. From [314].

used to obtain the results of figure 5.10 implicitly assumes a pure reaction process in which two well-defined sources emerge from the first instances of the collisions. We have already observed kinematical characteristics of light particles in figure 5.8 which point out more complicated patterns. Such features (which are discussed in the next section) suggest either the occurrence of strong deformations or the onset of a participant–spectator scenario, as discussed in section 5.2 in the case of relativistic heavy-ion collisions.

This dominance of binary-type collisions has, nevertheless, been observed

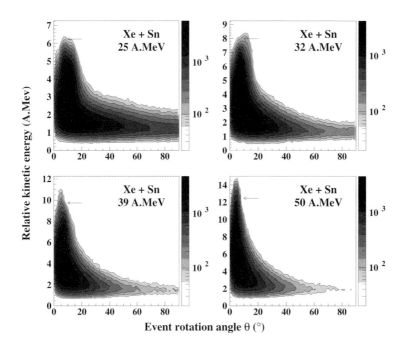

Figure 5.10. Correlations in Xe+Sn collisions at several incident energies between the relative kinetic energy (ϵ_{rel}) of the reconstructed primary QP and QT and the rotation angle (θ) of the whole system with respect to the beam axis. This angle was denoted θ_{flow} in section 4.2.3. The arrows in each panel correspond to the relative kinetic energy for elastic collisions (this is just the centre-of-mass incident energy per nucleon). Such a plot is reminiscent of the so-called Wiczynski plots obtained at low energy (see figure 5.2). Indeed, the relative kinetic energy is a measure of the dissipation and plays the same role as the centre-of-mass energy in figure 5.2. The same is true for θ. The quantities θ and ϵ_{rel} are generalizations of the variables used at lower energy for collisions in which several massive products are emitted either by the QP or the QT. From [314].

not only in several heavy- or medium-mass systems (Pb+Au [280], Kr+Au [430], Xe+Bi [24,294], Ar+Th [373], Xe+Sn [306] or Mo+Mo [121]), but also for lighter ones (Ar+Ag [258,400], Ar+Al [362], Ar+KCl [313] or Zn+Ti [440]).

5.3.2.2 *Formation of neck-like structures*

The most spectacular signature of new reaction patterns at incident energies around the Fermi energy is certainly the formation of neck-like structures. From

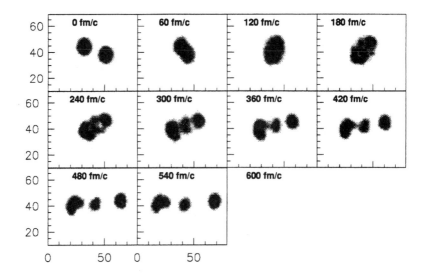

Figure 5.11. Density plots projected onto the reaction plane in semi-peripheral Pb+Au collisions at 29 MeV/u and an impact parameter of 6 fm, as predicted by a Landau–Vlasov simulation based on microscopic transport theory (see section 3.2.3.2). The beam axis is along the abscissa in the figure. After a contact phase starting arbitrarily at 60 fm/c, the relative motion between the two partners of the reaction is damped and the two nuclei strongly overlap (120 fm/c). Still, the relaxation is not enough to drive the system towards a compact configuration and the two partners tend to separate again (240 fm/c). However, in the overlap region, a piece of nuclear matter is produced, nearly at rest in the centre-of-mass connecting the two main bodies of the reaction: this is the formation of a neck-like structure around 300 fm/c. After 360 fm/c, neck fragmentation occurs, resulting in a final stage with three massive fragments. From [96].

a theoretical point of view, strong deviations from a pure DIC scenario have been observed in the simulations of nuclear collisions at intermediate impact parameters in the framework of semi-classical transport theories discussed in section 3.2.3.2 (see, for instance, [129, 130, 230]). Figure 5.11 is an illustration of this new phenomenon in the case of semi-peripheral Pb+Au collisions as simulated by the semi-classical Landau–Vlasov model (section 3.2.3.2). In the overlap region of the two incoming nuclei, a highly dense and hot piece of nuclear matter is produced, which, in some circumstances, may become an independent third source of emission.

In parallel to the theoretical results shown in figure 5.11, we discuss the experimental data obtained with the same system (Pb+Au at 29 MeV/u). Mid-peripheral events with three massive fragments were selected and analysed in terms of angular correlations as shown in figure 5.12. A clear alignment of

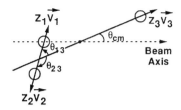

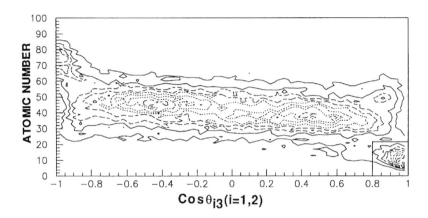

Figure 5.12. Lower part: Bi-dimensional plots of the cosine of the relative angle $\cos(\theta_{i3})$ versus the atomic number of the fragment labelled i in which i takes the values 1 and 2 according to the notations and the conventions given in the upper part of the figure. The selected events correspond to semi-peripheral collisions in which three massive fragments (labelled from 1 to 3) were emitted in coincidence with light-charged particles. The system is Pb+Au collisions at 29 MeV/u. The cosine distribution for fragments with atomic numbers close to about half the projectile or the target charge (associated with the symmetric fission of one of the two partners of the reaction) is flat, which is a signature of an isotropic emission. In contrast, those fragments which have either a high Z close to the projectile or the target or a small charge (typically lower than 20) have an emission angle $(\cos(\theta_{i3}) > 0.8$ or $\cos(\theta_{i3}) < -0.8)$ picked along the axis connecting the velocities of the two partners of the reaction. From [281].

the three fragments is observed when one of them has a size of the order of the volume of the overlap zone. This suggests a dynamical process producing a neck similar to the one observed in figure 5.11. A detailed study (not shown here, see [281]) of the relative velocity distributions of the three fragments reveals that the smallest fragment of the configuration is indeed essentially emitted along the axis connecting the two main bodies of the reaction, with a velocity close to the mid-rapidity (i.e. close to zero for this almost symmetric system). Such departures from pure binary mechanisms have already been observed in heavy-[500] and medium- [117] mass systems at lower bombarding energies around 20 MeV/u where strong anisotropic fragment angular distributions were analysed in terms of deformations. Similar trends have also been reported at higher incident energy (see for instance [63, 64]). Such features concern mainly heavy IMF's ($Z > 5$) and, to a lesser extent, lighter ones.

Neck fragmentation is a general feature of nuclear collisions in the Fermi energy range. Medium-mass systems as well as heavy-mass systems exhibit very similar trends to those shown in figure 5.13. A sizeable fraction of the emitted fragments (up to 80% for the light ones—IMF) may thus be concentrated in the intermediate velocity region. The corresponding products are mostly light IMF's as demonstrated by the rise and fall of the fragment yield Y_{neck} emitted in the neck region as a function of the charge of the fragment Z_{IMF} (figure 5.13). It is established that neck emission is strongly connected with the projectile–target geometrical overlap during the collision [296]. Thus, it can be tentatively interpreted as the first manifestation of the formation of a participant zone as observed at relativistic energies (section 5.2). Of course, one has to be cautious on this point because the overlap zone cannot be geometrically well defined for a given impact parameter due to the fact that, on the one hand, the Fermi and beam velocities are comparable and, on the other hand, the nucleon mean free path in nuclear matter is comparable with the nucleus size. It should also be noted that the results discussed earlier do not mean that all products emitted in the neck region have been emitted on a short timescale. Deformation mechanisms which are governed by the geometry of the entrance channel may involve long timescales as may be concluded from the time evolution of the matter density shown in figure 5.11. They indicate that, owing to fluctuations, the overlap region between two colliding nuclei may either lead to a third piece of nuclear matter (besides the QP and QT) or may be released attached to one of the partners [63].

What is the physics behind the formation and decay of neck-like structures? A possibility discussed in [322] is the occurrence of surface instabilities. In particular, in the course of the reaction, the overlap zone may be strongly stretched thus giving rise to possible mechanical ruptures. Indeed, the production of elongated structures can lead to strong surface–surface interactions such as those occurring in the fission process, namely the interaction between the surfaces of the two nascent fission fragments. The surface-surface interaction can be described by a 'proximity' force with the help of a finite-range interaction [62]. It is then possible to derive criteria for which an instability can occur in the system (see [322] for details). The main conclusion of this work is that such an instability happens when the 'thickness' of the neck is of the order of the range of the proximity force: a situation which may be achieved in mid-central collisions.

Another interesting point is the influence of the N/Z degree of freedom on the chemical composition of the matter produced in the neck region. Recent experimental results [154, 257] support the statement that the emitted matter

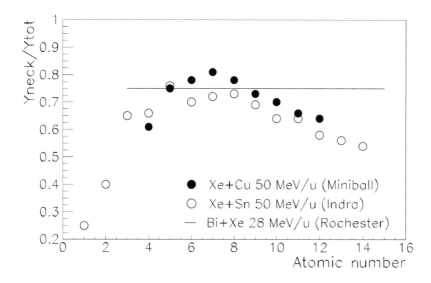

Figure 5.13. Percentage of matter emitted in the mid-rapidity region defined as $Y_{\text{neck}}/Y_{\text{tot}}$ and thus associated with the formation and decay of neck-like structures for medium- and heavy-mass systems in mid-peripheral collisions. In the Xe+Bi case, only a mean value integrated over charges ranging from 4 to 20 is indicated as a full line. Miniball data from [319], Rochester data from [465] and INDRA data from [296]. From [171].

in the neck region is more neutron rich than one would expect from simple considerations on the N/Z ratios of the projectile and the target. Such findings are obtained thanks to a very promising technique which consists in studying reactions with different projectile–target combinations differing by their neutron numbers, such as for instance ^{124,136}Xe+^{112,124}Sn in the present case. A similar method has been used at higher energy to study the degree of equilibration in central collisions [380].

The neutron richness of the matter emitted in the neck region observed experimentally has been theoretically investigated (for the same systems as discussed earlier) in the framework of standard transport theory in [434]. Of the two possibilities explored to explain experimental features, namely isospin equilibration and the production of symmetric light clusters (deuterons and alphas) leaving the remaining matter very neutron-rich, the authors of [434] favour the second hypothesis. Due to the uncertainty in the description of light composite particles in the framework of transport theories (see section 6.5), such a question deserves further investigation. Nevertheless, these first results on the formation and decay of neck-like structures in nuclear collisions with various

N/Z values are very promising and they will certainly constitute severe tests of the transport models.

5.3.3   Central collisions

We now come to the selection and characterization of the most central collisions. By 'central' we mean those collisions which lead ultimately to the formation of a *single* fused system, in contrast to the spacetime patterns depicted earlier, in which several sources were produced. Central collisions are the key reactions which really probe matter in its extreme states since they correspond to the largest dissipated energies and hopefully the largest compression. But, for obvious geometrical reasons, the cross-section corresponding to single source events is quite small (a few tens of millibarns). This means that efficient selection methods have to be used.

5.3.3.1   Evolution of complete fusion with E_{lab}

At low incident energy, fusion reactions are easily selected by searching for the signature of a compound nucleus. Since this latter is not exceedingly excited, its decay products (see chapter 7) will be easily recognized: these are either an evaporation residue or two fission fragments (if the fusion nucleus is heavy enough). Both processes are accompanied by the emission of a few light particles. However, in the fission case, there is a possible ambiguity because fission can also occur in binary dissipative collisions if the target is sufficiently heavy. Then the observable used to separate both contributions is the so-called folding angle (hereafter called θ_{AB}) between the two fission fragments. This angle is related to the recoil velocity of the fusion nucleus by simple kinematical considerations: the larger the recoil velocity is, the smaller θ_{AB} is. Indeed, low values of θ_{AB} correspond to a focalization of the two fission fragments in the laboratory frame and thus to a large momentum transfer. Thus, events with small θ_{AB} are associated with central collisions. The distribution of the folding angle has been studied as a function of the incident energy for N+U reactions as can be seen from figure 5.14.

At low E_{lab} (7.4 MeV/u), a single component is observed, peaked around 160°: it corresponds to a complete momentum transfer and thus to fusion. Increasing E_{lab}, three features are observed. First, a second contribution sets in, close to 180°: this is fission at small momentum transfer induced in peripheral collisions. This is due to angular momentum transfer from the projectile to the target, which enhances the fission probability by lowering the fission barrier. Second, the intensity of the peak at low θ_{AB} reduces progressively. This can be due either to a spreading in the component as a result of the successive evaporation steps of the fission fragments or to the fact that (for such highly excited fused systems) the fission probability decreases at the benefit of other decay modes such as fragmentation (see section 7.3.1). Finally, the maximum of the low θ_{AB}

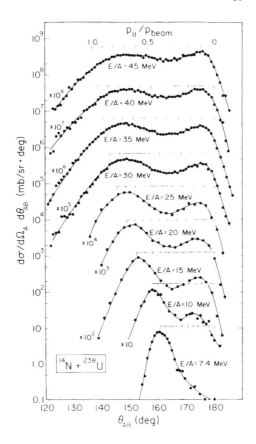

Figure 5.14. Folding angle distributions for binary fission fragments detected at various bombarding energies for N+U reactions. For each incident energy, the inner scale gives the amount of linear momentum transferred, with respect to the projectile momentum, to the composite system decaying into two fission fragments. A value close to zero is associated with very small transfer and thus presumably to peripheral collisions while a value close to one is a signature of complete momentum transfer leading to a fusion nucleus. From [180].

component is shifted towards larger and larger values. This fact is interpreted in terms of incomplete momentum transfer even for the most central collisions. The reason for this lies in the so-called pre-equilibrium emission of light particles (see chapter 6). This means that, on average, not all particles from the projectile are trapped inside the target during the interaction: some of them may escape the system because elastic nucleon–nucleon collisions and mean-field effects are not strong enough to slow down the motion of the incoming nucleons. This process is called *incomplete fusion*. However, the interpretation of experimental data in terms of incomplete fusion requires some caution. In very asymmetric systems, the kinematical properties (mass and velocity) of a QT produced in a DIC reaction may actually show features very similar to the ones of an evaporation residue. The only experimental way of distinguishing between both processes is to look for the QP which will be present only in the case of binary processes. This is not easy because the QP is emitted a long way forward (and thus difficult to detect) and it may also have undergone a long and complex decay chain. Conclusive results can be obtained only if the QP may be reconstructed from its decay products, which is possible if a 4π detection has been achieved. At the time at which most of the data of Figures 5.15 and 5.16 were taken, this was not a usual feature.

The transition from fusion to incomplete fusion is displayed in the compilation shown in figure 5.15 in which the percentage of linear momentum transferred to the target in asymmetric collisions has been plotted as a function of the incident energy corrected by the Coulomb barrier. A rather universal behaviour is observed: fusion is complete below 10 MeV/u incident energy and more and more incomplete above. It is worth noting that the general trend of the data is correctly bracketed by simple calculations taking into account either one-body or two-body dissipation. Incomplete fusion as well as the formation of neck-like structures thus turns out to be a specificity of the Fermi energy regime in which the dissipation mechanisms originate from various microscopic processes.

The results shown in figure 5.15 suggest a strong evolution of complete fusion cross-sections as a function of the incident energy. This is clearly put in evidence in figure 5.16. A large body of data shows the fast decrease and the disappearance of complete fusion as a function of the incident energy around $E_{\text{lab}} \sim 30$–40 MeV/u. This trend has been confirmed recently in a systematic study using argon beams from 17 to 115 MeV/u [128]. Complete fusion is therefore a very rare process above 40 MeV/u.

5.3.3.2 *Identification of fused systems in highly fragmented events*

In the previous section, the characterization of compound nucleus reactions, or at least of fused systems in central collisions associated with incomplete fusion, was achieved by identifying either a heavy evaporation residue or two fission fragments. As soon as one increases the beam energy, more and more kinetic energy is dissipated into heat, thus resulting in the production of more and more excited species. Anticipating the results discussed in chapter 7, we can say that new decay modes are expected as the excitation energy reaches values close to the binding energy. In particular, fragmentation (i.e. the disassembly of hot nuclei into at least three fragments with $Z > 3$, see section 7.3.1) becomes a competitive process.

For such a process, the simple techniques, discussed previously, to isolate fused systems become inappropriate. A solution consists of correlating the flow angle θ_{flow} defined in section 4.2.3.2 with the total kinetic energy (E_{TKE}, equation (4.16)) carried away by fragments. This is somehow similar to the

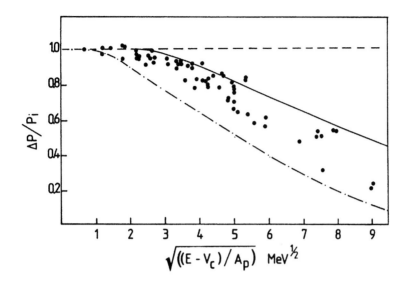

Figure 5.15. Percentage of linear momentum transfer in asymmetric nucleus–nucleus collisions as a function of $\eta = \sqrt{(E - V_c)/A_p}$ in which E is the total incident energy, V_c the Coulomb barrier in the entrance channel of the considered reaction and A_p the mass number of the projectile. For instance, in the N+U case, $\eta = 1.7, 4.7, 6.4, 7.7$ for 10, 30, 50, 70 MeV/u beam energies, respectively. The onset of incomplete fusion thus roughly corresponds to 10–15 MeV/u incident energy. The full line is the result of a geometrical calculation taking into account both the one-body and two-body dissipation, while the chain line accounts for one-body dissipation only. Finally, the broken line corresponds to full linear momentum transfer. From [290].

method described in section 4.2.3.2 and illustrated in figure 4.8. Figure 5.17, which corresponds to highly fragmented events detected in Pb+Au reactions at 29 MeV/u, shows the correlation between θ_{flow} and E_{TKE} (instead of E_{trans12}). The data display an evolution comparable with that shown in figure 5.10. A first contribution (associated with regions labelled 1 and 2 in the figure) is observed at small angles close to the grazing angle: it corresponds to mid-central collisions for which a strong memory of the entrance channel is still present. A second contribution, associated with a minimum value of E_{TKE}, and extending towards large values of θ_{flow}, corresponds to the most central collisions.

It is instructive to study the evolution of event shapes with help of the global variables defined in section 4.2.3.2: mid-central events are associated with elongated shapes aligned along the beam axis while a gradual progression towards more and more compact shapes is observed for more central collisions. Indeed,

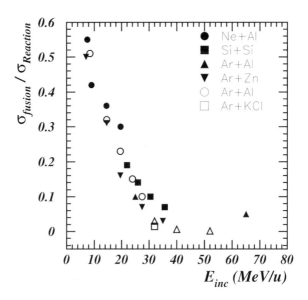

Figure 5.16. Evolution of the complete fusion cross-section compared to the total reaction cross-section as a function of the incident energy for a variety of systems. Compilation from [404].

at high E_{TKE} (zone 1), events are very elongated (rod shape) and the rotation of the matter measured by θ_{flow} is small. When decreasing E_{TKE} (zones 2 and 3), events become more and more compact evolving towards a disc-like structure, while rotation of the matter sets in. For the last two considered regions (labelled 4 and 5), matter has strongly rotated, while E_{TKE} reaches a minimum: events are very compact showing structures in-between a disc and a sphere.

It is finally interesting to examine the physical conditions (in terms of compression and temperatures) that could be reached in such violent central collisions. Figure 5.18 shows the predictions of a semi-classical transport model, namely BUU (section 3.2.3.2). The maximum density as well as the maximum temperature and entropy attained in central Au+Au collisions are shown as a function of the incident energy. In the energy range we are interested in, sizeable temperatures as well as entropies can be reached. Also interesting are the values of the compression: hot and compressed (up to $1.5\rho_0$) pieces of nuclear matter are thus predicted to be formed. Therefore, the rather large domain of compression and excitation energy stored in systems with very different masses makes central collisions in the Fermi energy range a unique tool to investigate the properties of hot nuclei which will be extensively discussed in chapters 7 and 8.

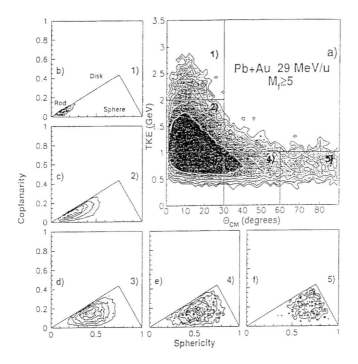

Figure 5.17. Correlation between TKE (denoted E_{TKE} in the text) and θ_{flow} for Pb+Au collisions at 29 MeV/u. Only those events in which five or more fragments with $Z \geq 5$ have been selected. The evolution of the shape of the events as a function of the impact parameter (from the peripheral (zone 1) to the most central ones (zone 5)) are displayed in the sphericity (S) (equation (4.11))–coplanarity (C) (equation (4.12)) plane. Generic shapes (rod, disc and sphere) in this plane are indicated in the first panel. A rapid evolution of the event shapes is observed from the rod-like structure for peripheral collisions (zone 1) (high E_{TKE} values and low θ_{flow} values) to more spherical shapes for the central ones (zone 5) (low E_{TKE} values and high θ_{flow} values).From [282].

5.4 Conclusion of the chapter

In this chapter we have described the various reaction mechanisms occurring in the Fermi energy range with short excursions at lower and higher incident energies. It appears that, in this energy range, heavy-ion collisions are very good tools with which to investigate the properties of hot and dense nuclear matter, as far as the various spacetime configurations achieved in the course of the reaction can be correctly identified and characterized with the help of the analysis tools

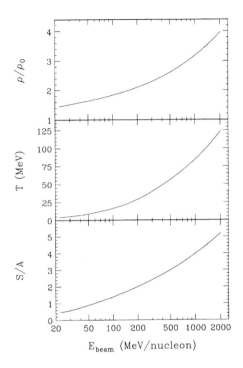

Figure 5.18. Maximum density, temperature and entropy reached in central Au+Au collisions as a function of the incident energy per nucleon as predicted by a BUU calculation (see section 3.2.3.2). From [148].

described in chapter 4. After a brief survey of the collisions in the relativistic regime and also close to the Coulomb barrier, we have introduced reactions in the Fermi energy domain by discussing general trends on the basis of simple estimates concerning energies and timescales. Such numbers show that this energy range is a transition region in which both low- and high-energy features coexist. We have also analysed the various storage modes of the energy during the reaction and emphasized the role of time in the coupling between the collective and the intrinsic modes.

   A schematic picture of the different processes and their evolution as a function of the incident energy is shown in figure 5.19. In peripheral and mid-central collisions, a smooth transition is observed from pure dissipative collisions close to the Coulomb barrier towards the onset of strong deformations culminating in the production of neck-like structures around the Fermi energy. This signals the onset of new instabilities (Rayleigh instabilities) [322] and opens

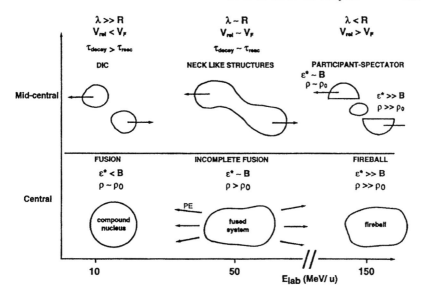

Figure 5.19. A schematic illustration of the various reaction mechanisms occurring in the Fermi energy range in mid-central and central collisions. In mid-central collisions, the transition from 'pure' deep inelastic binary processes at low energy towards a geometrical participant–spectator mechanism is shown with an intermediate behaviour corresponding to the formation of neck-like structures. Central collisions are characterized by the evolution from the formation of a compound system associated with complete fusion towards a fused system resulting from incomplete fusion and finally to the production of a very hot and compressed piece of nuclear matter, the so-called fireball. Also indicated are the various timescales, density and excitation energy regimes.

up the possibility of studying a highly dynamical process, namely thinning and stretching of pieces of nuclear matter at velocities close to the sound velocity. Finally, beyond about 100 MeV/u, geometrical aspects start to play a dominant role essentially because the mean free path becomes shorter, or at least of the order of magnitude of nuclear sizes, and also because the relative velocity becomes significantly larger than the Fermi velocity. In this energy region, the participant–spectator scenario prevails.

In central collisions, as E_{lab} increases, the ideal situation consisting of the production of a compound nucleus becomes highly improbable and is progressively replaced by the production of a fused system whose shape and excitation energy may not be fully equilibrated. We have seen that pre-equilibrium processes (to be discussed in more detail in chapter 6), as well as incomplete stopping power, were the main reasons for the observed transition from complete fusion to the so-called incomplete fusion process. These deviations from an idealized situation are unfortunately the price to be paid for reaching excitation energies as large or even larger than the total binding energy of the system.

Chapter 6

Fast processes towards thermalization

By nature heavy-ion collisions are truly dynamical processes. In the nucleonic regime, a sizeable fraction of collisions leads to the formation of a, possibly short-lived, composite which exhibits patterns of a more or less thermally equilibrated object. The question of the degree of thermalization attained in such collisions actually turns out to constitute one of the major open questions in the field (see also chapter 8). A key aspect is thus to access and possibly analyse the behaviour of the system *before* equilibration. It turns out that such studies provide valuable clues on the dynamics of the collision as a whole, and on the way to equilibrium. This is what we discuss in this chapter.

We begin by introducing some general considerations concerning the observation of fast processes in nuclear collisions in section 6.1.1. Next, the path towards equilibrium is described by discussing the results of microscopic transport models in section 6.1.2. We then proceed by discussing in more detail the two main classes of signatures of this early phase of the collision: flow (section 6.2) and particle production (section 6.3). In contrast to the statistical emission from an equilibrated system, the many particles emitted during the early phase of the collision keep a memory of the details of the conditions in which they were produced. An analysis of their collective kinematic properties such as flow, or the mere appearance of unexpected species, such as meson production, thus provides an almost direct link to the dynamics. One may then access, via comparison with transport models, a phase space localization and/or a timing of the production mechanisms. Here, hard photon production offers a particularly efficient tool, which has reached a high degree of accuracy with the results obtained from devoted photon spectrometers.

Still, this optimistic view has to be moderated by the many effects which may interfere in such processes. This requires experimental efforts to find the best suited observables as, for example, balance energy in the flow measurements (section 6.2). In turn, the comparison of data with transport models only allows us to point to the few most relevant quantities, such as momentum dependence of the interaction in the case of balance energy. In fact only the simplest observables, particularly the collective ones like flow, are easily accessible from the theory. In addition it turns out that they are often experimentally ill defined. Rare-particle production (section 6.3) is conversely not easily computable at energies well below the elementary threshold, and this is probably one of the aspects where the theory is the most deficient. The production of composite particles (section 6.5) is even less understood (see also chapter 8).

6.1 From contact to mixing

6.1.1 General considerations and experimental signatures

The evolution from the entrance channel configuration towards a thermalized object consists of energy transfer from the projectile–target relative motion to internal and collective degrees of freedom of the (possibly temporary) composite. If the interaction time is sufficiently long, the shape of the system may reach an equilibrium configuration, under the constraints of conservation of global quantities such as the total linear momentum, the total energy and the total angular momentum. The internal degrees of freedom are then relaxed and one may consider that the system is thermalized. This is the situation corresponding to the simple (at least in terms of reaction mechanism) formation of a compound nucleus. However, deviations from this idealized picture can occur.

- First, shape equilibration is not always reached. In many cases matter and energy exchanges lead to a final-state configuration with two excited and deformed fragments or even with neck-like structures (see section 5.3.2).
- Second, strong deviations, with respect to a full equilibration of the internal degrees of freedom, can be observed. These non-equilibrium features are related to the early emission of light particles and/or composites, which leave the system before thermalization.

Such processes do not occur at bombarding energies less than about 15 MeV/u because nucleons are then retained by the mean field inside the system. In this case, potentially energetic nucleons are slowed down *inside* the nucleus before leaving it, by energy exchange with the nuclear mean field. The dissipation mechanism is then one-body in nature. Above 15 MeV/u, mean-field effects are no longer strong enough, because the relative velocity between the two partners in the entrance channel is too large and other mechanisms start to play a leading role.

Figure 6.1 schematically shows the influence of the different velocities in the case of a central collision of a symmetric system which can lead to fusion. The two full circles correspond to the original Fermi spheres of the two nuclei in momentum space. They are separated by the relative velocity corresponding to the bombarding energy. The dotted circle corresponds to the Fermi sphere of the fusion nucleus. It is located at mid-distance between full circles if the projectile and target are symmetric in mass. The broken circle corresponds to

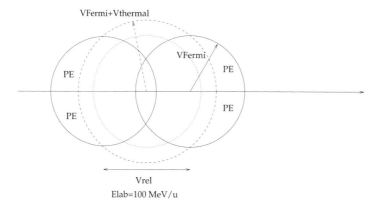

Figure 6.1. A schematic representation of the Fermi spheres (full curves) for two colliding nuclei at a relative incident energy of 100 MeV/u for a symmetric entrance channel. The dotted curve is the Fermi sphere of the fusion nucleus and the broken one corresponds to an extra thermal velocity. In this simple picture, nucleons located in the regions denoted PE can escape the system before fusion.

nucleons in the fusion nucleus with an extra average velocity associated with the thermal motion inside the fusion nucleus. Some nucleons, initially in the target and the projectile, belong to this domain: they are thus likely to be trapped in the nascent fusion nucleus. But those nucleons which are located outside (those regions are called PE in the figure) can escape the system. However, if they suffer efficient nucleon–nucleon collisions, their velocity can be sufficiently reduced so that in the course of the reaction, they can also be slowed down and trapped in the fusion nucleus. The particles that can escape from the system are often called pre-equilibrium particles. In the case of a reaction leading to two main bodies in the final state, the particles that have suffered hard nucleon–nucleon collisions can also escape the system by being emitted at a velocity in-between that of the projectile-like and the target-like: these particles are sometimes called mid-rapidity particles.

Experimentally, pre-equilibrium emission is a very commonly observed process in nuclear collisions. It is well known and has been studied for a long time now. Figure 6.2 is an illustration of this process in the case of neutron emission in reactions close to the Fermi energy. The kinetic energy spectra displayed in figure 6.2 clearly show two components. The low-energy part of the spectra is associated with the evaporation from the compound nucleus produced by the fusion of the two partners of the reaction as discussed in section 5.3.3.1. The high-energy part is associated with the neutrons that have escaped the system before full thermalization [167,407]. These particles have kept the memory of the entrance channel through their kinematic characteristics: they are mainly emitted

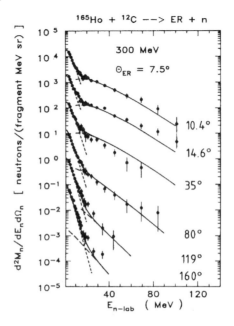

Figure 6.2. Differential neutron multiplicities in coincidence with evaporation residues (detected at 7.5°) following C+Ho fusion-like reactions at 25 MeV/u for different neutron detection angles. For clarity, the spectra at 10.4, 14.6, 35, 80, 119 and 160° have been multiplied by 10^n with n being 5, 4, 3, 2, 1, 0, respectively. The two broken curves are the results of a least-square fit with an evaporative (short broken line) and a pre-equilibrium (long broken line) component. The full line is the sum of the two components. Note the presence of a (weak) pre-equilibrium contribution even at very backward angles. From [245].

in the forward direction. However, one should note that they can also be observed at very backward angles.

Such a behaviour is observed not only for neutrons but also for light-charged particles and for composite clusters. Around 100 MeV/u beam energy, it is clearly observed even in more 'elementary' collisions such as proton–nucleus reactions. This is illustrated in figure 6.3. Composite particles (α-particles in the present case) can either be evaporated or be produced by coalescence of the incoming proton with several nucleons of the target during the proton's travel inside the medium. As already mentioned, such a coalescent process has a characteristic signature in terms of kinetic energies and angular distributions of the emitted particles. In particular, a high energy component is clearly visible, mainly at forward angles. We will come back to this question of composite particle production in section 6.5.

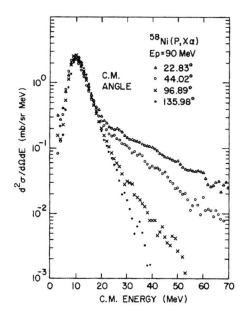

Figure 6.3. Kinetic energy distributions of α-particles emitted at several angles (indicated in the figure) in the centre-of-mass frame in p–Ni collisions at 90 MeV. Two components are clearly observed. The first one (the more intense) centred around 10 MeV, which is the same for all angles, is associated with the evaporation of the compound nucleus. The second contribution depending heavily on the detection angle is associated with a fast pre-equilibrium process. From [499].

6.1.2 Theoretical access to overlap and heating

In the previous section, we briefly discussed some experimental probes to trace back the path from the first contact to the complete mixing (if any) of the two partners of the reaction. We now address the question from the point of view of the microscopic transport models.

What are the leading physical mechanisms dominating the behaviour of the system in the first phase of the collision, before the formed composite system has actually reached a more or less equilibrated state? The answer to this question is both simple and complicated: simple because the causes of the various phenomena observed are essentially well known; complicated because the observed effects are numerous, sometimes hard to analyse (in particular, to disentangle from each other) and generally result from several causes acting together. In the beam energy domain in which we are interested, most of the physics of the entrance channel reflects the viscous nature of nuclei: when the two nuclei start to overlap, their relative motion is progressively slowed down

139La+12C at 50 MeV/A and b= 0.0 Fm

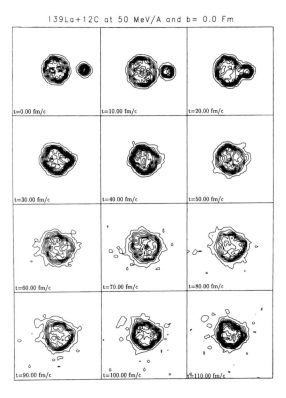

Figure 6.4. Density snapshots (in the reaction plane) for a central ^{139}La on ^{12}C collision at 50 MeV/u beam energy (BUU simulation, section 3.2.3.2). This reaction is a typical incomplete fusion reaction for such a highly mass asymmetric system. Each box represents a 40 fm × 40 fm surface and time runs in fm/c. It is enlightning to note how the carbon nucleus 'dissolves' inside the lanthanum nucleus. This provokes a heating of the system. Here thermalization takes about 80 fm/c. Note that thermalization is accompanied by sizeable energetic proton emission (pre-equilibrium emission), first along the beam axis and, later on, more isotropically. In this case the formed 'hot' nucleus has a temperature of order 3.2 MeV, corresponding to an excitation energy of order 1.3 MeV per nucleon. From [368].

and turned into internal agitation of the nucleonic degrees of freedom. This dissipative effect has its origin in nucleon–nucleon collisions not accounted for in the mean field (see chapter 3). It is illustrated in figure 6.4, for an *incomplete fusion* collision between ^{139}La and ^{12}C at 50 MeV/u beam energy and zero impact parameter. It is interesting to see, in this figure, how the ^{12}C nucleus is 'dissolved' inside the ^{139}La nucleus. The actual 'mixing' of the two colliding nuclei results precisely from friction forces. If nucleon–nucleon collisions had not been accounted for in this simulation, the carbon nucleus would have crossed the lanthanum nucleus, without having been significantly affected. Another interesting feature of figure 6.4 concerns the nucleon emission, primarily along beam axis, and, later on, more isotropically. This emission corresponds to the pre-equilibrium particles which have already been discussed from an experimental point of view in the previous section. They were emitted *before* the system had actually reached a thermalized state, and they carry away a sizeable fraction of the available energy. Beyond about 100 fm/c, in the particular example of figure 6.4, it turns out that the composite system does indeed reach a thermalized state which fixes an upper bound to the thermalization time in this case. In this example the actual mass of the incomplete fusion residue is around 140 and the temperature is of order $T \sim 3$ MeV. Note that the results of the simulations are in reasonable agreement with experimental data for this reaction, in particular, concerning the mass and charge of the incomplete fusion residue: $A_{exp} \simeq 146$, $A_{th} \simeq 137$ and $Z_{exp} \simeq 60$, $Z_{th} \simeq 56$ (with obvious notation) [368].

The density snapshots of figure 6.4 show the mixing of the matter constituting the two nuclei but they give no quantitative information on the degree of thermalization of the system. This is no surprise as thermalization corresponds to a rearrangment of the momentum distribution (section 6.1.1). This rearrangment first takes place locally (in real space), thus leaving possible temperature gradients in the system, and, next, globally (in real space), which corresponds to a uniform temperature all over the system. In order to follow the thermalization process in time one should thus study the time evolution of the momentum distribution. A first indicator of the degree of thermalization is thus provided by the amount of anisotropy of the momentum distribution. We use here the phase space one-body distribution $f(\boldsymbol{r}, \boldsymbol{p}, t)$, which is the basic ingredient of kinetic descriptions of heavy-ion collisions (section 3.2.3). One can then consider the quadrupole moment of the momentum distribution (averaged out over real space for the sake of simplicity)

$$Q_p = \int (2p_z{}^2 - p_x{}^2 - p_y{}^2) f(\boldsymbol{r}, \boldsymbol{p}, t)\, \mathrm{d}\boldsymbol{r}\, \mathrm{d}\boldsymbol{p} \tag{6.1}$$

which provides a simple and global indicator of thermalization. It is plotted as a function of time for a central ^{40}Ca+^{40}Ca reaction at 60 MeV per nucleon beam energy in figure 6.5. This reaction typically leads to the formation of a hot nucleus. Also indicated in this figure, for illustration, are the corresponding shapes of the momentum distribution $n(\boldsymbol{p}) = \int f(\boldsymbol{r}, \boldsymbol{p}, t)\, \mathrm{d}\boldsymbol{r}$ at some instances. The quadrupole moment Q_p is a basic observable for tracing relaxation in heavy-ion collisions in the nucleonic regime [34]. At initial time the two nuclei are separated in momentum space by the relative momentum associated with the beam energy. Overlap in real space allows two-body collisions to take place (typically in the 'free' open momentum space outside the Fermi spheres in figure 6.1) and leads to a thermalization of the system. This shows up as a decrease in Q_p towards zero. When thermalization is reached $Q_p \approx 0$, which

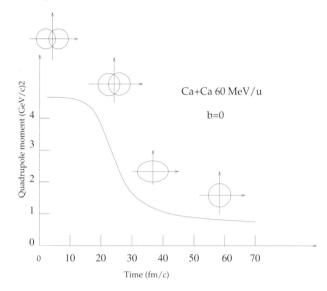

Figure 6.5. Time evolution (in fm/c) of the quadrupole moment Q_p of the momentum distribution (in $(\text{GeV}/c)^2$) in a $^{40}\text{Ca}+^{40}\text{Ca}$ collision at 60 MeV/u beam energy and zero impact parameter. Also indicated, for the sake of completeness, are snapshots of the momentum distribution at some instances. The decrease in Q_p towards zero reflects the relaxation of the system towards 'thermal' equilibrium.

reflects the sphericity of the momentum distribution.

A more detailed account of the way to thermalization is provided by the pressure tensor, which can be obtained following the usual definition of fluid dynamics:

$$\Pi_{ij}(\boldsymbol{r}) = \int \mathrm{d}\boldsymbol{p} \, (\boldsymbol{p}_i - \langle \boldsymbol{p}_i \rangle(\boldsymbol{r}))(\boldsymbol{p}_j - \langle \boldsymbol{p}_j \rangle(\boldsymbol{r}))f(\boldsymbol{r},\boldsymbol{p},t) \qquad (6.2)$$

where $\boldsymbol{p}_{i,j} = \boldsymbol{p}_{x,y,z}$ and where the local average momentum is defined as

$$\langle \boldsymbol{p} \rangle(\boldsymbol{r}) = \int \mathrm{d}\boldsymbol{p} \, f(\boldsymbol{r},\boldsymbol{p},t)\boldsymbol{p}. \qquad (6.3)$$

The pressure tensor Π_{ij} quantifies the local anisotropy of the momentum distribution. As soon as it becomes proportional to the unity tensor the system may be considered to be thermalized. In addition it provides a more detailed picture of the thermalization of the system than Q_p, which may easily wash out details in the averages it contains.

Indeed, a system may have a vanishing value of Q_p, while still a not perfectly isotropic pressure tensor, indicating a not yet fully thermalized system. The pressure tensor has been little considered in simulations of heavy-ion collisions as it is quite an involved quantity to compute, which can be easily polluted by numerical effects. Still, the few attempts made in this direction have shown that 'local' effects, which can possibly be traced back by $\prod_{ij}$, do not invalidate the information provided by global observables such as Q_p. For a first guess estimate the quadrupole moment of the momentum distribution may thus suffice to provide the correct orders of magnitude, in particular, in terms of timescales. Furthermore, one should keep in mind the fact that $\prod_{ij}$ is also an *a priori* difficult quantity to evaluate experimentally.

6.2   Sidewards flow and squeeze-out

In section 6.1.1 we introduced fast particle emission but we restricted ourselves to 'inclusive' observables such as angular or kinetic energy distributions. The advent of powerful multidetectors has allowed us to study the behaviour of such particles on an event-by-event basis and has thus given us access to their global characteristics. We have seen previously that particles emitted after a short time have experienced very few nucleon–nucleon collisions. It is generally admitted that about three nucleon–nucleon collisions are at least necessary to 'thermalize' an incoming nucleon. Nucleons experiencing less collisions are emitted preferentially at mid-rapidity, i.e. between the quasi-projectile (QP) and the quasi-target (QT). The kinematic properties of these particles emitted early are usually studied by considering the velocity distribution of the particles with respect to the rest of the matter. In particular, the study of the emission pattern with respect to the reaction plane, for a finite impact parameter, carries a lot of information about the dynamics of the collision.

From an experimental point of view, such processes have been explored and studied by different groups in the 1980s in relativistic nuclear collisions [153, 228, 229]. They were predicted earlier in hydrodynamical calculations (see section 3.1.4) and in microscopic transport model calculations (see [444]). Then, the exploration of these phenomena in the Fermi energy range was undertaken at MSU and GANIL.

The study of the in-plane emission of light particles is called sidewards flow while the out-of-plane emission is called squeeze-out. Before discussing these two processes, it is necessary to give some definitions and notation.

6.2.1   Definitions of flow measurements

A schematic picture of the entrance channel of a collision is displayed in the upper part of figure 6.6. Particles interacting in the overlapping region experience a deflection, at a mean angle θ_{flow}, while those having rapidities close to the projectile or the target are less deflected (this phenomenon is called the rebound). The distribution of transverse momentum per nucleon p_x/A *in the reaction plane* can then be plotted as a function of the particle rapidity y in a bi-dimensional

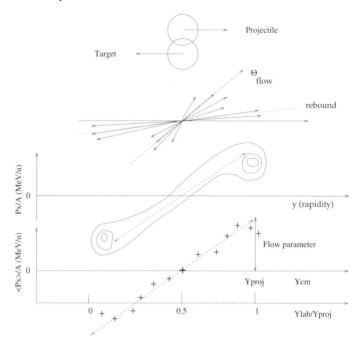

Figure 6.6. A schematic picture illustrating the various steps in the definition of the flow parameter F, as discussed in the text. In the upper part, after contact between the two partners of the reaction in real space, the kinematics of the emitted particles is schematically described in the reaction plane. The two contributions associated with the decay of the QT and the QP are displayed as two bumps located near the respective rapidities of the QT and the QP. The emission at mid-rapidity associated with processes occurring in the overlap region smoothly joins the two bumps. An averaging procedure for the transverse momentum gives the result in the bottom panel, resulting in a linear evolution of p_x/A as a function of the rapidity. The slope at mid-rapidity defines the flow parameter F.

plot $(y, p_x/A)$. Averaging p_x/A for various bins of the normalized rapidity, one obtains an S-shape curve whose derivative around mid-rapidity ($y_{\text{lab}}/y_{\text{proj}} \sim 0.5$ which corresponds to the nucleon–nucleon rapidity y_{NN}) gives the value of the flow parameter F, defined as

$$F = (y_{\text{proj}} - y_{\text{NN}}) \left(\frac{\mathrm{d}\langle p_x \rangle/A}{\mathrm{d}y} \right)_{y=y_{\text{NN}}} \tag{6.4}$$

where y_{proj} is the projectile rapidity. This procedure, however, requires a careful determination of the reaction plane which is defined by two vectors: the velocity of the projectile and the impact parameter vector.

From an experimental point of view, the reaction plane must be reconstructed from the particles emitted in the course of the reaction. Different methods have been developed so far:

- a method based on the study of transverse momentum distribution [147]; and
- a method based on azimuthal correlations [496].

Let us detail the first method. Estimating the reaction plane in the transverse momentum analysis is based on constructing the following quantity:

$$Q_\nu = \sum_{i=1}^{\nu-1} \omega^i p_{\text{perp}}^i \qquad (6.5)$$

where ω^i is a weight depending on the rapidity y_i of particle i. A usual procedure consists of taking $\omega^i = 1$ if $y_i - y_{\text{CM}} \geq 0$ and $\omega^i = -1$ if $y_i - y_{\text{CM}} \leq 0$ (see figure 6.6). Note that the summation holds for all detected particles but the particle of interest, labelled ν. Indeed, to avoid autocorrelation, it is necessary to construct a reaction plane *per particle* (Q_ν) instead of a reaction plane *per event* (Q in which the summation in equation (6.5) includes particle ν) because the definition of p_x/A requires a scalar product between the momentum p_{perp}^ν of the considered particle and the vector defining the reaction plane. Therefore, because of the autocorrelation, p_{perp}^ν should not be included in the summation of equation (6.5). Nevertheless, in order to preserve the total transverse momentum, a 'recoil correction' is implemented in the method. It should be noted that this correction does not affect the value of F, since it only generates a constant shift in the value of p_x/A [147].

6.2.2 Experimental results and comparisons with transport models

The sidewards flow has been studied in a variety of reactions at various incident energies. The flow parameter F can be evaluated for each impact parameter once the events have been sorted according to the methods described in section 4.2.3. F starts from zero in very peripheral collisions, then increases to reach its maximum value F_{max} in mid-central reactions. For symmetry reasons, it vanishes in central collisions. In figure 6.7, the evolution of the maximum value F_{max} of F as a function of beam energy E_{lab} is displayed. Generally speaking, the global trends exhibited by the data are better reproduced in semi-classical transport models using momentum-dependent interactions (MDI) [199] or a soft equation of state (EoS) [490].

Below 100 MeV/u, the flow parameter nevertheless remains close to zero. In order to be able to make quantitative comparisons with the predictions of transport models, the balance energy E_{bal} is then defined as the incident energy E_{lab} for which the flow parameter F vanishes: E_{bal} is hence associated with the change of sign of F. Indeed, at low E_{lab}, the nuclear force is attractive producing a negative deflection of the particles, hence leading to negative values of F, while at high

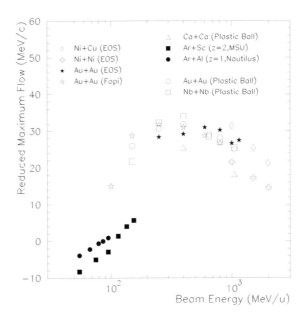

Figure 6.7. Compilation of F_{max} in symmetric collisions from 20 MeV/u up to a few GeV/u. The evolution from negative values to positive values is associated with a transition from an energy region where the nuclear force is attractive to a region where it becomes repulsive. After a strong increase of F_{max}, the decrease observed above 1 GeV/u is interpreted as a softening of the equation of state and could be a first hint for a transition from the hadronic matter to the QGP. From [172].

E_{lab}, the force becomes repulsive, leading to positive values of F. In fact, it is not possible to obtain, experimentally, the sign of the flow parameter, except if one measures the associated γ's [287,469]. When only charged particles are detected, it is implicitly assumed that the flow parameter is negative at low incident energy. But the balance energy can, nevertheless, be identified. The 'experimental' flow parameter F exhibits a cusp at E_{bal}, reflecting the actual change of sign of the physical F. This inversion of flow was predicted in [318] using transport theory.

A careful analysis of the results of transport models shows that the balance energy results from a subtle interplay between mean-field and two-body collision effects [44, 152]. In this respect it is an observable which typically emphasizes the various leading effects in the dynamics of collisions in the nucleonic domain. The balance energy thus constitutes a demanding test for microscopic dynamical models such as BUU. The measure of the sidewards flow, or equivalently of the balance energy, thus indeed puts constraints on the microscopic models,

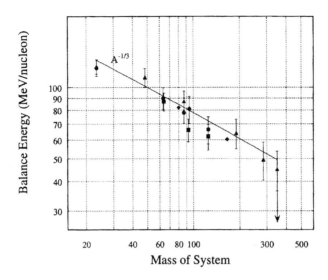

Figure 6.8. Mass dependence of the balance energy E_{bal}. The full line corresponds to a $A^{-1/3}$ mass dependence as predicted by geometrical arguments (see text). The two points without error bars are the predictions of transport models. It should be noted that the $A^{-1/3}$ is predicted by such models on the whole mass range. From [106, 490].

particularly in terms of the effective force and the in-medium nucleon–nucleon scattering cross-section entering these calculations.

Figure 6.8 shows a compilation of the balance energy as a function of the mass of the system. The data show a mass dependence in $A^{-1/3}$ [490]. Two computed points (results of a transport model) are also added to the experimental data in this figure. They coincide nicely with the data systematics.

Considering the impact parameter dependence of the balance energy (see [490] for a review), it was possible to show its relation with the momentum dependence of the mean field as reported in [435]. Last but not least, recent works at both experimental [357] and theoretical [30] levels have explored the role of the isospin degree of freedom in the sidewards flow process. They show promising results concerning the extraction of the isospin symmetry term of the nuclear force.

6.2.3 Squeeze-out and azimuthal distributions

So far, we have only considered the directed flow *in* the reaction plane. We shall now concentrate on the determination and the physical interpretation of the *out-of-plane* emission, often denoted as 'squeeze-out'. The squeeze-out is presumably related to the stiffness of the equation of state (EoS): the larger the

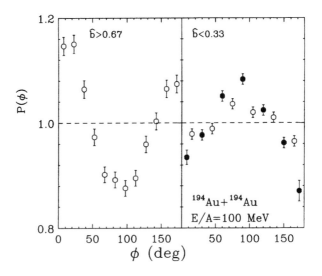

Figure 6.9. Azimuthal distribution of the emitted particles for Au+Au collisions at 100 MeV/u. The left-hand panel is for peripheral collisions while the right-hand panel is for central collisions. The reduced impact parameter estimation follows the method described in section 4.2.3. From [470].

effect is, the 'harder' the EoS is. One method of representation is to plot azimuthal distributions. The azimuthal angle ϕ is defined as the direction between the transverse momentum of the particle and the direction of the reaction plane Q. Such azimuthal distributions are illustrated in figure 6.9 for two different impact parameters in Au+Au collisions at 100 MeV/u [470]. In peripheral collisions, data exhibit a typical V-shape distribution which corresponds to 'soft' collisions for which most of the particles are emitted in-plane (minimum at $\phi = 90°$). In contrast, in central collisions, the distribution is maximum at 90° putting in evidence the strong enhancement of out-of-plane emission of particles.

A convenient way to reduce the information provided by the azimuthal distributions is to fit them with two cosines:

$$\frac{\mathrm{d}Y}{\mathrm{d}\phi}(\phi) = a_0 + a_1 \cos(\phi) + a_2 \cos(2\phi) \qquad (6.6)$$

where a_0 is a normalization constant and a_1 is a measure of the importance of the 'in-plane' flow. This expression thus provides another way to quantify the sidewards flow, as discussed in section 6.2.1. The anisotropy of the particle flow pattern is finally accessed through the a_2 term.

In the Fermi energy range, the experimental value of a_2 is always positive, whatever the rapidity of the considered particles, which means that the emission is preferentially in the reaction plane. This is in contrast to the results obtained at relativistic energies in which a_2 becomes negative. The transition ($a_2 = 0$) from a preferentially 'in-plane', to 'out-of-plane' emission is observed for Au+Au collisions around $E_{lab}/u \sim 100$ MeV/u. But a precise knowledge of this quantity would require new experiments.

It is worth noting that a_2 reaches its maximum negative value around 400–500 MeV/u (for Au+Au collisions) and then increases to become positive again around 5 GeV/u. This may be interpreted as a 'softening' of the EoS in this energy range and could be the first hint for the transition between hadronic matter and the QGP [438].

6.3   Particle production

Excited nuclear matter may decay in various ways. Up until now we have mainly discussed physical processes involving nucleons and their mutual interactions. In the course of heavy-ion collisions, even in the nucleonic regime, it turns out that not only nucleons are produced. For example, photons have long since been traced back as signals of collective motion. Here we aim to consider particles produced in less 'gentle' physical conditions, in particular, during the 'violent' overlap phase of heavy-ion collisions. This is typically the case of the so-called 'hard' photons, which because of their specificity are discussed in greater detail in section 6.4. In fact, many 'exotic' particles can be produced in the collisions in which we are interested, and it turns out that they provide valuable clues on the dynamical, and possibly 'thermodynamical', conditions prevailing in the system, which actually created them. In this section we shall thus discuss the production of several particles: baryons such as Δ's, mesons like π's, η's or even strange mesons like K$^+$'s, and of course the previously mentioned hard photons. And we shall try to understand what information they bring on the dynamics of the collision.

6.3.1   The role of beam energy

6.3.1.1   On production thresholds

Hard photons (up to detector thresholds) can be produced whatever the beam energy, but this is not the case for other particles ($\Delta, \pi, K, \ldots$), for which there exists a threshold, corresponding to the minimum energy necessary for producing this particle. Let us consider the Z-particle production $X + Y \rightarrow Z + \cdots$ from X and Y, where X and Y may be constituents of the system, nucleons, pions or even the colliding nuclei. The Z production threshold is defined as the minimum energy needed to create particle Z at rest, in the frame where Y is itself at rest:

$$E_Z = \frac{1}{2m_Y}[(m_Z + m_{...})^2 - (m_X + m_Y)^2] \tag{6.7}$$

with obvious notation.   It should be noted that, as particle production is
a consequence of the relativistic equivalence between mass and energy, the
threshold energy has to be defined relativistically. The threshold energy depends
on the production mechanism. In the following we shall mainly consider two
types of thresholds.   The *elementary* (or *free*) *threshold*, hereafter denoted
E_Z^{NN} corresponds to the dominant elementary channel (often N − N, hence the
notation). For example, the elementary threshold for Δ production (N − N →
NΔ) is $E_\Delta^{NN} \simeq 625$ MeV. In heavy-ion collisions the *absolute threshold*,
for a given projectile (A_p)–target (A_t) pair (in this case $m_X = m_{A_p}$ and
$m_Y = m_{A_c}$), corresponds to the case in which all the available beam energy
is actually consumed to create the particle. As nuclei, in the energy range we
consider here, behave as collections of nucleons, it is clear that the absolute
threshold is by construction smaller (and sometimes much smaller) than the
elementary threshold. For example the elementary threshold for K$^+$ production
from nucleons is $E_{K^+}^{NN} = 1582$ MeV (N − N → K$^+$ + Λ + N), while it is only
112 MeV in a ^{12}C+^{12}C collision.

6.3.1.2   *Systematics of meson production*

One elegant and telling way to compile the numerous data concerning meson
(hereafter labelled Z) production cross-sections σ_Z in nucleus–nucleus collisions
is to consider the reduced probability of production per participant:

$$P_Z^{red} = \frac{\sigma_Z}{\sigma_{\text{reac}}\langle A_{\text{part}}\rangle} \tag{6.8}$$

in which σ_Z is the measured Z-meson cross-section, σ_{reac} the measured total
reaction cross-section and $\langle A_{\text{part}}\rangle$ the impact parameter-averaged number of
participant nucleons. This last quantity can be estimated by means of a simple
geometrical *ansatz* by calculating the overlap volume $V(b)$ of the two interacting
nuclei at impact parameter b. Assuming normal density ρ_0, one then obtains:
$A_{\text{part}}(b) = \rho_0 V(b)$. Note that this procedure implicitly assumes that energetic
particles are produced in hard nucleon–nucleon collisions and thus essentially in
the overlap region of the two incoming nuclei.

The reduced production probabilities P_Z^{red} are plotted in figure 6.10. The
striking feature of this figure is that production probabilities for π's and η's fall
close to a single curve over about 11 orders of magnitude. In turn, systematic
deviations from this behaviour are observed for kaons, which may be down by as
much as a factor of five. This effect is usually attributed to the non-equilibration
of the strangeness (kaons are strange particles) degree of freedom.

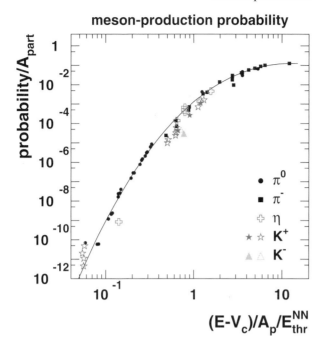

Figure 6.10. Meson production probability per participant nucleon as a function of $E - V_c$ (the incident energy minus the Coulomb barrier) normalized to the meson production threshold E_Z^{NN} in elementary free nucleon–nucleon interactions. In such a representation, points with an abscissa less than one correspond to the so-called sub-threshold regime. From [312].

6.3.2   Particle production and collision dynamics

6.3.2.1   Why to study particle production?

The study of particle production raises two classes of questions, once we assume that the elementary mechanism is well known. The first one concerns the actual, effective, production mechanism *in* the nuclear medium. Similarly to the (elastic) nucleon–nucleon interaction, inelastic channels are also sensitive to neighbouring nucleons (medium effects, see section 2.2.1.2). These effects have been particularly studied for Δ's and π's which have sizeable interactions with nucleons. This aspect is presumably less important in the case of strange particles like K^+ or with γ's, which interact more weakly with the surrounding medium. But altogether the questions raised by medium effects on particle production remain, in many cases, not fully solved.

The second class of questions, once effective medium effects are accounted for, concerns the information brought by particles about the reaction itself. For a given type of particle, the answer depends on energy. The study of particle production in a kinematic regime *above* the elementary threshold (for example kaons K$^+$ at 2 GeV/u beam energy) does not tell much about the underlying static or dynamical properties of the source, as particles are then produced by the simple, direct (usually well known) elementary process. When a nucleon from the projectile hits one from the target, there exists (even in the early stages of the collision) a probability of forming a particle equal to $\sigma_{N-N\to Z}/\sigma_{N-N|tot}$. In contrast, *below* threshold, this direct process is no longer possible, at the first collision. One needs an extra mechanism for understanding how pairs of nucleons may gain a relative kinetic energy above production threshold. Intuitively, it is thus clear that such conditions will likely provide some clues on the dynamics of the collision, in terms of timescales and/or in terms of 'thermodynamical' conditions. Furthermore, different particles, corresponding to various thresholds, will bring different pieces of information, depending on beam energy.

6.3.2.2 *Which particles to choose for characterizing the dynamics?*

Many particles have been studied in the context of heavy-ion collisions. We start with pions, which are abundantly produced compared to other massive particles, as their free threshold is relatively low (290 MeV in the direct $N - N$ channel). Beyond about 100 MeV/u they are mainly produced by disintegration of the short-living Δ's (NN $\leftrightarrow$ Δ $\leftrightarrow$ Nπ, with $\tau_\Delta \sim$ 1–2 fm/c), while below 100 MeV/u direct production dominates (NN $\to$ Dπ). Pions were originally expected to provide a clue to the stiffness of the equation of state [442] but, once produced, they interact strongly with the nuclear medium, so that the actually measured pions may have quite different characteristics compared to the ones originally produced. Kaons offer an appealing alternative here. Because of their strangeness, they interact little with the surrounding nuclear medium, which makes them a 'clean' probe[1]. They are mainly produced in BB $\leftrightarrow$ BYK$^+$ collisions (where B is a baryon, nucleon or Δ, and Y is a hyperon) and in πB $\to$ K$^+$ + Y processes. Kaons are presently studied actively both from the theoretical and experimental points of view. Note finally that an active field of research has also started around e$^+$e$^-$ pairs production, but at high beam energies (see for instance [169] and references therein).

Hard photons are especially interesting at low beam energy (typically below 100 MeV/u) where they result mainly from early neutron–proton collisions ('Bremstrahlung' radiation, section 6.4). At higher energies the γ signal is polluted by the electromagnetic disintegration of π_0's which start to be sizeably produced at these energies. Light mesons, such as π's, η's or K$^+$'s, allow us to consider a wider range of beam energies. For a given type of meson one should consider a beam energy high enough to provide a decent multiplicity, but well below the free threshold so that production cannot directly result from first chance

[1] Etas are also an interesting 'theoretical' possibility, but they raise problems in the experimental identification.

elementary collisions. For example, pions are *a priori* adapted to an energy range around 100–200 MeV/u, while kaons would be much better adapted to an energy range of order 500 MeV/u.

6.3.2.3 Theoretical approaches

Numerous models have been developed over the years to study particle production in heavy-ion collisions. Here we shall focus on microscopic approaches based on Boltzmann-like kinetic equations (section 3.2.3.2), because a realistic description of the early stages of collisions is to be based on microscopic dynamical approaches and Boltzmann-like kinetic equations offer here a proper tool of investigation.

In BUU calculations, particle production is described as an incoherent production mechanism, by summing up incoherently the contributions of elementary processes. The cross-section for Z production then is given by:

$$\sigma_Z \propto \int_0^\infty \mathrm{d}t \int 2\pi b \, \mathrm{d}b \int \mathrm{d}1 \int \mathrm{d}2 \, f(1) f(2) v_{12} \sigma_Z^{\mathrm{elem}} Q_\mathrm{P} \qquad (6.9)$$

where d1, 2 refers to a phase space integration over particles 1, 2 and b labels the impact parameter. In equation (6.9), Q_P is the Pauli blocking factor after creation of Z, and σ_Z^{elem} labels the elementary Z production cross-section. Note also that in general one should sum up the contributions of all the possible production channels: baryon–baryon, baryon–meson...

Figure 6.11 shows how particle production may give indications on the timing of production and thus on the dynamics of a heavy-ion collision, and illustrates the importance of the choice of the particle for a given beam energy. Here we compare π and η production in ^{40}Ca+^{40}Ca collisions at two beam energies, as a function of time. The 1 GeV/u beam energy is above both the π and η thresholds and production starts from the beginning in both cases. Particle production does not then bring much information on the dynamics of the reaction. In contrast, at 0.5 GeV/u beam energy (still above π, but below η threshold), while π production again starts from the beginning, η production is delayed so that a 'hot' zone has time to form (this requires five to ten nucleon–nucleon collisions, thus a time of the order of thermalization time, sections 3.1.1.2 and 6.1.2). In this case it is thus clear that particle production allows us to select a given time interval of production.

The question of linking particle production to the properties of nuclear matter, such as the incompressibility modulus K_∞, has long been debated. For the time being no clear conclusions have been obtained [118]. Of course the details of the production somewhat depend, for example, on K_∞, but they may as well depend on N–N cross-sections in a quantitatively comparable way. Furthermore, such effects usually remain relatively small (typically a factor of two or three in integrated cross-sections) compared to experimental error bars, so that in reality

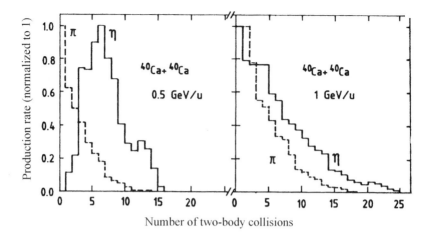

Figure 6.11. Timing of particle production (η and π) in a ^{40}Ca+^{40}Ca collision at 0.5 and 1 GeV/u. The production rate (normalized maximum value 1) is plotted as a function of the number of collisions encountered by the nucleons. Hence the abscissa is a timescale. At 500 MeV/u η's are produced after a well-defined time unlike π's for which the nucleon–nucleon threshold is well below this energy. In contrast, at 1 GeV/u η's and π's are produced at the same time. One can hence see that a timing is possible in a region kinematically adapted to a given particle, and impossible otherwise. These results have been obtained in a VUU-like calculation, including surface effects. From [119].

it is often hard to access them. One should also note that uncertainties in some elementary cross-sections may lead to further theoretical error bars.

6.3.3 Sub-threshold particle production at very low beam energy

Up to now we have considered particle production in a kinematic regime below the elementary threshold, but well above the absolute threshold (section 6.3.1.1), for which all the available energy from the beam is used to form the desired particle. The study of close-to-absolute threshold particle production is a hard task, from the experimental *and* theoretical points of view. Up to now no commonly admitted mechanism explains such a phenomenon. This very low energy production corresponds to very rare events, but the mere fact that it does exist is already quite surprising. Note that such a phenomenon is very different from usual particle production in high-energy physics, where a pair of possibly point-like particles (like e^+e^-) annihilates to produce a more massive particle. The major differences lie in the fact that energy here is spread over virtually *all* the degrees of freedom of an extended system (the nucleus!). Hence one has to invoke 'collective' effects, which still remain to be understood.

Close to the absolute threshold π production has been extensively studied over the years [93, 328]. More recently, sub-threshold kaons have been observed in the ^{36}Ar+^{48}Ti system at 92 MeV/u beam energy leading to a cross-section of the order of 2.5×10^{-7} mb [259, 286]. In a similar system (^{42}Ca+^{42}Ca at 92 MeV/u) Boltzmann–Langevin simulations lead to a comparable cross-section while BUU simulations miss the experimental point by typically 10 orders of magnitude [40]. It is likely that at very low beam energy the Boltzmann–Langevin results would also depart from experimental data, as the production mechanism in these calculations remains, to a large extent, of incoherent nature. Nevertheless, these results seem to indicate the importance of fluctuations in an intermediate energy domain between the energy range associated with a 'hot zone' and the absolute threshold. Furthermore, one may expect that, at these low energies, equation of state effects should be more sizeable than at several hundreds of MeV/u, somewhat similarly to the case of flow effects below 100 MeV/u (section 4.3.3). This aspect has still to be investigated in more detail.

6.4 Hard photon production

6.4.1 Systematics of hard photon production

As already mentioned, hard photons constitute very good probes of the reaction mainly because they interact weakly with the surrounding medium, once produced. The photon spectrum can furthermore be arbitrarily divided into soft and hard photons. Soft photons are associated either with the last steps of the decay processes of an excited nucleus or with the decay of collective modes such as giant resonances or superdeformed bands. The photons resulting from the decay of hot nuclei are usually called statistical photons and have energies not exceeding a few MeV. The typical energies of photons associated with the decay of giant resonances is around 15–20 MeV (see section 7.2.2). Hard photons are thus usually defined as photons of energies larger than 30 MeV.

Hard photons are produced in nucleon–nucleon collisions inside the medium through the so-called Bremstrahlung process. Similarly to the case of massive mesons one can perform a reduction of the total hard photon yield in terms of the elementary probability P_γ to produce a hard photon in an individual n–p collision. This leads to the following expression:

$$P_\gamma = \frac{\sigma_\gamma}{\sigma_{\text{reac}} \langle N_{\text{pn}} \rangle} \qquad (6.10)$$

in which σ_γ is the measured photon cross-section, σ_{reac} the total reaction cross-section and $\langle N_{\text{pn}} \rangle$ is the mean number of n–p collisions averaged over the whole impact parameter range. This last quantity is estimated, as before, by assuming that γ's are produced in the overlapping zone of the two nuclei. Figure 6.12 shows the systematics of reduced hard photon production (according to equation (6.10))

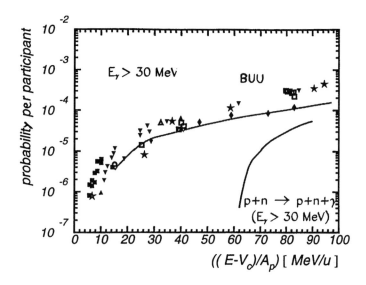

Figure 6.12. Systematics of hard photon production: the ordinate is the probability per individual nucleon–nucleon collision of producing a hard photon while the abcissa is the incident energy corrected for the Coulomb barrier. The full line labelled BUU is the prediction of a microscopic semi-classical transport model while the other full line, on the right, is the probability to produce a hard photon in an elementary nucleon–nucleon collision in the vacuum. From [419].

in the Fermi energy range. The huge difference between the probability for the elementary process in the vacuum and the in-medium probability emphasizes the role of the collectivity in producing energetic photons. Microscopic transport models (BUU in this case) are in reasonable agreement with the data over the whole energy range.

A detailed experimental study of hard photon production reveals (at least for sufficiently heavy systems) the existence of two distinct 'sources' of emission. The presence of two distinct contributions has been observed in the differential energy cross-sections of the γ spectrum: a first contribution, corresponding to the high-energy part of the spectrum, can be tentatively associated with the early instances of the reaction when the system is dense (this is the main contribution), while a second contribution (less intense and less energetic) could be attributed to a later stage of the reaction and could be associated with a thermalized source.

These two contributions are predicted by transport models. For instance, the time dependence of photon production in central Ta+Au collisions has been studied in the framework of the BUU model. A result is displayed in figure 6.13. Two different photon 'bursts' are indeed predicted by the model. The first one, which is rather insensitive to the value of the incompressibility modulus

K_∞ used in the simulation, is associated with direct nucleon–nucleon inelastic scatterings (one then speaks of *direct* photons) and constitutes a direct probe of the phase space occupancy of the nucleons in the early instances of the reaction: it corresponds, on average, to the most energetic photons, while the second burst is associated with later times, when the system has reached thermalization (one then speaks of *thermal* photons). The intensity of the second contribution is strongly dependant on the value of the incompressibility modulus K_∞: as K_∞ increases, so does the number of emitted photons. What is the microscopic origin of this second contribution? A study of the time dependence of the density reached in the centre of the system shows that, after a first compression phase associated with the early instances of the reaction, an expansion of the system is observed during which the system becomes dilute and the number of produced photons decreases because there are very few efficient hard nucleon–nucleon collisions. But for heavy systems, the expansion is stopped at a given turning point and a second, less intense, compression phase sets in, producing a new burst of γ's. Furthermore, a study of the correlation between photons emitted during the first step and the second step gives access to the spacetime characteristics of the collision as shown in the following (section 6.4.2).

6.4.2 Hard photon intensity interferometry

The study of the interference between two photons, each emitted by the two previously-mentioned sources (direct and thermal photons), can help in characterizing the spacetime extent of the matter in central nucleus–nucleus collisions. The experimental techniques of intensity interferometry have already been described briefly in chapter 4, so we will not discuss them any further as such. Two experiments have been devoted to the study of hard photon interferometry. Both were performed at GANIL: the first one used the MEDEA detector [21], while the second one used the TAPS device [307]. In both cases, the first step of the data analysis consists of evaluating the signal induced by the decay of the π_0 in the $\gamma\gamma$ channel, which induces a strong enhancement of the correlation function around the pion mass. Then, it is possible to obtain the correlation function free from the pion decay in the whole range of the relative four-momentum Q of the two photons. Note that Q is also the invariant mass, since photons are massless particles. Figure 6.14 shows the result obtained by the TAPS collaboration for two systems: Kr+Ni at 60 MeV/u and Ta+Au at 40 MeV/u. The two-photon correlation function can be tentatively described by the following expression:

$$C_{12}(Q) = 1 + \lambda \cdot \exp(-Q^2 R_Q^2) \cdot I_{\text{interf}}(Q) \tag{6.11}$$

where λ gives a measure of the coherence of the source: the larger λ becomes the more coherent the source becomes. In equation (6.11) $I_{\text{interf}}(Q)$ is the interference term between the two sources of emission intensity I_d for the 'direct'

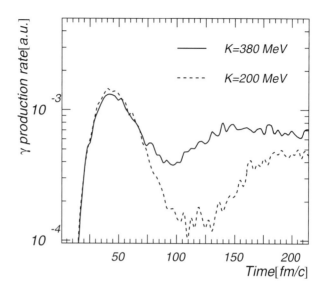

Figure 6.13. Time dependence of hard photon production as predicted by transport calculations in central collisions for heavy systems. A first 'burst' is observed in the early instances associated with a hot and dense phase. A second one, coming later, is associated with a second compression phase corresponding to a 'breathing' mode of the matter. The intensity and timescale of the second compression strongly depend on the value of the incompressibility modulus used in the calculation. From [419].

Table 6.1. Parameters of the fitting procedure of the correlation function of figure 6.14.

System	λ	R_Q (fm)	I_d (%)	Δ (fm)
Kr+Ni	0.52 ± 0.17	3.3 ± 0.9	79 ± 10	15 ± 3
Ta+Au	0.60 ± 0.03	4.5 ± 1.5	50 ± 16	37 ± 3

first burst; and I_t for the second, 'thermal', contribution, respectively. The interference term I_{interf} is given by

$$I_{\text{interf}}(Q) = I_d^2 + I_t^2 + 2I_dI_t \cos(Q\Delta). \tag{6.12}$$

The source geometry is characterized by its size R_Q (one assumes that the source size is the same during the two bursts). Finally, Δ is a measure of the spacetime distance between the two 'bursts'.

Fits of the correlation functions of figure 6.14 give the parameters in table 6.1

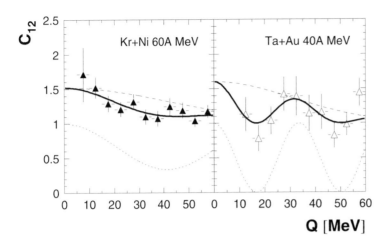

Figure 6.14. Two-photon correlation function for Kr+Ni (left) and Ta+Au (right) reactions at, respectively, 60 and 40 MeV/u. The full line is the result of a fit using equations (6.11) and (6.12). From [307].

for the sources. Note the evolution of the relative importance of the direct contribution when one goes from a medium-mass system (Kr+Ni) to a heavy one (Ta+Au): for a heavy system, the second photon 'burst' is more intense because the associated second compression is more important. The interpretation of Δ is more delicate: one finds values larger than the 100 fm/c predicted by the calculations (see figure 6.13). Up to now, no definitive interpretation of these numbers is available.

6.5   Composite particle production

Up to now, we have only discussed the production of fast nucleons and 'elementary' particles. We have shown briefly in section 6.1.1 that composite particles (say from deuterons to lithiums) could also be produced in fast processes. It is worth noting that the first attempts to theoretically describe such composite particle production in fast and non-equilibrated processes in nuclear collisions were initiated a long time ago in phenomenological models based either on cascade calculations or on the so-called exciton approach. It is not our aim to decribe in detail these pionering works and we refer the reader to standard textbooks [185, 198] or reviews [59]. It was already recognized at that time [59] that such processes are extremely difficult to address from a theoretical point of view. Several possibilities were explored ranging from preformed clusters in nuclei, which can scatter with an incoming nucleon, up to groups of nucleons whose correlated momenta match the cluster characteristics. Nevertheless, most of these approaches contained a large number of free parameters and thus were not really satisfying.

Although new developments have been proposed in the framework of the exciton model [60], most of the recent works are based on microsocopic transport theory. Composite particle production has been explored in the framework of the transport models, in particular, in QMD [361] and classical MD [79, 166] calculations and also in BUU simulations [148]. In molecular dynamics calculations, the mechanisms by which fragments are produced, or at least 'recognized', is by inspection of the phase space: nucleons which are connected in phase space are grouped together to form clusters. A condition based on energy considerations is applied to limit the amount of excitation energy stored in the reconstructed fragments. In [148], the production is explicitly taken into account by calculating the vertex that can lead to the production of deuterons, tritons, etc. To this end, a transport equation is derived for each species taking into account the formation as well as the dissociation cross-sections. Transition matrices are calculated as well as Pauli correlations. This promising attempt is, however, limited to very light composite particles. One is forced to recognize that a genuine quantum theory for the production of composite particles in nuclear collisions is far from being achieved.

6.6   Conclusion of the chapter

In this chapter, we have discussed the properties of the system on the way to thermalization by studying the first instances of the reaction. The two main features discussed are the mechanisms of production of early-emitted particles (nucleons, composites, mesons and hard photons) and for some of them their kinematic characteristics, through the study of the sidewards flow and the squeeze-out. From a theoretical point of view, the dissipation process can be traced back with the help of the microscopic semi-classical transport models. Encouraging results have been obtained in various aspects of these questions, concerning both the analysis of collective motion (flow in particular) and particle production. Still, standard transport models cannot account for the whole bunch of observed phenomena, as, for example, in the case of well below threshold rare-particle production.

Experimentally, the study of in-plane flow puts strong constraints on the transport models mainly by the evolution of the balance energy: they suggest the use of momentum-dependent forces in the theory and put limits on the variation of the in-medium nucleon–nucleon cross-section. The squeeze-out is not a dominant process in the Fermi energy domain but its evolution as a function of the impact parameter can also provide interesting tests for the theory.

Particle production remains an important subject in nuclear collisions. Various kinds of produced particles have been discussed: their production rate can be reduced with help of simple geometrical assumptions. Cooperative or at least collective processes possibly connected with high-order fluctuations in the medium are responsible for particle production close to the absolute threshold. A proper treatment of these fluctuations requires extensions of one-body theories and constitutes a future challenge for the theory. Weakly interacting particles such as γ's or kaons are excellent probes of the first instances of the collision. Particle–particle correlations (in particular $\gamma\gamma$ coincidences) can give valuable information on the spacetime pattern of the reaction. Finally, composite particle production far from equilibrium is still out of the scope of most of the microscopic approaches used up to now. Although several attempts have been made in nucleon–nucleus collisions, a truly microscopic quantal theory for the production of small clusters in out-of-equilibrium processes is still unavailable.

Chapter 7

Decay modes of hot nuclei: from evaporation to vaporization

In this chapter, we concentrate on the study of hot nuclear species produced in the nuclear collisions described in chapter 5. The methods for estimating the degree of thermalization and for measuring the excitation energy deposited in the system as well as the temperature, T, have been discussed in chapter 4. The study of the decay modes of hot nuclei is at the heart of nuclear thermodynamics. It can provide useful information on the instabilities present in the system. The latter will lead to particle, fragment or photon emission. Such studies also reveal the nature of the motion of the nucleons inside the system at finite excitation energy and thus are intimately related to nuclear dynamics.

The chapter is organized as follows. In section 7.1.1, the general features of the decay of hot nuclei are outlined introducing the lexicology used in what follows. Although hot nuclei are metastable objects a theoretical description based on a nuclear mean field is conceivable. It is described in section 7.1.2 where we introduce the concept of a limiting temperature, beyond which a static mean-field description of the system becomes definitively irrelevant. For high energies specific models are then necessary: they will be discussed separately in the next chapter in the framework of nuclear fragmentation. In the present chapter, the discussion about the decay properties of hot nuclei, particularly from an experimental point of view, has been arbitrarily divided into two parts: the low-energy part (section 7.2) and the high-energy part (section 7.3).

The low-energy part covers evaporation, decay by giant resonances, the production of evaporation residues and fission. Evaporation is associated with the 'chaotic' thermal motion of the nucleons at the surface of the nucleus. The correlation between thermal energy and temperature provides information on the heat capacity of hot nuclei. The evolution of the so-called level density parameter (section 7.2.1) reflects the various transitions occurring inside the nuclei as the temperature increases. Small collective motions such as giant resonances can be studied at finite temperature and their characteristic decay properties have been established as a function of excitation energy (section 7.2.2). In contrast, fission is a large amplitude collective motion. The evolution of its probability with excitation energy reveals the typical times needed to strongly deform a nucleus. It depends on the corresponding nuclear matter viscosity and its dependence on temperature (section 7.2.3). The properties of giant resonances, fission and evaporation residues are intimately related to the question of nuclear dissipation (section 7.2.3.3).

The section devoted to high-energy processes starts with the study of the rise of three-body decay (fragmentation). The general trends of nuclear fragmentation are then discussed without much detail. Indeed, it turns out that nuclear fragmentation is the key process that opens up the study of new instabilities in nuclear matter. Since it is currently a matter of intensive research, we have dedicated chapter 8 to this question. The chapter finally ends up with the description of the vaporization of the system into constituents with charges lower or equal to two (section 7.3.2).

7.1 Some experimental and theoretical properties of hot nuclei

7.1.1 The decay of hot nuclei: general experimental features

A strong evolution of the decay modes of hot nuclear species is expected as one goes from very moderate excitation energy deposits, corresponding to the vanishing of shell effects ($T \gtrsim 2$ MeV), up to values close to or even larger than the total binding energy of the system ($E^*/A = \epsilon^* \sim 8$ MeV/u), for which a complete dissolution is expected. Such a strong evolution is examplified in figure 7.1 showing the rapid evolution of the decay of a quasi-projectile (QP) nucleus produced in dissipative Pb+Au binary collisions at 29 MeV/u as a function of the neutron multiplicity M_n.

As previously mentioned, M_n is a good measure of the excitation energy deposited in the system (section 4.3.1.2). For the lowest values of M_n (less than 10 neutrons typically) the charge distribution is dominated by the production of *evaporation residues* with an atomic number close to the projectile charge ($Z_{\text{proj}} = 82$ in this case). When M_n and thus also the excitation energy E^* increases, the contribution associated with *symmetric fission* sets in (these are the events with $Z \simeq Z_{\text{proj}}/2$). For M_n reaching values around 40–50, *fragment emission* with atomic numbers between 3 and 30 becomes competitive with symmetric fission despite the difference in the barrier heights. When raising E^* up to a significant fraction of the binding energy, a transition from two-body decay to multi-body decay (i.e. the emission of several nuclear species with $Z \geq 3$) is then evidenced, indicating the onset of *nuclear fragmentation*. For the highest achievable excitation energy in such a collision, one ends up with many intermediate mass fragments (IMFs) associated with as many as 70 neutrons emitted coincidently. This corresponds to a total dissolution of the system. But before going into the detail of these various

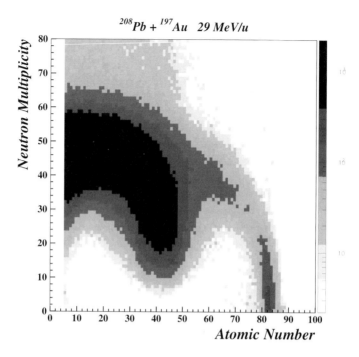

$^{208}Pb + {}^{197}Au$ *29 MeV/u*

Figure 7.1. Two-dimensional plots of the neutron multiplicity versus the atomic number of the products emitted by the QP in Pb+Au collisions at 29 MeV/u and detected in the forward direction. Neutrons are detected in the neutron-meter ORION in the whole space and thus correspond to the decay of both the QT and QP (this is why the multiplicity may be as large as 80). From [96].

decay modes, it is worthwhile discussing some general theoretical issues about hot nuclei.

7.1.2   On theoretical descriptions of hot nuclei

7.1.2.1   How to describe a metastable hot nucleus?

By nature, hot nuclei, once formed, are metastable objects, the major tendency of which are precisely to disappear, in one way or another. This obviously raises difficulties if one aims at a reasonable theoretical description of such systems. It is interesting to consider here a mean-field dynamical approach, which provides a first access to the time evolution of hot nuclei. The system is prepared at a given finite temperature, by giving each level of the single-particle spectrum a Fermi factor occupancy (section 2.3.1.4). An example of such a calculation is displayed in figure 7.2 for a ^{40}Ca nucleus at two temperatures: $T = 3$ MeV (left panel) and

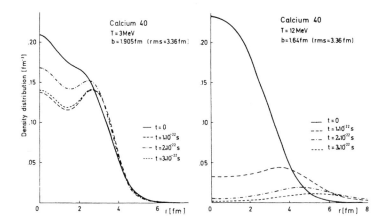

Figure 7.2. Density profiles of calcium nuclei at $T = 3$ MeV (left) and $T = 12$ MeV (right) as predicted by TDHF calculations for the different times indicated in each panel. From [479].

$T = 12$ MeV (right panel) [479]. In the low temperature case, one can see that the nucleus loses a small fraction of its nucleons, but still subsists as a nucleus on a relatively long timescale ($t \sim 3 \times 10^{-22}$ s ~ 100 fm/c), compared to the characteristic times of the system (section 3.1.1). In the high temperature case, in contrast, the system is vaporized within the same time interval. This means that at such a high temperature, only a dynamical description may make sense, while at lower temperature a static picture keeps some relevance. In such cases, one can then envision a static picture of the metastable hot nucleus, for example, as a nucleus immersed in the gas of emitted nucleons.

This static picture of hot nuclei has been the basis of numerous works in the mid 1980s, aiming to study, systematically, several properties of hot nuclei. Still, it has to be realized that a properly founded static description of hot nuclei, even at moderate temperature, requires some caution. The basic problem to be solved is to find a way to 'stabilize' the hot nucleus, for example, by exerting an external pressure which counterbalances the flux of emitted nucleons. Indeed, without such a stabilization mechanism, to build up a hot nucleus is impossible. This can be illustrated in a simple Fermi gas picture. In a Fermi gas at finite temperature (spin–isospin degenerate) the density can be written as

$$\rho = \int_0^\infty n(\epsilon)\omega(\epsilon)\,\mathrm{d}\epsilon \qquad (7.1)$$

where $n(\epsilon)$ is the Fermi occupation factor (2.63), $w(\epsilon)$ the single-particle level density (section 2.4.2) of the Fermi gas (equation (2.66)) and ϵ the single-particle energy, measured from the bottom of the potential well. If we now introduce the chemical potential μ and potential well V we obtain for the density ($\epsilon_F = \mu - V$):

$$\rho = \frac{1}{\pi^2}\left(\frac{\hbar^2}{2m}\right)^{-3/2}\int_0^\infty \frac{\sqrt{\epsilon}\,d\epsilon}{1+e^{(\epsilon-\epsilon_F)/T}} = \frac{1}{\pi^2}\left(\frac{\hbar^2}{2mT}\right)^{-3/2} J_{1/2}(\epsilon_F/T) \quad (7.2)$$

where we have introduced and defined the Fermi integral $J_{1/2}$. We now assume that the potential well is position-dependent $V \to V(r)$ (which amounts to making a so-called local density approximation) and we evaluate the resulting density at large distance (for $r \to \infty$, $V(r) \to 0$), which provides:

$$\rho(r) = \frac{1}{\pi^2}\left(\frac{\hbar^2}{2mT}\right)^{-3/2} J_{1/2}((\mu - V(r))/T)$$

$$\xrightarrow[r\to\infty]{} \frac{1}{\sqrt{2\pi^3}}\left(\frac{\hbar^2}{2m}\right)^{-3/2} T^{3/2}e^{\frac{\mu}{T}}. \quad (7.3)$$

The latter expression shows that the density does not vanish at large distance, which reflects the metastable nature of the system. This effect becomes practically dramatic beyond moderate temperatures of order 2–3 MeV. Any static description then has to accommodate this effect of finite asymptotic density of matter.

We shall not discuss here the various proposals made to solve this metastability problem (truncation of the space of accessible states, artificial external pressure, etc). Among the various approaches, the so-called subtraction method [73] probably gives the best picture in terms of a statistical physics justification and thus we shall briefly recapture its essence. The idea is to define a specific thermodynamical potential $\tilde{\Omega}_N$, which is neither the free energy nor the grand potential but the difference of two grand potentials associated with a liquid + gas Ω_{LG} (nucleus + evaporated nucleons) and gas Ω_G (evaporated nucleons providing the stabilizing pressure) phases, respectively:

$$\tilde{\Omega}_N(\rho_{LG}, \rho_G) = \Omega_{LG}(\rho_{LG}) - \Omega_G(\rho_G) + E_c(\rho^p_{LG} - \rho^p_G) \quad (7.4)$$

where $\rho_{LG,G}(r)$ label the nucleon densities in the two phases and where E_c is the Coulomb energy which is computed from the proton density of the nucleus $\rho^p_{LG}(r) - \rho^p_G(r)$. In this picture the difference between the two phases does represent the nucleus, for which the number of nucleons is supposedly fixed, while nucleon transfers (which amounts to adjust the nucleus mass) are allowed between the two coexisting phases. This subtraction method was primarily proposed at quantal level [73]. It has then been used in systematic calculations of limiting temperatures of hot nuclei (section 7.1.2.2) both at quantal level [74] and in a schematic approach based on the liquid drop model [46, 76]. Semi-classical approximations have also been developed [447] and used for evaluating the temperature depedence of level density parameter [446] and of giant resonances [202, 448].

7.1.2.2 Some properties of hot nuclei

- *Low temperature behaviours*
  For temperatures around 1 MeV, pairing effects disappear in nuclei. One can show that the gap strongly depends on temperature and actually vanishes when temperature becomes larger than typically 1 MeV. While zero temperature nuclei usually have a well-defined shape, tepid nuclei may explore various shapes. Temperatures around 1–2 MeV are thus the realm of spontaneous shape transitions. It should be noted that such transitions can just as well be triggered by angular momentum. Both effects have actually to be considered on the same footing, which leads to a type of a phase diagram for the various shapes, in the temperature/angular momentum plane [12]. Beyond about a temperature of 2 MeV, quantal calculations of hot nuclei show that shell effects (which play a major role in nuclear structure problems at zero temperature, section 2.1.2.3) disappear, so that beyond such temperatures all nuclei are expected to become spherical. This limit of 2–3 MeV thus signals the onset of the relevance of semi-classical methods for describing static properties of hot nuclei.

- *Fission instability*
  The Coulomb/surface ratio sensitively depends on the temperature. The stability of nuclei with respect to fission is thus strongly dependent on the temperature as this ratio fixes the gross properties of fission barriers. Calculations show that fission barriers $U_B(T)$, at least from a static point of view, disappear at temperatures of order 4 MeV. Systematic calculations of hot liquid drop nuclei have, for example, led to simple expressions for barrier heights, of the form $U_B(T) = U_B(T = 0)(1 - x(T))^3$ where $x(T)$ is a temperature-dependent fissility parameter $x(T) \simeq x(0)(1 + x_F T^2)$, with $x_F \simeq 7 \times 10^{-3}$ MeV^{-2} [225]. Note also that, here again, angular momentum plays a role similar to temperature, lowering the fission barrier with increasing angular momentum.

- *Limiting temperature*
  Nuclei cannot be heated up indefinitely. There are many reasons for this. Apart from the already largely discussed effects connected to timescales (see figure 3.1), the basic reason, in a static picture, lies in the strong temperature dependence of the Coulomb/surface ratio. One thus expects that beyond some temperature, surface tension becomes too weak to resist repulsive Coulomb effects. One then speaks of a limiting temperature T_{lim}, which was first identified in the previously discussed 'subtracted' calculations of hot nuclei [74]. The limiting temperature would be a critical temperature if one were to consider nuclei bound inside a box. It would then correspond to passing from a dense (nucleus) + dilute (gas) to a single gaseous phase, which indeed corresponds to a phase transition. Calculations of hot nuclei give values of the limiting temperature typically between 6 and 10 MeV. Such temperatures are significantly lower than the critical temperature of

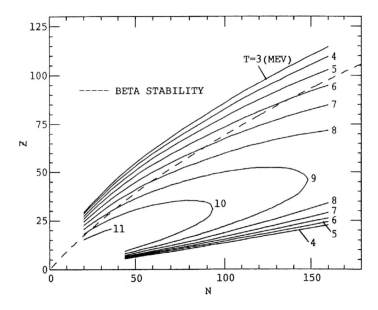

Figure 7.3. Chart of limiting temperatures $T_{\lim}$ (in MeV) of nuclei, in a schematic model of hot nuclei based on the subtraction method, in the N, Z plane. Isothermal lines are drawn every MeV. One should note the strong variations of $T_{\lim}$ as a function of mass and proton number. Light nuclei are altogether better able to resist temperature than heavy ones, because of the relative weakness of coulombic effects. The broken line corresponds to the stability valley. From [46].

infinite nuclear matter (section 2.3.2.1): this comes from the coulombic origin of the appearance of a limiting temperature in hot nuclei. Correlatively the value of $T_{\lim}$ strongly depends on the actual nucleus; light nuclei have larger limiting temperatures than heavy ones, and neutron-rich ones also have larger limiting temperatures than proton-rich ones, as illustrated in figure 7.3.

7.2   Low-energy processes

7.2.1   Nuclear thermodynamics at low temperatures: particle evaporation and the determination of $a(A, T)$

Evaporation is the standard decay process of a thermally equilibrated system in the low-energy regime. This results in the emission of light particles and of statistical photons. The theory of evaporation has been developed in section 2.4, where it was shown that an important ingredient of the theory is the level density parameter $a = A/K$ (A being the mass number). The level density parameter is defined in equation (2.93), valid at moderate temperatures ($T \lesssim$ 3–4 MeV), and is directly related to the density of states in the vicinity of the Fermi level.

From an experimental point of view, estimates of $a = a(A, T)$ are mainly provided by the study of evaporation spectra (see section 2.4). Kinetic energy distributions of the emitted particles are fitted with maxwellian-like distributions in order to extract both the emission barrier and the nuclear temperature (see section 2.3). The excitation energy is obtained by the standard 'calorimetric' methods described in section 4.3.1. A compilation of $K = K(T) = a(A, T)/A$ for systems with $A \simeq 160$ is shown in figure 7.4 (the same systematics also exists for $A \simeq 40$ [142]).

It is well known that, at low temperature, the value of K provided by the non-interacting Fermi gas model ($K = 16$ MeV) is too large by a factor of 2 compared to the empirical value obtained at low energy ($K = 8$ MeV). Of course, a realistic approach has to account for finite-size effects (surface, deformation, etc) as well as for structure effects (pairing, shell effects, etc). Hartree–Fock calculations may account for most of these aspects but they still do not allow for the recovery of proper experimental values. They typically lead to values of K of order 12 MeV. To go down to $K \sim 8$ MeV necessitates a proper account of correlations beyond the mean field, a step which can be performed by including the energy dependence of the effective mass [83, 238, 301]. These correlation effects turn out to be particularly large around the Fermi energy, which explains their impact on the level density parameter. As is obvious, the various (mean field and beyond mean field) ingredients entering the evaluation of K may sensitively, and differentially, depend on temperature, which may induce a non-trivial dependence of K on the temperature T. This is indeed what was shown in [83, 238, 479]. In particular, it was found that the 'collapse' of correlations for temperatures around $T \sim 4$ MeV (especially in the vicinity of the Fermi energy) is likely to explain the rapid evolution of the level density parameter $a(T)$ in this temperature region (see figure 7.4).

In order to make this discussion more quantitative we briefly describe here the model developed in [425] which accounts, in a simple semi-classical but transparent way, for most of the key effects of temperature on a. In the Thomas–Fermi approximation, the single-particle level density reads:

$$\omega(\epsilon) = 4 \int \frac{d^3p\, d^3r}{h^3} \delta(\epsilon - h(\boldsymbol{r}, \boldsymbol{p})) \tag{7.5}$$

in which the factor 4 accounts for the spin–isospin degeneracy, h is the classical single-particle Hamiltonian of the system and ϵ the energy. Integrating over momentum leads to

$$\omega(\epsilon) = \frac{1}{\pi^2} \int d^3r \, \frac{2m}{h^2} \left(\frac{2m}{h^2} (\epsilon - V(\boldsymbol{r})) \right)^{1/2} \tag{7.6}$$

where m is the free nucleon mass. Correlation effects, particularly the key energy dependence, can be included in the picture by introducing a model effective mass $m^* = m^*(T)$. Finally, one has to take care, at finite temperature, of the continuum part of the single-particle level density [73, 446, 447]. This can be simulated by subtracting the contribution of a free Fermi gas, which leads to

$$\omega(\epsilon) = \frac{1}{\pi^2} \int d^3r \left(\frac{2m^*}{\hbar^2} \right)^{3/2} \left(\left(\epsilon - \frac{m}{m^*} V(r) \right)^{1/2} - \theta(\epsilon)\epsilon^{1/2} \right) \qquad (7.7)$$

where $\theta(\epsilon) = 0$ for $\epsilon \leq 0$ and $\theta(\epsilon) = 1$ for $\epsilon \geq 0$.

Inserting phenomenological temperature dependences of the various parameters, as extracted from more sophisticated calculations, leads to the results shown in figure 7.4. This simple model thus accounts for the increase of K as a function of T indicating that the basic effects are correctly taken into account. At much higher temperature T the present theory is no longer valid because the nucleus becomes unstable when it reaches the limiting temperature previously discussed. The behaviour of the system is then no longer dominated by evaporation. This question will be specifically discussed in chapter 8. However, without leaving the moderate energy domain, we know (section 2.4) that evaporation does not exhaust, by any means, all the possible decay channels for hot nuclei. Evaporation is by nature a thermal process which does not imply a general motion of the nucleons inside the nucleus. We now consider the collective behaviours of nuclei at finite temperature, starting with a discussion of giant resonances.

7.2.2 Small amplitude collective motion: giant resonances

Almost half a century ago, actually very soon after the first experimental identification of nuclear giant resonances, it was suggested that such collective modes could also be built in excited ('hot') nuclei. Experimentally, the first indications of a 'hot' GDR were reported in the early 1980s (see e.g. [173, 197, 220, 285, 433, 476] for reviews on this topic). Up to now, only the GDR has been convincingly observed in hot nuclei. Systematics of both the GDR energy and width are now available, up to temperatures of the order $T \sim 4$–5 MeV. Note that beyond such temperatures one actually starts to face conceptual problems in terms of timescales, as typical emission times become sizeably smaller than the GDR period (see figure 3.1). Up to $T \sim 4$ MeV, all measurements show that the resonance energy essentially keeps its 'cold' value, within about 1 MeV. This is illustrated in figure 7.5 where an example of the systematics of GDR centroid energies is displayed. In contrast to the centroid energy, the GDR width Γ_{GDR} (see also figure 7.5) strongly depends on temperature. At moderate temperatures Γ_{GDR} increases rapidly with increasing temperature. This reflects the fact that the hot nucleus may explore various shapes, as a result of thermal agitation, as illustrated in figure 7.6. This figure clearly exhibits the role

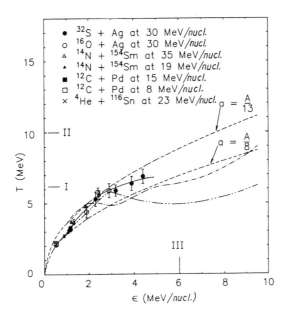

Figure 7.4. Evolution of K as a function of T for A close to 160. Data (points) evolve from the standard $K = 8$ MeV value at temperatures below 3 MeV to values as high as 14 MeV close to the value given by the non-interacting Fermi gas model. The broken lines correspond to the Fermi gas model limit for two values of the level density parameter a. The full line is the result of the model mentioned in the text. The chain lines include a contribution from shell effects which is not discussed here. From [337] and [425].

played by the finite temperature in smearing out the GDR cross-section. Indeed, with increasing temperature, thermal fluctuations become larger and larger and it becomes increasingly difficult to associate *one* single mean field, and thus one cross-section, with the hot nucleus. The total, observed, cross-section is the Boltzmann-weighted ($e^{-E(\epsilon,\gamma)/T}$), hence structureless, superposition of these various cross-sections corresponding to the various shapes (figure 7.6). At even higher temperatures, extracting the width becomes increasingly difficult. Still, measurements indicate a saturation of Γ_{GDR} with temperature, as illustrated in figure 7.5.

From the theoretical side, the description of hot giant resonances should be approached cautiously because of the metastability of hot nuclei. At low temperatures $T \lesssim 2$ MeV the extension of standard zero-temperature formalisms is possible, but beyond such temperatures the statistical occupation of a continuum cannot be treated simply [478]. In this respect the use of sum rules approaches, which provide simple and compact expressions for centroid energies

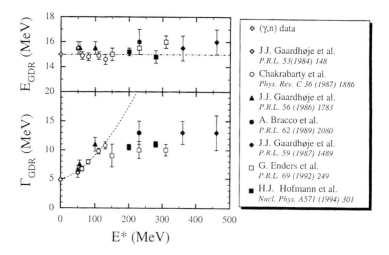

Figure 7.5. Systematics of the centroid energy E_{GDR} and width Γ_{GDR} of the GDR in nuclei of mass $A \simeq 110$, as a function of excitation energy. From [369].

[65], has allowed efficient and systematic calculations. One can, for example, show that the Goldhaber–Teller-type surface vibration (dipole mode) follows a ('weak') quadratic law in temperature: $\hbar\omega(T) \simeq (39.6 - 0.183T^2)A^{-1/6} = \hbar\omega(T = 0)(1 - 0.004\,62T^2)$ which is in good agreement with experimental data [202]. As in the zero temperature case, an accurate estimate of the GDR width is delicate and no systematic calculations are, to our knowledge, yet available. It should finally be noted that the study of giant resonances at finite temperature constitutes an important issue for our understanding of the properties of hot nuclei. Indeed, these collective effects are among the few properties of hot nuclei which 'resist' temperature well. This as such has to be understood, beyond the mere fact that it should allow us to trace back the existence of hot nuclei themselves, even at high temperature.

7.2.3 Large amplitude collective motion: nuclear fission

Heavy-mass (and also, but with smaller probabilities, medium-mass) systems can experience fission because of the competition between Coulomb forces that tend to repel the protons from each other, thus inducing strong deformations and surface effects which tend to restore the spherical shape of the nucleus. The balance between these two effects sets limits on the border of the nuclear chart. Fission can be induced by means of nuclear collisions. In the following, we will essentially discuss the fission of hot and possibly rotating compound nuclei produced in fusion reactions of asymmetric systems around 10 MeV/u incident

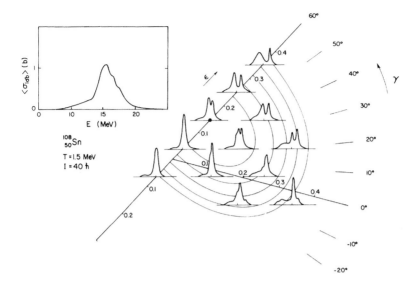

Figure 7.6.   Photoabsorption cross-section of the GDR in ^{108}Sn at temperature $T = 1.5$ MeV and angular momentum $40\hbar$. One can see from this figure the various patterns exhibited by the GDR cross-section as a function of the shape of the nucleus. The finite temperature allows the hot nucleus to explore these various coexisting shapes (here labelled by the Bohr and Mottelson ϵ, γ parameters), each of which leads to a possibly different photoabsorption cross-section. Strongly deformed shapes lead to geometrically fragmented GDR cross-sections, which, once averaged with the proper Boltzmann weights lead to an overall structureless, but wide, one-peak structure (see insert). From [201].

energies (for reviews see, for instance, [42, 242]). This energy range is somehow below most of the reactions that we have discussed up to now. Nevertheless, these collisions are discussed here in view of their interest for the study of the decay modes of 'tepid' nuclei and because they give access to estimates of the viscosity of nuclear matter. It actually turns out that massive fragment emission can also take place at such low energies but with very low probabilities: this process may then be called asymmetric fission. In the following, we consider a diffusion process in which both fragment emission and symmetric fission are treated on the same footing. This model will be used in the fission case while its applicability to fragment production will be discussed in the next chapter.

7.2.3.1 A general formalism for massive fragment emission

The description of fragment emission or fission requires a description based not only on phase space considerations as in evaporation theory (section 2.4), but also on the fact that such large amplitude motions may be strongly dissipative. The reason is that such processes lead to a large deformation of the system during which friction has time to act and possibly to play a crucial role. Descriptions using diffusion equations such as the Langevin or Fokker–Planck formalisms are thus typically well adapted (section 3.1.2). Still, a limiting case, in which dissipation is not taken into account, is the transition-state method, which we now describe and illustrate in the case of fragment emission.

Fragment emission can be macroscopically described as the passage of a system above a barrier. This is illustrated in figure 7.7 where a schematic representation of the potential energy of the system as a function of a deformation variable is displayed. If the system is at equilibrium, the emission process can be thermally driven. The emission probability thus depends upon the potential landscape, the inertia of the system and the nuclear temperature T. A key observable is then the escape rate, i.e. the number of emitted particles per unit time.

Historically, this problem was first treated in the case of fission in the celebrated paper by Bohr and Wheeler [67]. To deal with such an issue, they used a very general method which is common to many fields of physics: the so-called transition-state method. It was, however, soon realized by Kramers [272] that dissipation along the deformation path could significantly affect the escape rate mainly because of the fact that the collective motion is damped. This damping is due to the coupling of the collective variables with the intrinsic degrees of freedom.

To make the discussion more quantitative, we now recall the main steps of the derivation of the escape rate in the presence of friction, keeping in mind that the no-friction limit will then correspond to the transition-state theory [320, 321]. For the sake of simplicity, we restrict the discussion to the case of one single collective variable labelled q (the two-dimensional case has been studied in detail in [483]). In the case of fission, q typically represents a collective deformation as expressed, for example, by the real space quadrupole moment along the deformation axis. Starting from the Fokker–Planck equation with no external field (section 3.1.2.3), we now consider the phase space distribution $W(p, q, t)$ in the (q, p) plane in which $p = M_q \dot{q}$ is the conjugate variable of q and M_q is the mass parameter associated with the variable q. Taking into account the fact that the system now moves in an external potential $U(q)$, one obtains an extension of the Fokker–Planck equation (3.9), the so-called Kramers equation, which reads:

$$\frac{\partial W(p, q, t)}{\partial t} = \left[-\frac{\partial}{\partial q} \frac{p}{M_q} + \frac{\partial}{\partial p} \frac{\partial U}{\partial q} + \frac{\partial}{\partial p} \left(\gamma \frac{p}{M_q} + \gamma T \frac{\partial}{\partial p} \right) \right] W(p, q, t).$$

(7.8)

Note that in contrast to equation (3.9), here q is a real space variable and not a momentum. Seeking a quasi-stationary solution in which most of the probability is concentrated around $q = 0$ (corresponding to the bottom of the potential well)

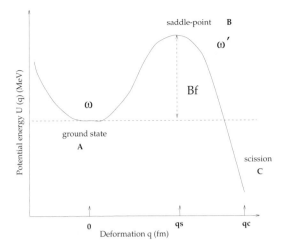

Figure 7.7. The potential energy of the system as a function of deformation along the emission path. The system is initially confined in the pocket ($q = 0$) around A where the curvature of the potential is ω. As q increases, the energy goes through a maximum at $q = q_s$: this is the fission or emission barrier B with a curvature corresponding to a frequency ω'. Then, the potential energy decreases until the deformation is so large that the two fragments separate: this is called the scission point (C at $q = q_c$). Adapted from [1].

with a small flow across the barrier, one can write $W(p, q, t) = W(p, q)$ as

$$W(p, q) = \phi(p, q)e^{-(p^2/2M_q + U(q))/T}. \tag{7.9}$$

A quadratic expansion around the saddle B (figure 7.7) then allows us to express $\phi(p, q)$ as

$$\phi(p, q) = C \int_{-\infty}^{u} e^{-((\alpha - \gamma)/2M_q T)x^2} \, dx \tag{7.10}$$

with $u = p - \alpha(q - q_s)$ and $\alpha = [\gamma + (\gamma^2 + 4M_q\omega'^2)^{1/2}]/2$ and where C is a normalization constant. The decay rate is then obtained by dividing the flux j across the barrier by the probability of presence n_A inside the pocket. Using the stationary phase approximation and a harmonic expansion of the potential around both the bottom of the well ($q = 0$) and saddle ($q = q_s$, figure 7.7) associated, respectively, with frequencies ω and ω', one obtains

$$j = C \left(\frac{2\pi M_q \gamma T}{\alpha - \gamma} \right)^{1/2} 2\pi \frac{T}{\omega} \tag{7.11}$$

and

$$n_A = CT \left(\frac{2\pi M_q \gamma T}{\alpha} \right)^{1/2} e^{-B_f/T}. \tag{7.12}$$

The Kramers decay rate then reads:

$$\Gamma_K = \frac{j}{n_A} = \frac{\omega}{2\pi}[(\beta_{red}^2 + 1)^{1/2} - \beta_{red}]e^{-B_f/T} \tag{7.13}$$

in which β_{red} is the reduced friction:

$$\beta_{red} = \frac{\beta}{2\omega'} = \frac{\gamma}{2m\omega'}. \tag{7.14}$$

In the absence of dissipation ($\beta_{red} = 0$) the Kramers decay rate reduces to the Bohr–Wheeler decay rate:

$$\Gamma_{BW} = \frac{\omega}{2\pi}e^{-B_f/T} \tag{7.15}$$

written here with $\hbar = k = 1$. In this non-dissipative case the decay rate is the product of the barrier assault frequency multiplied by a Boltzmann-like escape factor. Whatever the situation, namely whether the motion is underdamped or overdamped (corresponding, respectively, to $\beta_{red} < 1$ and $\beta_{red} > 1$), the Kramers factor in equation (7.13) is always less than one and thus induces a reduction in the emission rate compared to the non-dissipative case ($\Gamma_K < \Gamma_{BW}$). This is interpreted as the possibility for the system to go back and forth across the barrier due to the presence of thermal fluctuations, as in Brownian motion (section 3.1.2.1), while this is impossible in the non-dissipative case.

It is worth noting that, in their original derivation, Bohr and Wheeler furthermore assumed equal occupations of phase space around the barrier ($q = q_s$) and at $q = 0$, which leads to

$$\Gamma_{stat} = \frac{T}{2\pi}e^{-B_f/T}. \tag{7.16}$$

Thus, they did not consider the collective motion along the deformation path but only the thermal degrees of freedom (this is why we call it Γ_{stat}). Indeed, in equation (7.15), the typical frequency of assault is given by the energy ($\hbar\omega$) associated with the shape of the collective potential, while in equation (7.16) the typical energy is purely thermal and thus simply proportional to the temperature T.

As noted earlier, the inclusion of friction in the diffusion process tends to reduce the escape rate but it turns out that it also has a consequence in terms of timescales. Indeed, a careful analysis of the time dependence of the distribution function $W(p, q, t)$ shows that before W has reached the quasi-stationary value given by equation (7.9), there is a transient time needed for the flow to reach its asymptotic value [217, 218, 237]. This means that there are finally two possible ways to estimate the friction coefficient and its temperature dependence. A first indication can be given by a direct comparison between the experimental fission cross-sections (or its counterpart which is the production cross-section of evaporation residues) with the prediction of Kramers' formula. A second way is to estimate the transient time, and thus access friction, by measuring the fission timescales.

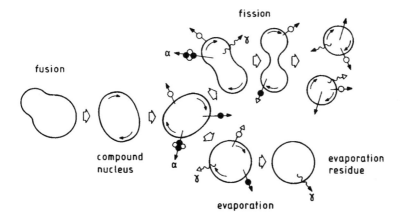

Figure 7.8. Sketch of the competition between fission and evaporation following the formation of a compound nucleus. Fission is a long and dissipative process during which particles may be evaporated, thus cooling the system and decreasing its fissility. Sufficient cooling can inhibit fission and lead to the production of an evaporation residue. The associated probabilities are governed by the fission timescale, this latter being essentially determined by potential landscapes and the temperature-dependent viscosity. From [195].

7.2.3.2 Competition between fission and residue production: fission timescales

We first discuss the fission rate in relation with the production of heavy evaporation residues. Experimentally, the detection of evaporation residues is difficult because of their small recoil velocity, which make them hard to identify. However, a large body of data is now available from low up to very high excitation energies of the order of several MeV per nucleon. A general feature of the production cross-sections is their unexpected large values compared to the prediction of a standard evaporation code. Strong deviations appear at such low temperatures as 1 or 2 MeV. In other words, the 'survival' probability p_s, which is nothing but $1 - p_f$, where p_f is the fission probability, remains rather large even for heavy, hence highly fissile, nuclei. The fission probability is defined as $p_f = \Gamma_f/\Gamma_{tot}$ in which Γ_f is the experimental fission width (to be compared with the theoretically expected value) and Γ_{tot} is the total decay width of the system. The sizeable production of evaporation residues thus suggests that fission is a long process, during which cooling, particularly via evaporation, may occur (competition between fission and evaporation).

The competition between fission and evaporation is schematically depicted in figure 7.8. We have already mentioned the possibility of measuring fission timescales 'directly' by means of crystal blocking techniques (see section 4.3.4.1). Here, we discuss another method for evaluating the fission timescale, deduced from neutron measurements, by the so-called neutron clock technique. The idea is to estimate both the number of neutrons (called hereafter pre-scission neutrons) emitted before, and the number of those emitted after (post-scission neutrons) the separation of the fission fragments. From these numbers and by estimating the frequency of emission with the help of a statistical model, one has access to the fission time defined as the time needed for the system to reach the scission point (point C of figure 7.7) when starting from the equilibrium point (point A) inside the pocket. Note that, in defining fission time this way, one makes the implicit assumption that the number of pre-equilibrium neutrons is negligible. Remember that pre-equilibrium neutrons are emitted in the early stage of the reaction before the occurrence of fusion (see section 6.1.1). One also assumes that the formation time of the compound nucleus is short compared to the evaporation time, so that no neutrons are emitted before full equilibrium is achieved. This thus implies that the method is only applicable to moderate excitation energies.

From a technical point of view, one takes advantage of the fact that neutrons do not experience coulombic interaction and are thus kinematically very sensitive to the velocity of the emitter v_{emitter} (protons are sensitive too, but this effect is masked by the additional velocity v_{Coul} due to the Coulomb acceleration which may be even larger than v_{emitter}). In particular, neutrons emitted by fully accelerated fission fragments are focused in a direction along the fission axis, connecting the two velocity vectors of the fission fragments, while the distribution of neutrons emitted before scission is isotropic. This means that a careful analysis of the neutron angular distributions allows us to disentangle between pre- and post-scission neutrons.

Numerous experiments have been devoted to such measurements. In figure 7.9, the evolution of pre- and post-scission neutrons is shown as a function of the excitation energy E^* of a compound nucleus with a mass number around 250. The two contributions (pre- and post-scission neutrons) evolve in a completely different way. The pre-scission component increases rapidly with E^* and becomes dominant around 200 MeV corresponding to nearly 1 MeV/u excitation energy per nucleon. This is associated with the rapid decrease of the lifetime of the system with respect to neutron evaporation (see section 2.4.3) and also to a possible saturation of the fission timescale due to an increase of the friction factor (section 7.2.3.3). In turn, the post-scission component hardly increases with E^*, showing that the scission point is reached late in the collision when the system is rather cold. This clearly indicates that fission is (i) a highly dissipative and (ii) a long process. With the help of the statistical model (section 2.4) it is possible to deduce the fission timescale by estimating the time needed for the system to emit an arbitrary number ν of pre-scission neutrons. Indeed, at each emission step, an estimation of the neutron decay width gives access to the mean neutron emission time (section 2.4.3). By considering all emission steps exhausting the total pre-scission neutrons, one obtains the fission

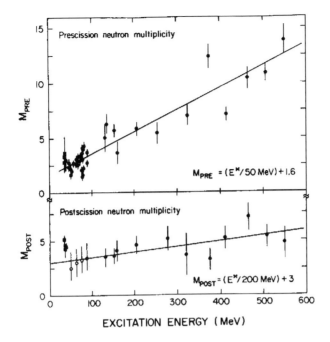

Figure 7.9. Pre- and post-scission neutron multiplicities measured as a function of the excitation energy deposited in a compound nucleus with a mass number around 250. The two multiplicities are obtained by a deconvolution of the neutron angular distributions as discussed in the text. Lines are drawn to emphasize the different evolution of the two distributions as a function of the excitation energy. From [416].

timescale [243, 244]. Such a procedure gives the result shown in figure 7.10 in which the estimated fission timescales based on pre-scission neutrons are displayed. Despite the large uncertainty observed in the experimental points, the general trends of the data exhibit a saturation of the fission timescales above 10^{-19} s. It is worth noting that recent estimations of the fission timescales based on the blocking technique in U+Si reactions at 24 MeV/u [212] lead to much larger values compared to those quoted in figure 7.10. This discrepancy is highly debated. One possible explanation [326] is the following: measurements based on neutron (and also γ) multiplicities do not cover the total fission 'dynamics' but, by definition, only the timescales accessible by the study of particle emission. In contrast, the measurement based on blocking techniques covers the full range of fission timescales, since then there is no dependence on a specific probe.

Nevertheless, when comparing the experimental fission timescales with those estimated with the help of the statistical model (full line in figure 7.10),

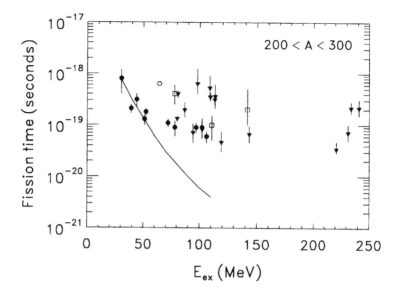

Figure 7.10. Systematics of the fission timescale as a function of the excitation energy of the compound nucleus in a mass number range between 200 and 300 units. The full line corresponds to the prediction of a statistical model in which no dissipation (i.e. no delay) is considered in the fission process. Strong deviation is observed above 50 MeV excitation energy demonstrating the importance of nuclear viscosity (see also figure 7.11). From [474].

a strong deviation is observed as soon as E^* reaches values around 50 MeV. This behaviour is easily understood in a statistical model where there is by definition no friction: the fission time decreases as a function of the excitation energy, as the fission flux across the barrier increases as a function of the temperature as suggested by equation (7.15) or equation (7.16). This deviation has been exploited in order to estimate the nuclear viscosity and, in particular, its temperature dependence from the study of nuclear collective motion. Such studies have triggered much experimental effort in the last decades. Two main signatures are now discussed here in relation to these remarks.

7.2.3.3 *Collective motion and nuclear dissipation*

As stressed in the two previous sections, collective motion is, to a large extent, governed by nuclear potentials and the reduced nuclear friction β_{red}. Estimations of β_{red} at $T \simeq 0$ have been obtained by comparing the predictions of the wall-and-window one-body dissipation model and the alternative two-body dissipation approach [428] with the total kinetic energy of the two fragments released in the fission process [480].

From a theoretical point of view, the temperature dependence of the nuclear viscosity is a matter of debate. In the theory of Fermi liquids, a T^2 dependence is expected, at least up to a certain temperature over which the viscosity could then drop down [37]. This behaviour has been observed in a typical Fermi liquid, namely ^{3}He [2]. There are, however, significant differences between liquid ^{3}He and atomic nuclei. Atomic nuclei are systems in which surface effects can play an important role as far as dissipation is concerned. In contrast, ^{3}He is a bulk medium. Maybe more importantly [359], ^{3}He is a strongly bound system while nuclear matter consists of 'weakly' interacting nucleons (2.1.3) (see [329] for an illuminating discussion on this point). Finally, medium effects lead to the opposite behaviour of the effective masses m^* (see section 2.1.3.3): m^* range is 3–$9m_{\text{free}}$ for ^{3}He while m^* is close to m_{free} for atomic nuclei. Therefore, the behaviour of atomic nuclei may show differences compared with macroscopic Fermi liquids.

Figure 7.11 summarizes various experimental observables used to evaluate the temperature dependence of friction, namely $\beta_{\text{red}}(T)$, from experiments performed in the O+Pb system. The first method is based on a study of evaporation residue (ER) cross-sections. The ER cross-section, following O+Pb fusion, is displayed in the top left-hand panel of figure 7.11 as a function of the incident energy E_{lab}. The general trend of the data shows a saturation of the cross-section at high E_{lab}, emphasizing the persistence of ER production even at large temperatures. This is incompatible with the prediction of a standard evaporation calculation, in which friction is not taken into account in the collective decay modes. Indeed, such a model predicts a strong decrease in the ER channel probability (broken lines in figure 7.11) as compared to the fission channel. This result is complementary and in full agreement with the one of figure 7.10 showing a saturation of the fission time and thus a saturation of the fission probability (middle left-hand panel). In the same experiment, pre- and post-scission neutrons were measured. Their multiplicities follow the same trends as a function of beam energy, as already pointed out in figure 7.9. Here again, a comparison between a model without friction (broken lines) and a model with a temperature dependent friction (full lines) clearly favours this latter. The introduction of dissipation into the model naturally increases the pre-scission neutron multiplicity (to be compared with the data (black triangles)) and it also improves the agreement with the post-scission neutron multiplicity.

Finally, a comparison is shown between calculations and data concerning the energy distribution of the γ's around the value expected for the GDR. Here, the main effect of including the dissipation is to delay fission and thus to increase the probability of survival of the compound nucleus and thus the probability of observing the GDR decay. It is worth noting that the three observables described here have been used 'simultaneously' to constrain the value of β_{red}. The resulting T-dependence of the viscosity is shown in the bottom panels of figure 7.11. It turns out that, unfortunately, the fit is compatible either with a linear or with a

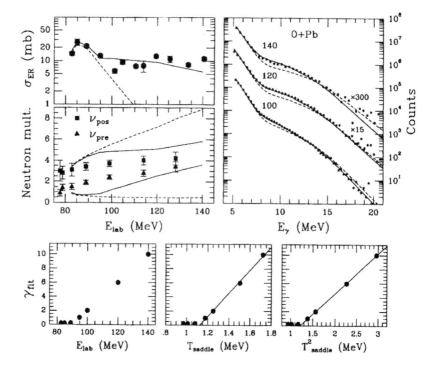

Figure 7.11. Experimental results showing three different signatures that can be used to measure nuclear viscosity and its temperature dependence: top left, evaporation residue cross-section; middle left, pre- and post-scission neutron multiplicities; top right, photon production in the energy range corresponding to the GDR for three different incident energies (100, 120 and 140 MeV). All data are for the same system $^{16}O+^{208}Pb$. Deviations are evident with respect to a statistical model with no dissipation ($\beta_{red} = 0$, broken lines) when $E_{lab} = 90$ MeV corresponding to an excitation energy close to 40 MeV and to a temperature T close to 1.2 MeV. The temperature dependence of β_{red} needed to reproduce the data above 50 MeV is shown in the lower figures. The characteristic T^2 dependence is in agreement with the predictions of the theory of Fermi liquids [37] but a linear temperature dependence is not to be excluded. From [359].

quadratic evolution of β_{red} with temperature. Experiments at higher temperatures would thus be highly welcome to see whether friction still increases or decreases above $T \sim 2$ MeV. However, one important problem then is the onset of strong dynamical effects associated either with entrance channel effects (pre-equilibrium emission), which make the analysis complicated, or with the advent of new decay processes that may become dominant as we will now discuss.

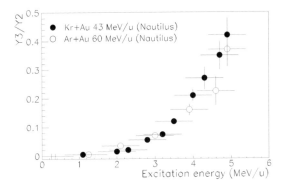

Figure 7.12. Evolution of the competition between fission (Y_2) and three-body (Y_3) decay as a function of ϵ^*. Due to experimental limitations, the atomic number of each detected fragment is ≥ 8. Selected impact parameters correspond to central and mid-central reactions so that the mass number of the excited source is around 200–220 mass units. The excitation energy was estimated by the 'subtraction' method (section 4.3.1.3). From [55].

7.3 High-energy processes

We now consider the decay modes of hot nuclei when the excitation energy per nucleon ϵ^* increases significantly above 1 MeV/u and may even reach values close to the binding energy per nucleon. In a somewhat arbitrary way, the transition from low-energy decay processes to high-energy processes is associated with the transition from fission to fragmentation, namely the transition from a decay process in which two and only two massive fragments are emitted, to a process with at least three massive fragments in the final state.

This transition has been studied quantitatively for systems with mass around 200 in [55]. A comparison of the two decay channels is made possible by counting the number of events in which two and only two massive fragments are observed and events in which at least three fragments have been emitted. In figure 7.12, the yield of three-body (Y_3, fragmentation) versus two-body (Y_2, fission) mechanisms is shown as a function of excitation energy per nucleon ϵ^*: a sharp increase is observed around 3 MeV/u. From this energy on, fragmentation begins to be a competitive process, although evaporation and fission are still present. It is worth noting that the ratio Y_3/Y_2 displayed in figure 7.12 is independent of the entrance channel (i.e. both of the projectile and the bombarding energy) and thus of the mechanism that led to the production of the hot nuclei considered in these experiments. This gives strong support to the fact that the transition is essentially governed by the excitation energy.

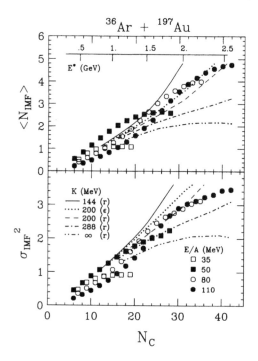

Figure 7.13. First and second moments of fragment multiplicity distributions as a function of the light-charged-particle multiplicity in dissipative Ar+Au collisions between 35 and 110 MeV/u incident energies. An excitation energy scale is shown in the upper part of the figure. Note the quasi-independence of the curves as a function of the incident energy. Lines are the result of the Friedman model [194] to be discussed in section 8.3.1. From [158].

7.3.1   Rise and fall of fragmentation

Figure 7.13 displays the evolution of the mean number of emitted fragments $\langle N_{IMF} \rangle$ as a function of the light-charged particle multiplicity N_c (section 4.2.3) for various incident energies and for the same system ^{36}Ar+^{197}Au. The number of emitted fragments $\langle N_{IMF} \rangle$ as well as the associated variance is an increasing function of N_c. An interesting point is the quasi-independence of this result with respect to incident energy, in perfect compatibility with the results shown in figure 7.12. This is a strong indication that the fragment multiplicity is, to a large extent, dominated by the excitation energy deposited in the system, independently from the incident energy. A statistical model (to be discussed in section 8.3.2) can account for the general trends of the data, in particular, when the variances are considered (bottom part of figure 7.13). For large values of N_c, the data begin

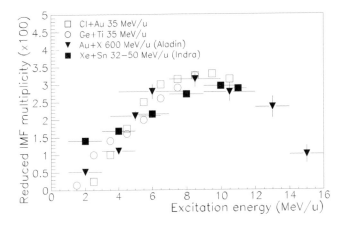

Figure 7.14. Rise and fall of the reduced IMF multiplicity as a function of excitation energy per nucleon ϵ^* for a variety of systems. From [171] and references therein.

to exhibit a saturation at $\epsilon^* \simeq B$ (the binding energy per nucleon). In those events, more and more light particles are produced. Consequently, due to charge conservation, fragment multiplicity decreases.

In order to account for finite-size effects and to allow comparisons between different systems, the number of emitted fragments can be scaled according to the total mass of the considered system. Results are shown in figure 7.14. The 'rise and fall' of nuclear fragmentation is demonstrated here. The maximum IMF production is observed for excitation energies around 9 MeV/u and then drops down, being progressively replaced for ϵ^* close to B by vaporization (i.e. the decay by particles with $Z = 1$ and 2 only). It is impressive to note the large variety of projectiles, targets and incident energies that lead to similar results once they are scaled according to the dissipated energy.

7.3.2 Nuclear vaporization

Nuclear vaporization has been observed in central Ar+Ni collisions in the Fermi energy range between 52 and 95 MeV/u [20, 80, 81, 401]. Even for medium-mass systems, the cross-section for such a process is very small (of the order of the picobarn) as it corresponds to very central collisions. A detailed analysis of the kinematic characteristics of the events reveals the binary character of the collision. A calorimetric study of the two sources gives access to the dissipated energy and thus to the excitation energy stored in each excited source. The chemical composition of the vaporized projectile-like (QP) source is shown in figure 7.15 as a function of excitation energy per nucleon ϵ^*. Low-energy events are predominantly composed of α-particles. As ϵ^* increases, neutrons and protons become increasingly numerous, while no significant evolution is observed for the other species. Note that the neutron ratio was not directly measured, but obtained by mass balance. In the bottom part of figure 7.15, the mean kinetic energies in the centre-of-mass frame of the QP have been plotted for all species. It turns out that these energies are all approximately equal as should be the case for a thermalized system. This suggests an analysis of the data in the framework of a statistical approach. The predictions of a quantum statistical model (QSM) (see section 8.3.3 for a detailed discussion of this model) are shown in the same figure (lines). One of the major parameters in such an approach is the density of the system. A good agreement could be obtained when assuming the formation of two dilute ($\rho \simeq \rho_0/3$) equilibrated sources (QP and QT) within a temperature range from 10 to 25 MeV. Note that complete equilibrium (both thermal and chemical) is assumed. This indicates that the results are mainly governed by the available phase space.

7.4   Conclusion of the chapter

In this chapter we have described the various possible decay modes experienced by hot nuclear species produced in dissipative collisions. The key quantity that allows us to follow the evolution of the process from low-energy-like phenomena (evaporation and fission) to high-energy-like processes (fragmentation and vaporization) is obviously the excitation energy per nucleon ϵ^* or the nuclear temperature T.

The relation between these two quantities has been discussed in section 7.2.1 by studying the emission properties of the evaporated light particles. The importance of the nucleonic correlations has been put forward in the determination of the level density parameter. This question will be addressed again in the next chapter in a more 'extreme' context associated with complete nuclear disassembly at high excitation energy.

Collective motions at moderate temperatures (GDR and fission) have been reviewed and their importance for the study of nuclear dissipation pointed out. Nuclear fission is a slow process: this is the only way to understand the large survival probability for observing evaporation residues. A successful comparison of the data with the predictions of the Kramers theory shows very nicely the evolution of nuclear dissipation with temperature which is compatible with the properties of Fermi liquids.

The 'transition' energy from fission to fragmentation has been measured as well as the rise and fall of fragment emission. This transition takes place around $\epsilon^* = 3$ MeV/u and becomes the dominant process up to excitation energies per nucleon close to the binding energy per nucleon. It is then replaced by the 'ultimate' decay process: vaporization. Anticipating the features of the next chapter, we have seen on an example that, surprisingly, nuclear vaporization

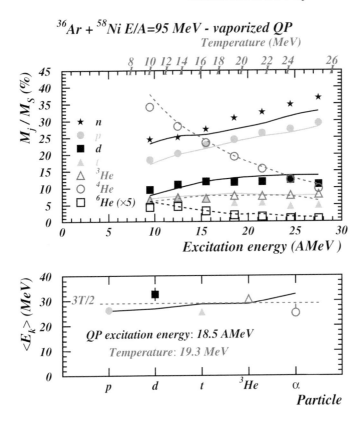

Figure 7.15. Upper part, ratios of various light species produced in the vaporization of the quasi-projectile produced in Ar+Ni central collisions at 95 MeV/u. The lines are the result of a quantum statistical model (QSM) (see text) [226]. Lower part, mean experimental kinetic energy for each light species in the centre-of-mass frame of the emitter (the QP in this case) for $\epsilon^* \sim 18.5$ MeV: the full line is a prediction of the QSM model; the broken line corresponds to $\langle E_{\text{kin}} \rangle = 3T/2$. From [81].

can be understood in the framework of nuclear thermodynamics, although the temperatures to be considered here are extremely high.

It turns out that low-energy decay modes probe nuclear matter at densities close to the equilibrium value, while it is plausible that high-energy processes occur in extended dilute systems. Evidence for this transition and its possible implications for our understanding of nuclear matter far from the equilibrium point in terms of a possible liquid–gas transition are the main motivations for the studies presented in the next chapter.

Chapter 8

Nuclear fragmentation and the liquid–gas phase transition

In the last chapter we described the various decay modes experienced by hot nuclei produced in dissipative reactions, which emphasizes the key role played by the deposited excitation energy per nucleon ϵ^*. The rise ($\epsilon^* \sim 3$ MeV/u) and fall (ϵ^* close to the nuclear binding energy) of fragment production has been pointed out. Fragmentation thus appears as an intermediate mechanism bridging the low-energy decay modes dominated by evaporation (and fission, for heavy systems) and the high-energy decay modes characterized by the complete vaporization of the system. The main physical issues related to nuclear fragmentation are associated with the properties of the nuclear equation of state (EOS) at finite temperature T and low-density ρ. In particular, the possibility of observing a liquid–gas phase transition has triggered numerous experimental and theoretical investigations justifying a whole chapter on this subject.

Section 8.1 is devoted to a preliminary discussion of some fundamental issues. In particular, the relevance of the concept of phase transition in open transient structures such as hot pieces of nuclear matter is certainly a key issue. In section 8.2, the dynamical aspects of fragmentation are discussed through their connection with the EOS (section 8.2.1) by an analysis of the dynamical path followed by the system in the T–ρ plane. New types of matter instabilities are described with a particular emphasis on the physics of the spinodale decomposition (section 8.2.1.3), which is expected to occur in central collisions in the Fermi energy range.

The study of the dynamics can shed some light on the occurrence of a global (or at least partial) equilibrium inside the system, allowing the study of dilute and heated systems with the help of thermodynamical concepts. The thermodynamical properties of fragmenting systems is a very exciting challenge in this field. The models dedicated to this subject are discussed in section 8.3. The statistical models can be schematically divided into two categories depending on the degrees of freedom considered. In the first ones (multifragmentation

220

statistical models, section 8.3.2), the concepts of nuclear droplets and heated vapour composed of light particles are used, while nucleons are considered in the second ones (lattice-gas models, section 8.3.4). Both classes of models have been used to investigate the important question of the experimental signals of a phase transition in a finite system, which would allow us to relate nuclear fragmentation to a critical phenomenon. Owing to their universality, the lattice-gas models can thus also be used to study the generic patterns of fragmentation and confront the properties of fragmenting nuclear systems with systems in other fields of physics and chemistry.

From an experimental point of view, several questions arise in the quest for the liquid–gas phase transition and its characterization. First, a proper spacetime analysis of the fragmenting matter is needed. In particular, fragmentation timescales must be evaluated (section 8.4.1.1). A long timescale is associated with a sequential 'low-energy-like' process, while a short timescale corresponds to a 'simultaneous' phenomenon which has been called *multifragmentation*. The transition between these two regimes is discussed in relation to the corresponding evolution of the charge distributions (the so-called partitions) of the fragments. Another important piece of information is the evidence for a collective motion associated with the disassembly process (section 8.4.2). This is an indication that the system might have reached a dilute state and thus constitutes a first clue for a liquid–gas phase transition. Experimental signals for this are discussed in the last part of the chapter on the basis of nuclear thermodynamics (section 8.4). The key words here are *reducibility* and *thermal scaling* (section 8.4.3.2) as well as the search for *phase coexistence* (section 8.4.3.3), *critical behaviour* (section 8.4.3.5), discontinuities or fluctuations in the nuclear specific heat in the so-called *caloric curves* (section 8.4.3.4) and finally the possible experimental evidence for *bulk instabilities* (section 8.4.4).

8.1 The issues of nuclear fragmentation

8.1.1 A new physics?

As mentioned in the introduction, nuclear fragmentation is the key process that bridges evaporation and vaporization. The first question then is whether fragmentation is reminiscent of fission (the low energy side of the bridge) or the onset of nuclear disassembly (the high-energy side of the bridge). An experimental answer is given in section 8.4.1.1 (see figures 8.6 and 8.7) by showing the evolution in terms of timescales and charge distributions from a slow process (the so-called sequential fragmentation) below 5 MeV/u excitation energy per nucleon up to a fast process (the so-called multifragmentation) above 5 MeV/u. Are these two regimes of fragmentation just a matter of words or are they conceptually different? This issue is related to the physical nature of fragmentation as either a process dominated by a statistical occupancy of phase space or a process essentially driven by the dynamics of the collision and the transport properties of nuclear matter [297, 322].

In the first case, shortening the timescales is just a matter of growth of the available phase space, but there is no conceptual difference between the two fragmentation processes. One should, however, note that the sequential fragmentation models fail to reproduce data at high excitation energies, suggesting that the sampling of phase space may require varying prescriptions (in particular the density of the decaying system) as the excitation energy varies.

In the second case, the two processes could be the manifestation of new nuclear instabilities. Close to the normal density, shape and Coulomb instabilities have been discussed in chapter 7. They are responsible for the fission process of hot nuclei. The deformation, which may be due to angular momentum, occurs around the normal density and is essentially driven by surface properties at finite temperature. There, the pressure is moderate and thus the collective motion is purely induced by coulombic forces. The process is thus governed by the transport properties of nuclear matter at normal density, particularly the viscosity. However, central collisions can produce matter at low density in a compression/expansion cycle: here, new instabilties related to the bulk properties (see section 8.2.1) are expected. They correspond to a situation in which part of the available energy is stored preferentially in collective degrees of freedom.

8.1.2 Phase transitions in the nuclear context

In the following, we will discuss the results of nuclear fragmentation in the context of a possible liquid–gas phase transition. Here, we would like to recall some of the basic concepts of phase transitions as well as their relevance in finite systems. Indeed, phase transitions are usually introduced in the case of infinite systems in the thermodynamic limit. Having defined two distinct phases in the system, the latter is characterized by functions (for instance the entropy) calculated with variables characterizing each phase (for instance their respective densities). Singularities in such functions signal the occurrence of phase transitions. The transition may be either first or second order. First-order phase transitions are characterized by a strong increase in the entropy thus leading to the existence of a latent heat as expressed by the Clapeyron equation. The derivative of the entropy with respect to the internal energy becomes constant leading to the divergence of the heat capacity at constant pressure. Second-order phase transitions do not exhibit a discontinuity in the entropy; but the heat capacity exhibits a discontinuity with no divergence.

Phase transitions can be thermally driven. In this case, the temperature is the so-called tuning parameter in that it controls the phase transition. However, phase transitions also exist in non-equilibrium phenomena (see, for instance, [371] and references therein). In the following site percolation will be discussed in the context of nuclear physics: this is a typical 'geometrical' phase transition.

As already discussed in section 2.3.2.1, the nucleon–nucleon effective interaction in the nuclear medium shows similarities with the Van der Waals intermolecular forces. Therefore, a liquid–gas phase transition is expected to occur in infinite nuclear matter. The characteristics of such a transition have been derived in section 2.3.2.1 in the framework of a simplified Skyrme interaction. In such a case, below the critical point, the transition is of first-order. In particular, a latent heat is produced to enable the evolution from the gas phase to the liquid phase. At the critical point, the transition is of second order. The densities of the gas phase, ρ_g, and liquid phase, ρ_l, are both equal to the critical density ρ_c: therefore, the order parameter of the transition, which is defined as $\rho_l - \rho_g$, vanishes at the critical point.

In infinite nuclear matter, the liquid phase is identified as the one in which nucleons are bound together at a density close to the normal density, while the gas phase is identified with free nucleons (similar to the monomers in the liquid–gas transition of macroscopic fluids); the corresponding density is much smaller. Unfortunately, the study of nuclear matter is only possible on earth by means of small excited pieces produced in nuclear collisions. Therefore, major difficulties arise due to the fact that phase transitions in finite systems are not as well defined as in infinite matter. In the case of excited nuclei, the liquid phase may be tentatively identified with fragments and the gas phase with light particles. The transition from a system in the liquid phase to the gas state, which corresponds to vaporization, has then been described in section 7.3.

A key question concerns the extent to which the transition occurs at equilibrium. In other words, is the decay process thermally driven? This question is extensively discussed in section 8.4.3. We will see there that phase space dominance seems to be a general feature in nuclear fragmentation. This may be due to the strong dissipative behaviour observed in nuclear collisions which rapidly drives the system towards equilibrium. However, we will also see that the estimation of a nuclear temperature still meets technical and also conceptual difficulties (section 8.4.3.4). This is linked to the fact that the systems under study are not stationary in the sense that they are open and have finite lifetimes. Thus, they exchange matter with the outside and their description with the help of static concepts may be problematic. It is clear that the physics of open and transient systems is probably not as simple as expectd from simple statistical approaches.

However, even when assuming that the transition is thermally driven, the question of the signatures of phase transitions in finite systems remains open. The understanding of finite-size scaling in the nuclear context has recently made significant progress (section 8.4.3.5). This is certainly a question in which the study of nuclear objects can shed light on this very general question and be instructive for other fields of physics. It should finally be noted that two other characteristic features make hot nuclear droplets rather unique physical objects, as far as liquid–gas phase transitions are concerned:

- Hot nuclei are charged objects. The role of a long range force mixed with a short range interaction makes hot nuclear droplets almost unique systems in the study of phase transitions in finite systems.

- Hot nuclei are furthermore made of two different species: neutrons and protons. Indeed, it turns out that finite two-component systems exhibit new 'chemical' instabilities [331]: thus they have a more complex phase diagram depending on the N/Z ratio. This leads to a richer phenomenology in the decay processes.

8.2   Dynamical description of nuclear fragmentation

8.2.1   Exploring the nuclear equation of state phase diagram

Among the two kinds of instabilities advertised for explaining fragmentation (surface (shape) instabilities at low energy and bulk (density) instabilities at high energy), bulk instabilities play a special role in the sense that they imply more or less direct connections with the equation of state of nuclear matter. They have thus motivated a lot of studies, either inspired by, or directed to, the phase diagram of nuclear matter. The basic reaction mechanism assumed is an expansion of the initially hot and compressed composite. This implies a 'preferred' exploration of the low-density regions of the nuclear matter phase diagram. In this section we thus briefly discuss what might occur to such an excited piece of nuclear matter exploring low-density regions of the nuclear phase diagram. We focus, in particular, on nucleation and instability growth in the spinodale region.

8.2.1.1   *The nucleonic matter phase diagram*

As discussed in section 2.2.1 the nucleon–nucleon interaction is repulsive at short range and attractive at medium range (at least in the dominant channel relevant for our discussions). This possibly causes the nuclear matter equation of state to exhibit a phase coexistence region in which nuclear matter phases of low (gas) and high (liquid) densities may be in equilibrium with each other. Along an isothermal line, namely at fixed temperature, phase coexistence defines the common pressure and chemical potential of the two phases in equilibrium, as in a typical simple gas, as described, for example, by a Van der Waals equation of state.

This phase coexistence is basically an equilibrium concept whatever time is hidden in the process of forming a liquid droplet inside a gas or a gas bubble inside a liquid. Thus in truly equilibrated situations the picture makes sense. A typical nuclear example concerns the nuclear matter equation of state in supernovae cores in which the slowness of the collapse (in terms of nuclear timescales) fully justifies an 'instantaneous' adjustment of equilibrium properties such as the temperature of the system [364]. In the case of heavy-ion collisions the coexistence picture is more difficult to justify as the typical timescales of the evolution of the system are comparable to the timescale needed for reaching a thermal equilibrium (section 5.3.1). Once again, one is facing competition between the timescales and thus a definite conclusion is not easy to reach. As discussed in section 8.2.1.2, it is therefore worth trying to attain some quantitative estimates of the effects at work.

The coexistence region of the nuclear phase diagram furthermore includes a specific region in which dynamics by its very nature plays a more direct role: the spinodale region. The spinodale region is the region of the phase diagram in which nuclear matter is unstable with respect to a small mechanical perturbation, which means that the derivative of the pressure with respect to the density is negative in this region (see figure 2.8). The spinodale region is bounded by the spinodale line defined as

$$\left.\frac{\partial P}{\partial \rho}\right|_Q = 0 \tag{8.1}$$

where Q labels the conserved quantity in the process. For a heavy-ion collision, in which the nuclear composite expands to enter the spinodale region, the (preferred) conserved quantity is the entropy per nucleon and one then speaks of the isentropic spinodale line. The spinodale regions (see figure 8.2) disappear at the critical point of the equation of state (section 2.3.2.1) above which matter is mechanically stable and phase coexistence meaningless.

The spinodale region is expected to play an important role in nuclear fragmentation. However, the *entrance* into this unstable region is not the only process leading to phase coexistence between gas (free nucleons) and liquid (nucleon clusters (IMF)). Inside the coexistence region of the phase diagram, but outside the spinodale region, the matter remains stable with respect to small perturbations but large amplitude perturbations are also likely to create a phase separation: the corresponding mechanism is called the nucleation process.

By nature, the spinodale region is associated with a dynamical process, namely the growth of instabilities. In this respect it hardly corresponds to an equilibrium process. Nevertheless, the quantitative difference with phase equilibrium lies mainly in differences in the timescales associated with nucleation and instability growth, respectively. It is thus worthwhile discussing these two aspects in some more detail.

8.2.1.2 *Nucleation*

Because of the more naturally dynamical picture underlying the spinodale instabilities, the question of nucleation, namely large amplitude fluctuations leading to phase separation, has been little considered in the context of heavy-ion collisions. Still it is worth discussing. A general introduction to the suject may be found in [50]. A kinetic approach to nucleation in the nuclear context may be found in [412]. Here we shall briefly summarize the results obtained by the Copenhagen group at the end of the 1980s [160]. The idea is to discuss the effects of large amplitude fluctuations (as originating from single-particle fluctuations), in particular, in the formation of bubbles inside a hot expanding phase of nuclear matter. Above temperatures of order 1 MeV quantum fluctuations are known to play a minor role in these questions [61] so that a purely classical picture becomes applicable. Starting from simple arguments of classical nucleation theory one can then estimate the manner in which, how many and how fast, bubbles form inside an initially hot and compressed nuclear phase.

A typical result of these calculations is given in figure 8.1 where various relevant quantities as a function of time are plotted. The system is initially prepared with a small amount of compression (10%) with respect to the saturation density and a (possibly large) temperature. Both quantities are followed in time, as well as the total number of bubbles n_b. A critical radius r_c which roughly separates bubbles into two classes is also plotted: on average, bubbles with a radius greater than r_c will tend to grow, while bubbles with a radius less than r_c will tend to shrink (we speak here of tendencies to keep in mind the crucial role of fluctuations in bubbles of radii around r_c). Figure 8.1 exhibits several interesting results for our purpose. First, it should be noted that in the 'low' temperature case ($T = 6$ MeV) the nucleation process is not very efficient. The $T = 9$ MeV case provides nucleation, in particular, beyond times of order 30 fm/c, which is comparable to a typical expansion time. A further indication of the role played by the nucleation process may be found in [160]: these authors have performed systematic calculations as a function of the initial temperature/density of the system. As a result, one can identify a region of the phase diagram which is likely to provide 'favourable' initial conditions for nucleation in the course of the expansion. Very roughly speaking, it corresponds, in the density/temperature plane, to regions with entropy between slightly less than 1 (per baryon) and about 2.4, which is of the same order of magnitude as the estimated values of the entropy per baryon in heavy-ion collisions in the nucleonic regime. Still, it is striking to note in the example in figure 8.1 that precisely beyond 30 fm/c, which is the typical time for nucleation, the system is very likely already inside the spinodale region of the nuclear equation of state ($\rho/\rho_0 \lesssim 0.6$, section 2.3.2.1). This means that standard nucleation probably competes with bubble/fragment formation stemming from small amplitude fluctuations. Even if these latter fluctuations will need some time to become 'macroscopically' visible (typically another 20 to 30 fm/c), it is possible that the instabilities due to the spinodale region play a significant role in the story, besides the large shape fluctuations dynamically induced in the entrance channel (see section 8.2.3).

8.2.1.3   *Spinodale instability*

The spinodale region is mechanically unstable: a small amplitude perturbation (hence corresponding to a long wavelength and/or little energy), which would be damped at (or close to) normal density is amplified in this region. This means that *any* density fluctuation may grow to become of the order of the average density itself and thus it will break the system into pieces. This mechanism of spinodale instability was initially proposed in [43], on the basis of the form of the nuclear equation of state at low density. This scenario is appealing as it integrates a dynamical component into an equation of state picture: an initially hot and

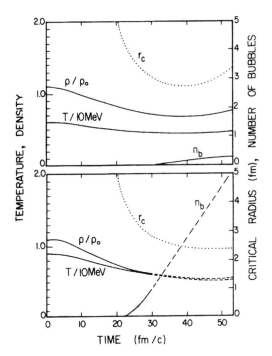

Figure 8.1. Time evolution of the density and temperature for a nucleus $Z = 50$ and $A = 120$. The critical radius r_c and the number of bubbles n_b are also indicated. From [160].

compressed nucleus expands to reach the spinodale region in which fluctuations are amplified, and in which, once a sufficient amount of time has elapsed, the system breaks into fragments. This picture of spinodale instability has been further analysed in detail in several papers [138, 241, 363]. We particularly refer the reader to the pedagogical reference [363] for details.

To make the discussion a bit more quantitative we study, in a static uniform piece of infinite nuclear matter of given density ρ, how small density fluctuations develop, following the line of [363]. This is most easily achieved when analysing density fluctuations in terms of their Fourier components, which decouple and have dispersion relations of the form

$$\omega_\nu^2 = c_s^2 q_\nu^2 \qquad (8.2)$$

where ω_ν is the angular frequency, q_ν the wavenumber associated with the multipolarity ν and where c_s^2 is the isothermal sound velocity

$$c_s^2 = \frac{1}{m}(\partial P/\partial \rho)_T.$$

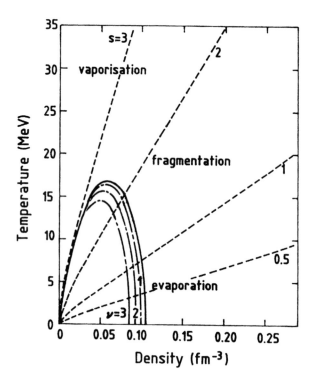

Figure 8.2. Phase diagram of nuclear matter in the density (in fm^{-3})–temperature (in MeV) plane. Characteristic lines of the equation of state as well as typical heavy-ion trajectories have been indicated in this figure. The chain lines inside the spinodale region correspond to the domains inside which a given multipolarity becomes unstable. Broken lines represent possible heavy-ion collisions trajectories and correspond to fixed entropy per baryon (from 0.5 to 3). Finally, the dominant de-excitation mechanism has been indicated for each region of the phase diagram. From [241].

The isothermal spinodale line thus corresponds to $c_s^2 = 0$. Outside the spinodale region, $c_s^2 > 0$ and ω_ν is real, leading to stable (sound type) density oscillations. Inside the spinodale region, $c_s^2 < 0$ and ω_ν is purely imaginary, which leads to damped or growing solutions. In the 'catastrophic' case of an exponential growth of the fluctuations ω_ν reads $\omega_\nu = -\mathrm{i}\lambda_\nu^2$, which leads, for a given multipolarity ν, to a time evolution of the form $\exp(\mathrm{i}\omega_\nu t) \propto \exp(\lambda_\nu^2 t)$. The exponential growth of small amplitude perturbations inside the spinodale region thus somewhat depends on the dominant multipolarity of the perturbation [363], through the instability thresholds (see figure 8.2).

The exponential nature of the growth mechanism does not by itself ensure that the system will break into pieces. As is obvious, there is a typical timescale associated with such a growth process, namely the time needed by the system to generate a density fluctuation with an amplitude of the order of magnitude of the average density itself. The key parameter thus turns out to be the accumulated growth, which is expressed, for a mode of multipolarity ν, as

$$G_\nu = -i \int \omega_\nu \, dt. \tag{8.3}$$

The system actually breaks into pieces for $G_\nu \gtrsim 3$ [363]. This corresponds to an accumulation in the ν mode of density fluctuations $\delta_\nu \rho$ such that $\sqrt{\langle \delta_\nu \rho^2 \rangle} \sim \rho$. Of course, in a realistic case, it is likely that several multipolarities will be excited simultaneously, so that a total density fluctuation of order ρ, which sums the contributions of these various multipolarities, will be attained for smaller values of the G_ν. Still, the picture of an accumulation of fluctuations in time remains valid and points to the importance of the time spent by the system inside the spinodale region. Note finally that one can, at least qualitatively, compare the various fragmentation channels (roughly speaking, a given multipolarity can be associated with a given number of fragments) by comparing the values of G_ν. This is a further justification of this analysis in terms of modes of various multipolarities.

These features are also illustrated in figure 8.3 which displays the spinodale decomposition of a finite piece of nuclear matter in a mean-field calculation including an account of fluctuations through a dedicated stochastic term, i.e. in solving or simulating a Boltzmann–Langevin-like stochastic equation. This is an approach similar to the one used to describe Brownian motion by a Langevin equation. In [224] the stochastic term is introduced as a local fluctuation of the potential which has been adjusted to reproduce the agitation of the most unstable modes. The calculations have then been performed for a heavy nucleus (Au) expanded to about half the saturation density and which is then let free to evolve under the combined influence of the one-body field and the residual two-body collision processes, with the effects of the fluctuations included whenever local spinodale instability occurs. The system quickly expands into a hollow. Such unstable configurations decay by the emission of several IMFs. The typical length of the most unstable mode is the order of 10 fm and thus leads (after about ~ 120 fm/c) to five to seven fragments with approximately the same sizes. Of course, the main defect in these descriptions is that they need 'preparation' of the system at a given temperature and density. In other words, it is assumed that an equilibrium step (at least partial) has been reached in the early phases of the reaction. The validity of this assumption is addressed in section 8.2.2.

8.2.2 From the phase diagram to heavy-ion collisions

The brief analysis we have just performed (sections 8.2.1.2 and 8.2.1.3) is interesting because of its link to the nuclear matter equation of state. Still, one

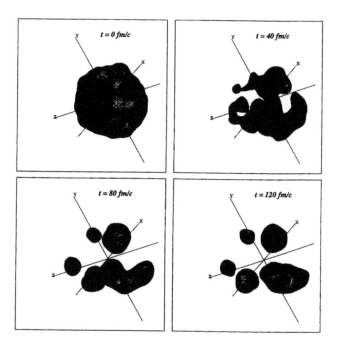

Figure 8.3. Simulation of the spinodale decomposition of a finite piece of nuclear matter ($A = 210$ and $Z = 90$) in the framework of a mean-field calculation with a stochastic component. Timescales are of the order of 100 fm/c. The initial density was set close to $\rho_0/2$ and the temperature $T \sim 3$ MeV. The size and the multiplicity of the fragments is directly linked to the wavelength of the most unstable mode as discussed in the text. From [223].

should keep in mind the fact that the underlying models are very schematic, which introduces some limitations on the impact of these conclusions. In particular, a key question here is to define the state of the system when it enters the coexistence or the spinodale region. But the primary evolution of the (possibly) already 'hot' and (likely) compressed system towards low densities is not so easy to predict in detail. Two-body collisions are likely to play a dominant role during this phase of the evolution, slowing down the expansion of the system and transforming some of the available energy into heat. Hence, it is not very clear how far these simple and appealing descriptions of nucleation or spinodale instability are actually involved in a real situation. Microscopic calculations such as BUU (section 3.2.3.2) or QMD (section 3.3.3) add some valuable pieces of information as two-body effects are included in these models. Indeed several calculations seem to indicate [79, 162, 214, 215, 265, 317] that the fragmentation pattern, as observed after the breaking up of the system, does indeed reflect some initially large fluctuations present in the system at the beginning of the expansion phase. The presence of such initially large fluctuations might then be explained by invoking the fluctuation–dissipation theorem: the larger the dissipation is, the larger the fluctuations are, and the most dissipative phase of the collisions is undoubtedly at the very beginning of the overlap during the compression and well before the possible expansion. It is, nevertheless, delicate to reach this conclusion at this stage, as this type of involved microscopic calculations is not necessarily fully robust against such details. Further work is certainly needed, in relation to the coming experimental results, in order to attain a more conclusive statement.

The weakness of the equation of state model does not lie purely in the difficulties related to following a reaction path. There are, in fact, deeper problems connected to the capability of the equation of state model itself to depict such a path. The first problem lies in the fact that an interpretation in terms of the equation of state requires the existence of the variables characterizing the system, such as pressure, temperature or density. And it is only marginally true that the system has enough time, for example, to allow thermalization. A second difficulty concerns the fact that, in collisions, one is dealing with finite systems, with all the specificities they carry. For example, although it presumably plays a minor role during the early phase of fragmentation, the Coulomb interaction, for example, cannot be totally overlooked as it will, even marginally, affect the motion of protons, and thus of the neutrons, even during the beginning of the expansion (the Coulomb interaction is not considered in any (infinite) nuclear matter analysis). Finally, it should be noted that plotting, for example, the pressure as a function of density definitively gives a particular weight to the monopole channel. Indeed, during the early phases of the collision, a sizeable amount of beam energy is converted into the collective monopole energy, which then shows up as a radial expansion. Still, the monopole is probably not the sole multipole to be excited, even if it is likely that it is the most robust one. To choose the density as the single 'abcissa' variable is thus certainly a bit schematic. One could even go further along this line of thinking and imagine that a homogeneous and isotropic variable such as the density overlooks the exploration of specific geometries in nuclear matter. Such questions were actually considered in the case of supernovae, where it was shown that various phases of low dimension nuclear matter can be made stable [279, 384, 497] (section 2.3.2.2). Of course the situation in these cases is quite different from the case of heavy-ion collisions, particularly because of the composition of the system which contains electrons responsible for a screening of the Coulomb interaction. Still, the question is probably worth also considering for heavy-ion collisions.

Altogether, we would thus like to conclude this analysis of the nuclear matter equation of state with a word of caution. The spinodale instability picture, which has so often been advertised in this field, may reflect some part of the reality but one has to bear in mind the fact that this picture is mainly an interpretation, which may actually serve as a guideline, but which by no means should constitute the end of the story.

8.2.3   Improving the description of the dynamics

The concepts discussed in the previous sections have been implemented in a variety of theoretical phenomenological models. Here, we do not aim at being fully exhaustive in the description of the large amount of work performed during the last two decades. Generally speaking, the theoretical description of nuclear fragmentation is a difficult task. The understanding of the formation of fragments at the microsocopic level must take into account the clusterization of a quantum fluid at finite temperature, which *a priori* implies solving the N-body quantum problem. Since this is a very far-reaching programme, most of the models rely on simplifying assumptions.

The dynamical aspects of nuclear fragmentation have already been schematically discussed in the context of the nuclear equation of state by considering nucleation and the spinodale decomposition of the matter in the low-density phase. As the nuclear transport models described in chapter 3 are basically one-body theories, they are not well suited to studying fragmentation processes because they cannot account for the high-order correlations necessary to produce fragments. The extension of such models to higher order correlations in the phase space distribution has been envisaged by including stochasticity on the transport theory (see section 3.2.5). However, in most models, this is not sufficient to describe fragment production self-consistently. In general, a cluterization procedure in phase space is realized by means of numerical algorithms. These 'build' fragments on the basis of energy considerations by means of minimization procedures (see for instance [377]), thus reintroducing statistical features into the dynamics.

One of the most promising phenomenological approaches is certainly the antisymmetrized molecular dynamics (AMD, section 3.3.4.2). A recent study of fragment production in the framework of such a model has shed some light on the importance of quantum effects in the process as illustrated in figure 8.4. A drastic change is observed in the time evolution of the nuclear density when one compares calculations with (AMD-V) and without (AMD) wavepacket diffusion. These calculations compare rather well with experimental data [356]. They are thus very promising and offer the possibility of following in detail the dynamical evolution of the matter towards disassembly. They also allow a comparison of the different effective forces that may be used in the simulation. Another example of the comparison of a dynamical approach with experimental data will be shown at the end of the chapter (section 8.4.4).

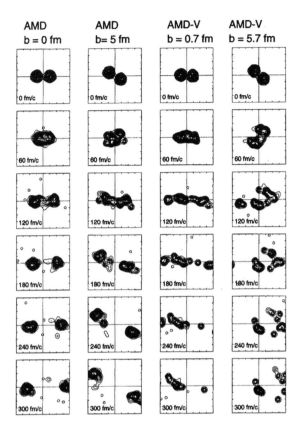

Figure 8.4. Time evolution of the density projected in the reaction plane for Ca+Ca central collisions at 35 MeV/u at two impact parameters $b \sim 0$ and $b \sim 5$ fm. The size of the area is 40 fm by 40 fm. Left-hand panels are the predictions of AMD without wavepacket diffusion while right-hand panels correspond to AMD-V calculations in which wavepacket diffusion is considered. The introduction of such a process strongly affects the dynamics of the collision and leads to more fragmented configurations. This result illustrates the importance of quantum effects in the production of fragments in nuclear collisions. From [356].

8.3   Statistical description of nuclear fragmentation

Since data are not easy to compare with dynamical models because of the previously mentioned difficulties, a direct contact between data and phenomenological approaches can be attained in the framework of statistical theory. Statistical models of nuclear fragmentation assume partial or global equilibrium. These models may be subdivided into those which treat fragment emission macroscopically such as the multifragmentation statistical models and those which treat the process microscopically by explicitly considering nucleons such as, for instance, the lattice gas models. We briefly discuss successively these various models.

8.3.1 Low-energy statistical models

Fragment formation is a process which is taken into account even at the level of the standard statistical model (section 2.4), although it is then treated under particular assumptions (section 2.4.4). More complicated situations may also be accounted for in more sophisticated models. For example, in chapter 7, the decay properties of hot nuclei for low-energy processes such as evaporation and induced fission were discussed. The theoretical description of such processes is achieved by means of the theory of the compound nucleus. The statistical competition between the various decay processes is governed by energy balance and by the available phase space. The theory of light particle evaporation has been described in chapter 2. The emission of complex fragments (at least in fission processes) has been discussed previously by considering the transition state method and its extension, the Kramers formalism, taking into account nuclear viscosity (section 7.2.3.1).

In the context of nuclear fragmentation the practical implementation of these ideas is realized in the statistical binary models (see for instance [120, 170, 397]): the decay is described as a sequence of two-body splittings ('fissions') at normal density, until the system is sufficiently cold enough to prevent particle emission. The emission of γ-rays can also be considered in such codes. The splittings are well separated both in space and time so that the system can achieve equilibrium between each emission. Matter stays at normal density: evaporation and fragment emission are then understood as surface processes. However, when the excitation energy deposited in the nuclei reaches and even exceeds the total binding energy, it is likely that the systems will expand under thermal pressure or under mechanical constraints (compression) generated in the early instances of the collision, during the overlap of the two partners of the reaction. Evidence for systems fragmenting during expansion will be shown in the following. Under such conditions, the system is dilute. Thus the very notion of surface emission may become questionable: fragmentation is no longer a surface process but has become a bulk process.

This latter transition between surface and bulk emission has been studied and quantified in the expanding emitting source (EES) model [194]. In this model the concept of evaporation is applied to sources in expansion. For densities close to the saturation density ρ_0, the formalism is equivalent to the standard theory of evaporation described in section 2.4. However, there exists a density (in fact one close to the critical density ρ_c of the nuclear matter equation of state, as defined in 2.3.2.1) for which the entropy variation associated with the emission process becomes more favourable for a volume emission than for a surface emission. This transition towards bulk processes is associated with the onset of multifragmentation, since it corresponds to a shortening of the timescales. However, for technical limitations, this model cannot account for the emission of massive fragments. A complete statistical description of fragmenting dilute systems is the programme of high-energy multifragmentation statistical models.

8.3.2 High-energy multifragmentation statistical models

8.3.2.1 *Context and constraints*

A historical account of the developments of statistical multifragmentation models may be found in appendix A of [222] and we refer the reader to this reference for details. The key assumptions of the statistical theory have been outlined in section 2.4. The transition matrix (see equation (2.76)) connecting the initial and final state of the decay process is constant. The system has no history or at least this history is so complicated that it is hopeless and useless to describe it. It is hence assumed that dissipation has driven the system towards equilibrium so that the phase space has been homogeneously populated. Therefore, the key quantity to be estimated is the density of states of the final configuration. The first simple approach along this line was proposed in the grand canonical ensemble in [381]. But a key question here concerns the definition of such a configuration in the case of a path towards multi (≥ 3) fragments. In chapter 7, fission has been described using a diffusion equation. The stationary solution of this equation provides the fission flux across the saddle-point, and hence the fission width. The saddle-point is thus the locus at which the fate of the system is decided. The extension of the concept of saddle-point to a fragmentation path in more than three fragments is, nevertheless, not straightforward and has thus been a matter of debate.

An attempt to extend the 'fission' picture to the fragmentation channel is described in [292, 383]. The problem is to consider a multi-dimensional saddle-point, but such a point is hard to define because many degrees of freedom are involved. Another major difficulty is the description of the evolution of the system in the vicinity of such a point. Indeed, the motion is not only driven by the multi-dimensional potential landscape but also by friction which in such complicated geometries is very difficult to take into account.

A simpler strategy consists of considering a 'scission' configuration (in the spirit of the Fong statistical model of nuclear fission [189]) in which the considered degrees of freedom are nuclear droplets (hereafter called pre-fragments) embedded in a vapour of light particles. Matter is confined in a fixed volume called the *freeze-out* volume, V_{fo}. This latter presumably results from the strong dissipative stage of a nuclear collision, during which thermal and mechanical pressures drive the system to low density. Equilibrium is then achieved due to the ergodicity of the process. At freeze-out there are no further

matter and energy exchanges between the pre-fragments. However, pre-fragments
may be excited and will decay (this is called secondary decay) 'in flight', during
their time evolution in their respective Coulomb fields. It is implicitly assumed
that the break-up occurs on a very short timescale, which means simultaneously
with respect to the typical expansion time of the system. Freeze-out is associated
with a density ρ_{fo} which corresponds to a mean distance between surfaces of
fragments of order 2–3 fm. This value corresponds to the typical range of the
nuclear force leading to values of ρ_{fo} in the 0.1–0.5ρ_0 range.

The most commonly used multifragmentation models are the Berlin model
[221] and the Copenhagen model [78]. Recently, new developments have also
been undertaken in quite a similar line [379]. The main differences between these
models lie in the chosen statistical ensembles. In the Copenhagen model, the
canonical description is used while in the Berlin model, the density of states is
calculated at the microcanonical level. These models show technical differences
but these are of minor importance as far as the predictions are concerned. For the
sake of completeness, but without going too much into the detail, we briefly recall
some of the key equations used in these models.

8.3.2.2 *Basic equations of statistical multifragmentation models*

We consider a parent nucleus of mass A_0 and charge Z_0 with an excitation energy
E^*, a linear momentum P_0 (usually taken to be equal to zero), and an angular
momentum L_0. At freeze-out, this hot nucleus decays inside a volume V_{fo} into
several light particles and clusters. The set of particles and clusters is usually
called a partition. Let us suppose that it is constituted of ν species with masses
$\{A_i, i = 1, \ldots, \nu\}$ and charges $\{Z_i, i = 1, \ldots, \nu\}$. The energy balance equation
between the energy of the parent nucleus and the energy of the partition E_{part}
then reads:

$$E_{part} = E^* + M(A_0, Z_0) = \sum_{i}^{\nu} M_i(A_i, Z_i) + E_{inter} + \sum_{i}^{\nu} \epsilon_i^* + \sum_{i}^{\nu} \epsilon_i^K \quad (8.4)$$

where E_{inter} is the interaction energy between the particles and ϵ_i^*, ϵ_i^K and M_i's
are, respectively, the excitation energy, kinetic energy and mass of species i.

The interaction energy E_{inter} is purely coulombic in origin since it is
assumed that species exchange neither energy nor mass at the freeze-out stage.
It can then be calculated exactly by explicitly taking into account the positions of
the constituents of the partition, inside the source at freeze-out or by averaging
using standard approximations. The kinetic energies ϵ_i^K may be purely thermal
and are thus simply related to the temperature T_{part} of the partition. However,
collective motion such as rotation or expansion may also be present and this
results in additional terms in the kinetic energies of the constituents.

The temperature T_{part} of a given partition is calculated self-consistently with
help of equation (8.4). To this end, a relation between ϵ_i^* and T_{part} is needed by

means of the level density parameter a (section 7.2.1). There is no universal prescription among the models to determine a. A Fermi gas prescription may be used as well as a full mass temperature dependence prescription taking into account, for example, the temperature dependence of the surface tension (see section 7.1.2.2).

From the values of the internal excitation energies ϵ_i^* of the constituents it is possible to calculate the statistical weight W_{part} of each partition. As already mentioned, the estimation of W_{part} depends on the statistical ensemble considered in the model. Let us just discuss the microcanonical case, in which energy, momentum, angular momentum, mass number and atomic charge are strictly conserved. For a given partition one then obtains:

$$W_{\text{part}}^{\text{micro}} = \frac{1}{\xi} \exp[S_{\text{part}}(E_{\text{part}}, V_{\text{fo}}, P_0, L_0, A_0, Z_0)] \tag{8.5}$$

with

$$\xi = \sum_{\text{part}} \exp[S_{\text{part}}]. \tag{8.6}$$

The entropy S_{part} is obtained (see section 2.3.1.3) by estimating the free energy F_{part} of each partition from the corresponding partition function. One introduces a free energy which may be decomposed into various terms: those depending on temperature, thus contributing to the entropy, are included in the kinetic term associated with the thermal motion $F_{\text{part}}^K = \sum_i^\nu F_i^K(T_{\text{part}})$ and the term associated with the excitation energy of each species $F_{\text{part}}^* = \sum_i^\nu F_i^*(T_{\text{part}})$. A detailed description of these terms may be found in [221] and [78]. Let us just detail, for instance, the free energy $F_i^*(T_{\text{part}})$ associated with the internal degrees of freedom of each species i. We have, in the framework of the liquid drop model,

$$F_i^*(T_{\text{part}}) = F_{A_i,Z_i}^{\text{bulk}}(T_{\text{part}}) + F_{A_i,Z_i}^{\text{surf}}(T_{\text{part}}) + F_{A_i,Z_i}^{\text{sym}} + F_{A_i,Z_i}^{\text{Coul}}. \tag{8.7}$$

The last two terms have, to a good approximation, no temperature dependence so that they do not contribute to the entropy. The first one is associated with the bulk and reads:

$$F_{A_i,Z_i}^{\text{bulk}} = -a_v A_i - a(A_i, T_{\text{part}})T_{\text{part}}^2 \tag{8.8}$$

in which a_v is the bulk energy per nucleon at zero temperature (equation (2.2)) and in which the level density parameter $a = a(A_i, T_{\text{part}})$ can be evaluated according to the prescriptions discussed in section 7.2.1. The surface term reads:

$$F_{A_i,Z_i}^{\text{surf}} = 4\pi R(A_i, Z_i)^2 \sigma(T_{\text{part}}) \tag{8.9}$$

in which the temperature-dependent surface tension σ is such that $\sigma(T_c) = 0$, T_c (~ 16 MeV) being the critical temperature as defined in equation (2.71).

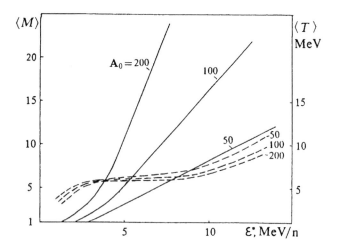

Figure 8.5. Evolution of the average temperature $\langle T \rangle$ (broken lines, right-hand scale, similar to $\langle T_{\mathrm{part}} \rangle$ in the text) and the fragment multiplicity $\langle M \rangle$ (full lines, left-hand scale) as a function of ϵ^* as predicted by the Copenhagen model (SMM) for three different sources with mass number A_0. From [78].

8.3.2.3   A caloric curve

The multifragmentation statistical models have met significant success in their comparison with experimental data in dissipative nuclear collisions as will be discussed in section 8.4. Here, we just want to present one of the most interesting predictions of these models, regarding the existence of a liquid–gas transition in the nuclear context. Figure 8.5 shows the predictions of the Copenhagen model concerning the evolution of the temperature T ($T = \langle T_{\mathrm{part}} \rangle$ where the average is taken over all possible partitions) and the particle multiplicity $\langle M \rangle$ as a function of the excitation energy ϵ^* for three different source sizes A_0. The evolution of $\langle M \rangle$ (full lines, left-hand scale) shows a strong increase around $\epsilon^* \sim 2$–4 MeV/u depending on the parent nucleus mass A_0. This is to be compared with the experimental rise of fragment emission, as shown in figure 7.12. It constitutes another clue for the strong dominance of the phase space in nuclear fragmentation. The broken lines in the same figure (right-hand scale) display the evolution of T with ϵ^* (this is usually called the 'caloric' curve in the literature). An interesting feature is the existence of a plateau around $T = 5$ MeV between $\epsilon^* = 3$ and 8 MeV/u, resulting in a deviation of the caloric curve with respect to the behaviour of a Fermi gas. This can be interpreted as a characteristic signal for a first-order liquid–gas phase transition. The temperature stays constant in an excitation energy range (corresponding to a latent heat), in which the energy is used to produce fragments, essentially at the cost of increasing the free surface of the system. This is thus characteristic of phase coexistence, in which the liquid phase is dominant at the beginning of the plateau around 3 MeV/u and is progressively replaced by the gas phase at the end of the plateau, around 8 MeV/u. This feature is also predicted in the framework of the Berlin model. It is worth noting that the plateau behaviour is still there despite the fact that the system has a finite-size. This finding has triggered numerous studies concerning the experimental determination of the caloric curve as will be discussed in section 8.4.3.4.

8.3.3 Quantum statistical models

The statistical multifragmentation models discussed previously have limited prediction power concerning the production and characteristics of small clusters because for these one has to take into account the very detailed structure effects which have not yet been fully implemented in the computer codes. This is an important issue, because we have seen the impact of the correct treatment of such small clusters for the determination of nuclear temperatures (section 4.3.2).

These features are properly taken into account in the quantum statistical model (QSM) originally proposed in [232]. In such a model, the population of each species i is given by

$$Y_i = \frac{V_{\text{fo}} g_i}{h^3} \int d^3 p \, n_i(p) \tag{8.10}$$

where g_i is the degeneracy factor of each species i and n_i the occupation factor in momentum space. At the canonical level for classical particles, assuming a temperature T, one obtains

$$Y_i = \frac{8\pi V_{\text{fo}} g_i (m_i T)^{3/2}}{2^{1/2} h^3} \int_0^\infty dz \, z^{1/2} \exp\left(-z + \frac{\mu_i}{T} - \frac{E_i}{T}\right) \tag{8.11}$$

in which $z = \epsilon/T$ is the integration variable over all possible kinetic energies (see section 2.4.1); E_i and μ_i are, respectively, the mass energy (taking into account a possible internal excitation energy) and the chemical potential of species i. This latter is determined by the law of mass action:

$$\mu_i = \mu_Z Z_i + \mu_N N_i \tag{8.12}$$

where μ_Z and μ_N are the proton and neutron chemical potentials. These quantities are constrained by the conservation of atomic and mass numbers of the source. In this formulation, each species is treated as an independent structureless particle, thus assuming neither a final-state interaction among the species nor internal discrete states.

The introduction of quantum statistics, internal structure and final state interactions in the formalism has been proposed by considering, for instance, excluded volume effects in [226]. Taking into account quantum statistics amounts to replacing the Boltzmann factor in equation (8.11) by a Fermi or Bose factor (see section 2.3.1.4), which leads to the following relation:

$$Y_i = \frac{8\pi V_{\mathrm{fo}} g_i (m_i T)^{3/2}}{2^{1/2} h^3} \int_0^\infty \mathrm{d}z \, z^{1/2} \left[\exp\left(z - \frac{\mu_i}{T} + \frac{E_i}{T} + f_i \right) \pm 1 \right]^{-1}$$

(8.13)

in which f_i is a suppression factor taking into account the proper 'volume' of each species.

The quantum statistical models have been extensively used in the study of nuclear thermometry but also in the description of nuclear vaporization. In particular, the predictions of such a model have been successfully compared with vaporization data in chapter 7 (figure 7.15). It is worth noting that such models have also been used at much higher energy by including mesons and resonances produced in central collisions in the relativistic domain. Under the constraints of reproducing the populations of various species, a freeze-out temperature as well as a density have thus been obtained up to the limits of the predicted QGP (see [172, 438]). This point will be discussed again in section 9.4.2.1.

8.3.4   Lattice-gas models

The degrees of freedom considered in the previously described models are the ones of hot nuclear droplets, possibly embedded in a nuclear vapour composed of light particles. In such models , special emphasis is thus put on liquid drop properties (section 2.1.2.1). The purpose of lattice gas models is to consider the nucleons themselves as the degrees of freedom. Practically, a very useful method consists of mapping the Ising model into a nuclear lattice gas model [113, 227, 358, 385, 409]. In such a model, a schematic Hamiltonian H represents the interaction between the nucleons. In the spirit of the Ising model, nearest neighbour interaction is assumed between nucleons located at the sites of a cubic lattice, which leads to an Hamiltonian of the form

$$H = \sum_i \frac{p_i^2}{2m_0} \tau_i + \sum_{\mathrm{lattice}} \epsilon \tau_i \tau_j$$

(8.14)

where the last sum is restricted to nearest neighbours and where τ_i is the occupation factor at site i (0 or 1), while ϵ is the energy coupling constant, which is chosen to reproduce the saturation energy of nuclear matter (section 2.2.2.1). Such models may be thermal in nature but some of them focus on geometrical properties, such as those based on percolation theory. Historically, these were the first to be developed in nuclear physics [33, 109, 110]. They were originally designed to search for critical phenomena in nuclear fragmentation. In particular, the study of the various moments of the cluster size distributions has lead to the use of the so-called Campi plots (section 8.4.3.5). In turn, one of the major interests of the thermal lattice gas models is their ability to address in a well-defined context the question of the liquid–gas phase transition in finite systems (see, for instance, [125, 227]). In particular, the question of the experimental signature of such a transition has been directly addressed in these models, as well as the influence of the N/Z degree of freedom. We will come back to these points in the following.

8.4 Experimental aspects: towards the liquid–gas phase transition

The high-energy statistical models predict the occurrence of a liquid–gas phase transition in hot and dilute pieces of nuclear matter such as hot nuclei, the signature being a plateau in the caloric curve as shown in figure 8.5. The quest for experimental signatures of this liquid–gas transition in the nuclear context was initiated a long time ago. Following the scheme developed in the introduction of this chapter, we discuss the possible signatures of the liquid–gas transition as a function of the 'complexity' of these signatures: from the simplest ones to the most elaborated ones.

But before proceeding, it is necessary to discuss the topological characteristics of nuclear fragmentation. Indeed, owing to the theoretical discussions in the previous sections, the applicability of the concepts useful in the study of phase transitions requires a number of experimental conditions to be achieved. First, the fragmentation timescale should be short enough so that the whole process may be described as a single-step process. Otherwise, sequential fragment emissions make the definition of a single well-defined temperature a questionable issue. Second, the system should decay at low density. Indeed, the critical density ρ_c (see equation (2.71)) is found to be close to about one-third of the normal density for typical Skyrme-like forces. Such a low value is also found in the case of the multifragmentation models. The key question is thus to determine to what extent the experimental data on nuclear fragmentation show these two characteristics (short timescales and low density).

8.4.1 Fragmentation timescales and charge distributions

8.4.1.1 Fragmentation timescales

The techniques of nuclear chronometry have been described in section 4.3.4. Emission times are estimated by analysing spacetime correlations between fragments taken two by two, taking advantage of the 'proximity' effects induced by the Coulomb interaction. This is the so-called fragment intensity interferometry. A number of experiments have been devoted to such measurements. They used either angular correlation or reduced velocity correlation functions. An example of such a study has already been shown in section 4.3.4.3 (see figure 4.15) using the relative angle variable to build the correlation function. Generally speaking, the mean fragment emission time is obtained with the help of simulations by direct comparison with the data as

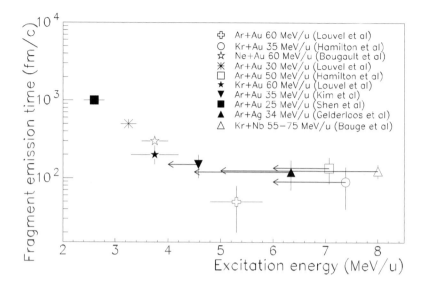

Figure 8.6. Systematics of the fragment emission times as a function of the excitation energy per nucleon ϵ^* for a variety of systems. The points with arrows correspond to the maximum available excitation energy in fusion reactions. From [171] and references therein.

described in section 4.3.4.3. Some of the data obtained in the last decade are shown in figure 8.6. They have been sorted according to the estimated excitation energy per nucleon ϵ^*. A strong decrease in the emission time with increasing ϵ^* is observed up to about 5 MeV/u beyond which a saturation of the emission time is observed around 100 fm/c. For such short times, fragments are emitted almost 'simultaneously' so that their emissions cannot be treated as successive splittings. This is the multifragmentation regime. This result thus justifies the use of statistical multifragmentation models for ϵ^* larger than 5 MeV/u.

8.4.1.2 *Charge distributions: Dalitz plots*

The onset of fragmentation around $\epsilon^* \sim 3$ MeV/u (section 7.3.1) is associated with the emission of low-Z fragments corresponding to very asymmetric fission-like processes. As ϵ^* increases, it is thus instructive to evaluate the evolution of the charge distribution of fragmented events with increasing ϵ^*. A usual and easy-to-visualize technique consists of considering the three largest fragments detected per event displayed with the help of Dalitz plots. Let Z_1, Z_2, Z_3 be the atomic numbers of the three considered fragments. One defines the following coordinates

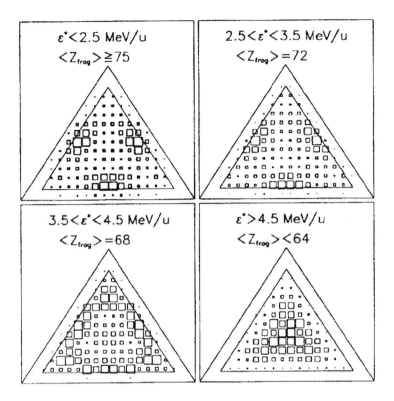

Figure 8.7. Evolution as a function of ϵ^* (in MeV/u) of the atomic numbers of the three largest emitted fragments (shown as Dalitz plots) following the decay of excited nuclei with mass close to 200. The variable $\langle Z_{\text{frag}} \rangle$ is the mean atomic number of the fragmenting source. Up to $\epsilon^* < 3.5$ MeV/u, most of the events populate the sides of the triangle. Therefore, according to equation (8.15), they are associated with two large fragments (similar to two fission fragments) and a smaller third fragment. As ϵ^* increases up to 4.5 MeV/u, the corners become more populated: they correspond to a large fragment and two smaller ones. In the last panel corresponding to the largest ϵ^*, most of the events are in the centre of the triangle, which corresponds to nearly equal atomic numbers for fragments. From [55].

in a Cartesian frame:

$$x = \tfrac{1}{3}(Z_2 - Z_3), \quad y = Z_1 - \tfrac{1}{3}S_{123} \qquad (8.15)$$

in which $S_{123} = Z_1 + Z_2 + Z_3$. Then, each point of coordinate (x, y) lies in a triangle, the distance d_i to each side i of the triangle being equal to Z_i. Therefore, in such plots, the corners of the triangle are populated by events with one large

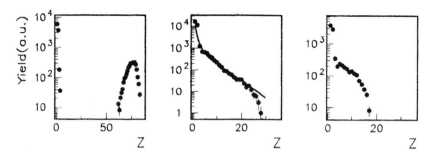

Figure 8.8. Evolution of the charge distributions observed for the Au+Au system at 35 MeV/u bombarding energy. On the left-hand plot, the involved excitation energy is limited and the decay corresponds to light-particle evaporation leading to a heavy residue (liquid nuclear matter). Conversely, when most of detected products are light particles, the deposited energy may lead to a complete vaporization of the system (right-hand plot: see section 7.3.2). In between, many light particles and fragments are emitted which may be interpreted as the coexistence of the liquid and gas phases (multifragmentation regime). From [309].

remnant and two small fragments. The sides of the triangle are associated with fission-like events (two large fragments and a small one), while the centre is populated with equal-mass fragment events. An example of the evolution of such Dalitz plots as a function of the excitation energy is displayed in figure 8.7, showing a clear evolution from a fission-like process towards more symmetric splittings as ϵ^* increases from 3 to 5 MeV/u.

It is instructive to note the similar evolution between the timescales and the charge distributions as a function of ϵ^* both suggesting a transition from sequential processes to very fast processes: multifragmentation is thus associated with a shortening of the timescales and with charge distributions which are more and more symmetric. Such charge distributions are coherent with a change from a process where most of the mass remains concentrated in a single heavy product (a piece of liquid nuclear matter surrounded by gas) towards an emission of several pieces of liquid nuclear matter as would be expected when the system explores a phase transition in the spinodale region. This point is illustrated in figure 8.8 showing the evolution of the charge distribution as a function of the dissipated energy in Au+Au collisions at 35 MeV/u.

In the description of the statistical multifragmentation models and also of the spinodale decomposition process, special emphasis has been put on the fact that the system should be at low density. We now consider possible signatures of such a situation by discussing collective motion in nuclear fragmentation.

8.4.2 Collective motion

8.4.2.1 Fragment kinetic energy distributions

In a collective ordered motion, and in contrast to the case of thermal motion, there is a correlation between the position of a particle (or fragment) and its velocity at the instant of emission corresponding to freeze-out. Collective motion may thus be studied with the help of the kinetic energy distributions of the emitted fragments [292]. This is valid if the coupling between intrinsic degrees of freedom and collective degrees of freedom is not strong enough to completely damp the motion. Within this picture, the kinetic energy of a *given* fragment of mass number A can be decomposed into two terms, a thermal one and a collective one:

$$\langle E_{\text{kin}} \rangle = \langle E_{\text{thermal}} \rangle + \langle E_{\text{collective}} \rangle \qquad (8.16)$$

where E_{thermal} takes the value $3T/2$ in the classical approximation if only translational degrees of freedom are considered. Since the relevant quantity for an expansion process is the expansion velocity, it is convenient to introduce the mean expansion energy *per nucleon* thus leading to

$$\langle \epsilon_{\text{kin}} \rangle = \left\langle \frac{E_{\text{kin}}}{A} \right\rangle = \frac{3T}{2A} + \langle \epsilon_{\text{collective}} \rangle. \qquad (8.17)$$

It turns out that the thermal contribution plays a sizeable role only for light emitted particles and becomes negligible for fragments, when compared to the collective energy. Therefore, a good variable to disentangle the thermal and the collective terms is to measure $\langle E_{\text{kin}} \rangle$ or $\langle \epsilon_{\text{kin}} \rangle$ as a function of the charge or mass of emitted products.

The collective part of the kinetic energy of the fragments can be decomposed into a Coulomb term plus an 'expansion' term. It is not easy to derive the 'expansion' term directly from the data and one very often relies on simulations [293, 306]. In such calculations, the kinematics of the multifragmentation of a single source is calculated with various possible assumptions concerning the geometry of the source and the initial outward flow of the fragments at freeze-out (section 8.3). The motion is supposed to be self-similar which means that the initial velocity of the fragments is proportional to their distance from the centre-of-mass of the configuration:

$$v(r_i) \propto r_i \qquad (8.18)$$

where $v(r)$ is the radial velocity of the particle located at distance r from the centre of the system. It is worth noting that this prescription is not a unique possibility [283]. However, equation (8.18) is well supported by the predictions of microscopic transport theories [402]. Therefore, in equation (8.4), one should add a contribution in the kinetic energy term which reads:

$$\epsilon_i^{\text{coll}} = \tfrac{1}{2} A_i m \alpha_0 r_i^2 \qquad (8.19)$$

and in which α_0 is a fitting parameter to be obtained by comparison with the data. It scales the velocity gradient inside the fragmenting matter. For moderate collective motion, this term has a small feed-back effect on the composition of the partitions. Generally, it is taken into account in the formalism by simply subtracting (on average) this collective term from the total available energy E^* in equation (8.4). But for large collective motion, the radial dependence of the term in equation (8.19) should be taken into account explicitly and not only on average.

8.4.2.2   Kinetic energy distributions of light particles: the blast model

In view of the preceding discussions, collective motion is best measured with the help of fragments. There are, however, situations (at high energy) where the dominant process is vaporization or quasi-vaporization leading to very few fragments. Then the kinematic characteristics of light particles must be used. The method then consists in analysing the kinetic energy distribution of light particles within the framework of the so-called 'blast' model [77]. This leads to the following expression for the kinetic energy distribution:

$$\frac{\mathrm{d}N}{\mathrm{d}E\,\mathrm{d}\Omega} \propto p e^{-\gamma_f E/T} \left[\frac{\sinh \alpha}{\alpha}(\gamma_f E + T) - T \cosh \alpha \right] \qquad (8.20)$$

in which E and p are the total energy and momentum of the particle in the centre-of-mass, $\gamma_f = (1 - \beta_f^2)^{-1/2}$, and $\alpha = \gamma_f \beta_f p/T$. The nuclear temperature T and the collective velocity β_f are adjusted to reproduce the kinetic energy distribution of the emitted particles. A typical example of such an analysis is shown in figure 8.9 in the case of Au+Au collisions at 1 GeV/u. All kinetic energy distributions are simultaneously fitted with the help of equation (8.20). In the case displayed in figure 8.9, a collective motion is clearly needed to reproduce the data.

8.4.2.3   Systematics of collective motion

A compilation of the collective velocity measured in a large range of beam energies is shown in figure 8.10. A threshold can be identified around $E_{\mathrm{lab}} \sim 30$–40 MeV/u, beyond which a rapid increase in the collective velocity is observed as a function of beam energy. A comparison with microscopic transport model calculations (BUU and QMD) is also displayed in figure 8.10. All in all, the models do a reasonable job, although it turns out that the BUU model predicts an anisotropic flow (more flow in the transverse direction than in the longitudinal one), while the data are compatible with an isotropic expansion. A possible interpretation lies in the treatment of the individual nucleon–nucleon collisions, which allows too much stopping in the simulations. Another interesting aspect of figure 8.10 is the apparent weak dependence upon the stiffness of the equation of state as shown by the results of the QMD simulations. One way to understand this feature is that a 'soft' EoS leads to high densities but to a low pressure gradient

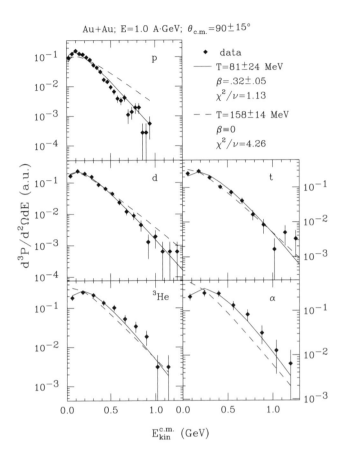

Figure 8.9. Energy spectra for light particles (p, d, t, He^3, α) emitted at 90° in Au+Au reactions at 1 GeV/u. The best fit using a collective expansion (equation (8.20), full line) is shown with the corresponding chi-square per degree of freedom χ^2/ν. Also shown is the best fit with a pure thermal emission (no expansion) which fails to reproduce the data (broken lines). The effect of the temperature is predominantly observed with the protons through the slope of the distribution. This is due to the fact that protons are very light and thus very sensitive to temperature T while the effect of the collective velocity inducing a shift in the energy distribution is mostly observed for α's due to their larger mass number. From [393] and references therein.

while a 'hard' EoS leads to the opposite, which altogether provides the same flow at the end of the reaction.

A significant amount of collective motion shows up beyond a beam energy

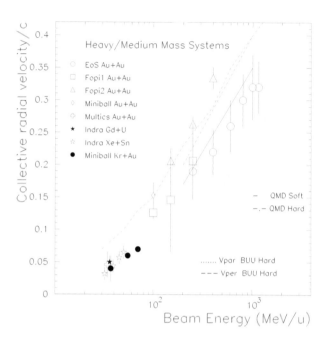

Figure 8.10. Systematics of the collective radial velocity as a function of the beam energy for medium- and heavy-mass systems in central collisions as measured by various collaborations. Lines correspond to the predictions of transport models. The labels 'Hard' and 'Soft' are associated, respectively, with a hard and soft equation of state. For BUU calculations, the collective motion is found to be anisotropic so that both v_{par} and v_{perp} contributions have been shown. From [172] and references therein.

of about 30 MeV/u for central symmetric collisions and for excitation energies around 5 MeV/u. Figure 8.10 shows that the proportion of the total available energy measured in the expansion mode evolves rapidly from a few percent below 50 MeV/u up to 30 to 50% around and above 100 MeV/u. The fact that the collective radial energy may be scaled with the beam energy does not, nevertheless, imply that this latter is the only relevant parameter. Indeed, the results shown in figure 8.10 refer to central collisions for systems with little or no asymmetry. Therefore, there is a direct link between the incident energy and the excitation energy deposited in the system. Could it be that the excitation energy is the most important parameter in the process?

A non-exhaustive compilation of the measured collective energy in nuclear collisions as a function of the excitation energy per nucleon is shown in figure 8.11. The onset of collective motion seems to be correlated with the onset

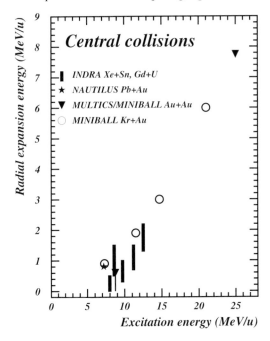

Figure 8.11. Systematics of the collective energy as a function of the excitation energy per nucleon in central collisions for a variety of heavy systems at incident energies in the Fermi energy range. Error bars have been indicated in the case of the INDRA data. Compilation from [405].

of multifragmentation (see previous section) for a value of ϵ^* close to 5 MeV/u. From the wealth of data displayed in figures 8.10 and 8.11, it is not clear whether the physical origin of the collective motion is related to a mechanical effect or to a thermally driven expansion. In the first case, the key quantity is the incident energy since the higher the bombarding energy is the larger the compression reached in central collisions. In the second case, the key quantity is the excitation energy. Then, the expansion should be independent of the impact parameter meaning that peripheral relativistic collisions and intermediate energy central reactions could lead to the same signal, provided the excitation energy deposited in the system is the same in both cases. If this is true, this would mean that the system has lost the memory of the initial compression (if any).

Another aspect concerns the respective roles of the collective and thermal motion in the fragmentation process. In a statistical approach, we have seen the importance of the concept of 'freeze-out'. Let us discuss this concept from the point of view of the timescales in the presence of a collective motion. In order for the system to reach equilibrium at freeze-out, the expansion of the system should not be too fast so that global equilibrium may be achieved and pre-fragments have time to be formed. Let R_{fo} be the radius of the system at freeze-out (with a volume V_{fo}) and let us suppose that the system expands with an average collective velocity v_f. The typical expansion time τ_{exp} is then given by the following relation

$$\frac{1}{\tau_{exp}} \sim \frac{1}{V_{fo}} \frac{dV_{fo}}{dt} \tag{8.21}$$

which leads to

$$\tau_{exp} \sim \frac{R_{fo}}{3v_f}. \tag{8.22}$$

For a system of mass around 200 and $\rho_{fo} = \rho_0/3$ one obtains, respectively, 100 fm/c, 50 fm/c and 20 fm/c for, respectively, $v_f = 0.03c$ (0.4 MeV/u), 0.06c (1.7 MeV/u) and 0.14c (9.1 MeV/u) mean radial expansion velocities. In the Fermi energy range, the measured collective velocities correspond to expansion times which remain larger (but not very much) than thermalization times. This is no longer the case at higher incident energies when details of the expansion dynamics should be taken into account.

8.4.3   Thermodynamical signatures

To summarize the results of the two preceding sections, we can say that there is experimental evidence for the existence of a fragmentation regime starting around an excitation energy per nucleon $\epsilon^* = 5$ MeV/u in which

- fragment emission timescales become very short: this is the so-called multifragmentation regime;
- the system experiences a (moderate) collective expansion;
- the atomic number distributions of the fragments evolve from asymmetric splittings towards equal-mass partitions.

All these features hint that the system experiences fragmentation through its passage in the low-density coexistence region of the EoS as discussed in section 8.2.1.1. There is, of course, the question as to the extent that such a process is either thermodynamic in origin or the result of a complicated non-equilibrated process. In the following, we discuss in more detail the characterization of multifragmentation from the point of view of the thermodynamics.

8.4.3.1   Comparison with statistical models

The characteristics of fragmentation events described in the previous section have been compared with the predictions of the statistical models outlined in section 8.3. Generally speaking, good agreement is obtained as far as fragment charge and multiplicity distributions are concerned. For instance, figure 8.12

shows the successful comparison between experimental charge distributions obtained in central Au+Au collisions at 35 MeV/u and the predictions of the Copenhagen statistical model [143]. It should however be noted that this model needs input quantities such as the excitation energy ϵ^* and the mass of the source and has at least one free parameter, namely the freeze-out density ρ_{fo}. In this respect, an interesting aspect of the results shown in figure 8.12 is the fact that the input quantities needed to fit the data are not unique. For instance, different pairs of the $\rho_{\text{fo}} - T$ variables (section 8.3) lead to the same agreement with the data. A high value of ρ_{fo} (upper part of figure 8.12) maximizes the interaction energy E_{inter} (see equation (8.4)) between the pre-fragments since they are closer in real space than for a lower value of ρ_{fo} (lower part of the same figure). Consequently, the available energy both for thermal motion and internal excitation energy of the pre-fragments is smaller for high values of ρ_{fo}. Thus, in this case, in order to obtain the same atomic number distributions, it is necessary to increase ϵ^* from 4.8 to 6 MeV/u in the inputs of the model. Finally, one should note the sensitivity of the predictions to a slight variation in ϵ^*. A change of 1 MeV/u strongly affects the charge distribution in the high Z region. This is why ϵ^* is very often fixed in the model by fitting the mean value of the charge of the heaviest fragment in each partition.

Despite these successes, one must be aware of the fact that the multifragmentation statistical models are not the only models that can adequatly reproduce the charge and multiplicity distributions in nuclear fragmentation. For instance, the ALADIN data were successfully reproduced using a percolation model [273]. But in the following we describe the approach of the Berkeley group which aims to 'extract' the statistical features of the fragmentation data directly by means of a reduction procedure.

8.4.3.2 Reducibility and thermal scaling

In section 7.2.3.1 the transition-state theory has been used to calculate the decay width associated with fragment production. The decay width for the emission of a fragment of atomic number Z reads:

$$\Gamma_Z = \hbar\omega_Z e^{-B_Z/T} \tag{8.23}$$

where B_Z is the emission barrier and ω_Z is the frequency of assault of the barrier (see figure 7.7). Assuming that the frequencies of assault are approximately equal whatever the fragment, the elementary emission probability is given by

$$p_Z \sim \frac{\Gamma_Z}{\Gamma_{\text{tot}}} \sim e^{-B_Z/T} \tag{8.24}$$

in which Γ_{tot} is the total decay width. Such a formalism has been successfully used to extract fragment emission barriers in [323]. This is achieved by identifying the decay widths with the measured experimental cross-sections in

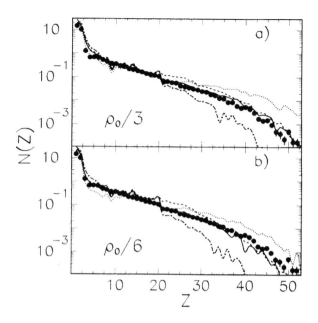

Figure 8.12. Comparisons of the measured charge distributions of fragmented events in central collisions with the predictions of a multifragmentation statistical model. The data (black points) have been compared with the results of the calculation (full lines) with two sets of parameters: the freeze-out density is indicated in each panel and $\epsilon^* = 6.0$ MeV/u in (a) and $\epsilon^* = 4.8$ MeV/u in (b). The dotted and chain lines are calculations with $\epsilon^* \pm 1$ MeV/u. From [143].

complete or incomplete fusion reactions:

$$p_Z^{\text{exp}} = \frac{\sigma_Z^{\text{exp}}}{\sigma_{\text{tot}}^{\text{exp}}} \tag{8.25}$$

where σ_Z^{exp} is the measured cross-section to produce fragments with atomic number Z and $\sigma_{\text{tot}}^{\text{exp}}$ the total cross-section. Estimating the temperature T with the methods of section 4.3.1, the barrier is obtained for each species and can be compared to the predictions of the models.

Experimentally measured cross-sections provide the probability $P^{\text{exp}}(n)$ to observe n fragments on an event-by-event basis: these are the branching ratios for binary, ternary, ..., n-ary decay shown in the right-hand part of figure 8.13. These probabilities have been measured in Ar+Au collisions at 80 and 110 MeV/u. Events have been sorted here according to the total transverse energy E_t (see section 4.2.3). A striking feature observed by the Berkeley group is the possibility of fitting these distributions with a binomial law (full lines in the left-hand part of figure 8.13):

$$P^{\exp}(n) = P_n^m = \frac{m!}{n!(m-n)!}p^n(1-p)^{m-n}. \tag{8.26}$$

where p is the elementary emission probability averaged over all fragments:

$$p = e^{-\langle B \rangle / T}. \tag{8.27}$$

The underlying interpretation of equation (8.26) is the following: if the system has the opportunity to try m times to emit an 'inert' fragment with an average probability p, then the probability P_n^m of emitting exactly n fragments after m tries is given by equation (8.26). For each transverse energy bin, it is possible to extract from the corresponding multiplicity distribution the value of $\langle n \rangle$ (the mean fragment multiplicity) and σ_n^2 (the corresponding width) in order to deduce p and m. Indeed, for a binomial law, one has the relations $\langle n \rangle = mp$ and $\sigma_n^2 = \langle n \rangle (1-p)$.

The identification of the measured p with its definition given by equation (8.27) provides a test of the statistical aspects of fragment emission in the framework of the transition state theory. However, since the temperature T is not directly measured, it has to be assumed that E_t is related to T according to the predictions of the Fermi gas model:

$$T \sim \sqrt{E^*} \sim \sqrt{E_t}. \tag{8.28}$$

Thus, the relation between p and E_t reads:

$$p \sim e^{-\langle B \rangle / \sqrt{E_t}}. \tag{8.29}$$

As shown in the right-hand part of figure 8.13, equation (8.29) is impressively verified by plotting $\log(p)$ as a function of $1/\sqrt{E_t}$. The same analysis has been performed for a variety of systems in the Fermi energy range [323] leading to the same conclusions.

The agreement of the data with a binomial law suggests a mechanism in which the whole decay process is reducible to a series of independent individual processes, hence the use of the term *reducibility*. Moreover, since the elementary probability is directly related to the temperature via equation (8.27), this result is interpreted as a signature of *thermal scaling*. This implies a strong dominance of phase space in fragment emission. Indeed, the probability of observing a given species is solely dictated by the temperature and barrier as in a variety of chemical and biological processes, hence the denomination of Arrhenius plots for the curves in figure 8.13. These results have been recently highly debated (see, for instance, [36, 86, 467, 472, 491, 492]). A crucial point is the question of the relation between E_t and ϵ^* or T as assumed in equation (8.28). In [467], it is claimed that E_t and ϵ^* are weakly correlated. The interpretation of these authors is the following: Fragment production is basically a dynamical process in the

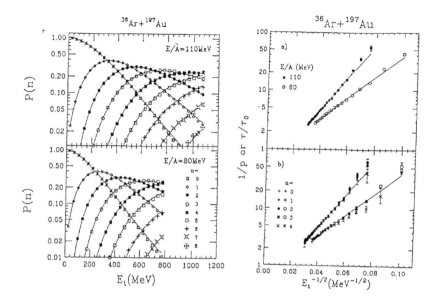

Figure 8.13. Left-hand side: probability for emitting n fragments as a function of the total transverse energy. Points are experimental; data curves are calculated from relation (8.26) with p values indicated in the right-hand part of the figure. The linear evolution of $1/p$ (in log scale) with the inverse of the square of the transverse energy is expected if the evolution of the system is governed by thermal equilibrium. From [323].

Fermi energy range. It thus corresponds to a new mode of energy dissipation [464, 466] as far as the dissipated energy reaches values close to 3 MeV/u. Therefore, there is no reason that equation (8.29) should remain valid if one were to use the 'real' excitation energy to estimate T instead of the transverse energy. Calculations using statistical models seem to support such a statement (see also [492]). Another point of debate is the observed autocorrelation between the IMF multiplicity and the transverse energy (see [472, 491, 492]).

Originally, reducibility had been interpreted as evidence for the sequentiality of fragment emission. This statement has been discussed in [86] in which it is shown that the Berlin statistical multifragmention model (see section 8.3.2) which assumes a simultaneous process can reproduce the observed trends of the data. In the framework of this model and from a general point of view, the reducibility of multifragmentation events to the product of a single fragment emission is a strong signature of the dominance of phase space (in the sense that transition amplitudes which contain all the dynamics seen to play a marginal role) but it is not a signature of sequentiality. The reason for this is that the fragment multiplicity distribution is not sensitive to the (small) distortions induced by fragment–fragment interactions in the final state. The only way to put such correlations in evidence is to build more sophisticated spacetime correlations such as those shown, for instance, in section 4.3.4.3.

To summarize, there is no doubt that the reducibility of fragmentation events is an experimental fact. This was first reported in [323] and confirmed in [492]. This shows up when one uses the definition of the transverse energy taking into account fragments and with the assumption that the transverse energy is proportional to the square of the temperature. Concerning this last point, it would be interesting to study the predictions of models in which fragments and particles are produced dynamically (for first attempts see [161, 461]). In particular, one may wonder whether the observed reducibility can be obtained in non-equilibrium incoherent processes (fragments could be produced independently in coalescence processes as suggested in [464, 466]). In such a picture, the transverse energy would be related to the impact parameter and thus to the size of the region in which fragments are formed (the larger the region is the larger the fragment multiplicity).

8.4.3.3 *Phase coexistence*

In the previous section, the concepts of reducibility and thermal scaling for fragment multiplicity distributions were introduced. The same formalism may be used for the analysis of charge distributions [204, 366]. In this context, one writes:

$$P_i(z) = p_Z \tag{8.30}$$

where $P_i(z)$ refers to the probability of emitting a fragment of charge z in an event with fragment multiplicity i. It turns out that equation (8.30) is not exactly verified by the data, which means that the fragment charge distributions do depend on the fragment multiplicity. The reason for this discrepancy is of combinatorial nature and reflects charge conservation constraints. It is possible to fit the data if we replace equation (8.24) by

$$P_n(z) \propto e^{-(Bz/T)-ncz}. \tag{8.31}$$

The parameter c is then a measure of the deviations with respect to a completely reducible distribution with no charge conservation constraints. The evolution of c with the excitation energy can be understood if one notes that for limited excitation energies charge conservation constraints are weak because the heavy remnant acts as a charge reservoir so that c should stay close to zero. Conversely, in the multifragmentation regime, this reservoir no longer exists and charge conservation induces an increase of c. The rise of c as a function of excitation energy can thus be interpreted as a transition from a monovariant system (phase coexistence, $c \simeq 0$) to a bivariant system (one phase, $c \geq 0$).

Such a behaviour has indeed been experimentally observed, as illustrated in figure 8.14 [323]. Simulations of finite nuclear systems based either on the

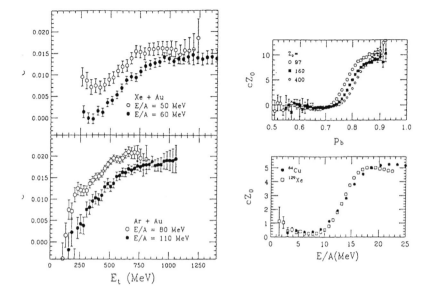

Figure 8.14. Left-hand side: evolution with the transverse energy E_t of the measured c parameter (see text) for the Ar+Au and Xe+Au systems. Right-hand side: corresponding simulations in a percolation calculation in which p_b is the probability to break a bond on the percolation lattice (this parameter plays a similar role as temperature or a total transverse energy) (top) and a binary sequential evaporation calculation (bottom). From [323].

percolation model or on the sequential binary decay model (left-hand part of the figure) support the idea of a transition from a saturated vapour in equilibrium with a liquid (the residue) at low temperature to an overheated unsaturated vapour evaporation at high temperature. The evaporation model places the transition around 10 MeV/u excitation energy per nucleon, which corresponds to the maximum fragment production (see figure 7.14).

8.4.3.4   Caloric curves and the study of the heat capacity of hot nuclei

Up to now, we have only considered the study of fragmentation partitions. We now come to a discussion of the caloric curves which connect the temperature T and the excitation energy per nucleon ϵ^*. We have seen in section 8.3.2 that one of the major predictions of the high-energy statistical models was the existence of a plateau in the caloric curves. The search for such a behaviour in the decay of hot nuclei has been the motivation of numerous experiments in the last decade, triggered initially in the fragmentation regime by the ALADIN group [372]. The methods to measure both ϵ^* and T have been described in chapter 4. Low-energy data have been discussed in section 7.2.1 (see figure 7.4). A non-exhaustive compilation of the experimental results obtained using the three methods to measure nuclear temperatures described in section 4.3.2 is shown in figure 8.15. The data cover different collision regimes in both the Fermi energy range and the relativistic domain

Within a given method of extraction of ϵ^* and T, the data from different collaborations do agree, while the application of the three previously mentioned methods do not. This means that the different methods provide, at best, *apparent* temperatures. Thus, they need to be either corrected for spurious effects or at least to be inter-calibrated. The 'kinetic' temperatures follow, approximatively, a Fermi gas law, while the 'excited state' temperatures seem to saturate (see also [372]). Finally, the temperatures based on a 'double ratio' seem to increase slowly.

These results are somewhat puzzling. This apparent contradiction can be solved in the framework of QSM [232, 269] by including excluded volume effects [226]. In particular, the saturation of the 'excited state' temperatures can be explained by the fact that the yield of excited clusters is not only determined by the temperature but also by the 'geometrical' volume occupied in the source by such excited species. Indeed, it turns out that most of them decay by particle emission, which somehow means that such excited clusters occupy a larger volume than in their ground states. However, other explanations are also possible [111, 338]. In particular, cooling [194, 246, 429] may be an important effect: nuclear species would be produced at different steps of the disassembly process, thus light particles could be expected to be produced first (even at the pre-equilibrium stage), while fragments would be emitted later, at lower temperatures, implying a hierarchy in the different temperatures experimentally observed. In [253] the authors have compared results using either the population of discrete excited states or the double ratios of isotope yields. The observed discrepancies are, nevertheless, not fully understood: they could reflect side feeding effects [372]. More simply, the precision with which the values of the 'double ratio' temperature have been obtained with the help of equation (4.29) presumably becomes questionable at large temperatures. For these temperatures, as the logarithm is a highly non-linear function, an uncertainty about the measured population yields can lead to an overestimation of the temperature by several MeV [299].

However, if one assumes that the 'double ratio' temperature is a correct measure of the real temperature, the resemblance between the results displayed in figure 8.5 and those of figure 8.15 suggests the observation of a liquid–gas phase transition in the decay of hot nuclei as claimed for the first time in [372]. This statement has been, and still is, highly debated even if the 'double ratio' temperature they obtained with He–Li isotopes seems to be relatively robust. However, even when taking the experimental value of the temperature for granted, there remains a question about the physical conditions prevailing in the system: is pressure or volume constant?

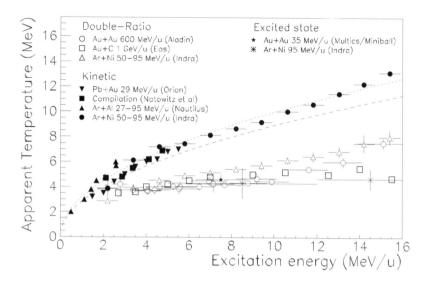

Figure 8.15. Systematics of measured nuclear temperatures with the three methods described in section 2.3 as a function of ϵ^*. From [171] and references therein.

These two situations lead to different ϵ^*–T plots [323]. Indeed, Moretto *et al* have pointed out that, for any real fluid, the observed plateau (or at worst the very slight increase) is expected only if the pressure in the system is constant: a situation which may not be reached in a finite open system such as a hot nucleus. In contrast, the evolution of the system at constant volume does not lead to a plateau in the corresponding caloric curve. This question is discussed in [408] and in [126] (see figure 8.16). It is seen in this figure that the correlation $T = f(\epsilon^*)$ (caloric curve) strongly depends on a third variable λ which is the Lagrange multiplier associated with the volume of the system (here 216 particles). This calculation has been performed in the lattice gas model approach discussed in section 8.3.4. According to this work, the experimentally observed caloric curve may be any cut in the plot of figure 8.16 since the volume or equivalently λ are not directly measured experimentally. For the most negative values of $\ln(\lambda)$ (associated with average densities close to $\rho_0/3$), a backbending is observed indicating that the heat capacity of the system becomes negative: this is associated with the liquid–gas phase transition in a finite system as discussed in the next paragraph. However, other paths (associated with larger densities) in the plot do exhibit a monotonic behaviour.

From the previous discussion, it appears that, up to now, there is no clear consensus on the interpretation of the caloric curves. It could even

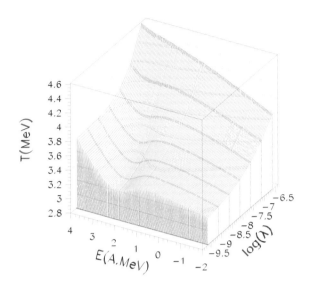

Figure 8.16. Three-dimensional plot showing the correlation between the temperature, the excitation energy (here labelled E) and the λ parameter. λ is the Lagrange multiplier associated with the volume for an isolated system (microcanonical description) of 216 particles described in the lattice gas model framework. From [126].

be, for instance, that out-of-equilibrium clusterization phenomena could lead to the production of species mimicking to some extent the experimental data. Dynamical calculations performed within the framework of FMD (section 3.3.4) [184] can indeed 'produce' a caloric curve with a plateau. However, it is not clear that the boundary conditions imposed in the calculation are those encountered in nuclear collisions. Thus, a truly microscopic interpretation of the results shown in figure 8.15 is still not at hand. This task represents a true theoretical challenge for future studies.

In any case, the most relevant phase transition signatures have to be found in an event-by-event analysis rather than from analysis involving many events. It is indeed worth noting that all temperature measurements imply an addition of the contributions of many *similar* events: for instance, a 'kinetic' temperature is obtained from the slope of an energy spectrum built from *many* particles detected in *many* events. It is far better to try to identify phase transition behaviour in a single event. Such a method implies the analysis of correlations inside each single event which is now possible with modern multidetectors (see section 4.1.2).

Recently such studies have been proposed in two directions: correlations between fragments of a single event [82] (see the next section) or correlations between the energies associated with independent degrees of freedom [125, 144]. In this last case, the idea is that a phase transition is associated with large fluctuations in the sharing of the available energy among the various possible degrees of freedom. It has been proposed [126] to look at the relative values of the potential and kinetic energy parts. The heat capacity of the system may then be calculated from the variance associated with the kinetic energy part. One obtains the following relation for the heat capacity of an isolated nucleus:

$$C = \frac{C_{\mathrm{K}}^2}{C_{\mathrm{K}} - \sigma_{\mathrm{K}}^2/T^2}. \tag{8.32}$$

In this expression, C_{K} is the kinetic microcanonical heat capacity and T the temperature. If the kinetic energy variance σ_{K}^2 is large compared with T^2 the total heat capacity becomes negative. Such a result is expected from the microcanonical description of a first-order phase transition occurring in a finite nucleus [227]. It has been obtained recently from experimental data ([90, 144, 145]) but has still to be confirmed from systematic measurements with, on the one side, systems for which the phase transition is expected and, on the other, systems for which it is not.

8.4.3.5 *Search for a critical behaviour in nuclear fragmentation*

At the critical point in infinite systems, both the specific heat and the isothermal incompressibility become infinite. Near the critical point, the behaviour of these quantitites (and also of other ones) can be described with the help of universal numbers called critical exponents. The search for critical exponents in the decay of hot nuclear matter follows the analysis developed in condensed matter physics for the study of critical phenomena. Fragmentation data were first analysed in this context in the mid-1980s in [109, 110, 374]. A percolation model (section 8.3.4) was used and its results compared with the behaviour of charge distributions obtained in the fragmentation of gold nuclei around 1 GeV/u incident energy [484]. One defines the various moments m_k of the charge distribution as

$$m_k = \frac{1}{Z_{\mathrm{tot}}} \sum_i n_i Z_i^k \tag{8.33}$$

where the sum runs over all fragments i with multiplicity n_i and with charge Z_i. The lowest order moment m_0 thus provides the total fragment multiplicity and m_1 gives the total charge Z_{tot}. But more interesting is the fact that the correlation between m_0 and m_2 provides a direct signal of a critical behaviour as shown in figure 8.17. Finite-size effects were clearly emphasized at that time as well as the importance of the dimensionality of the system. Such analyses have triggered numerous studies as more and more complete data became available.

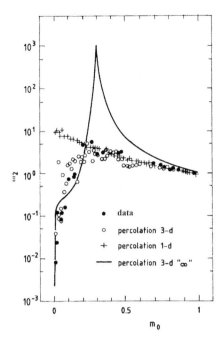

Figure 8.17. Experimental (black points) correlation between the two moments m_2 amd m_0 (equation (8.33)) observed in the fragmentation of Au nuclei [484] compared with the predictions of the percolation model for three different geometries. The full line corresponds to a very large three-dimensional lattice (125 000 sites): for such a very large system a critical behaviour is clearly observed around $m_0 = \frac{1}{3}$ (for an infinite system, one would have a divergence at this point). Going to a smaller three-dimensional lattice (open points) where the number of sites is of the order of the number of nucleons in the source (around 200), the divergence is unfortunately reduced to a broad maximum. This is a consequence of finite-size effects. Nevertheless, the model is in good agreement with the data. Finally, a one-dimensional lattice calculation (crosses) does not show any maximum in the distribution but a monotonic evolution. This is expected because it can be shown that no critical behaviour is expected in such a one-dimensional system. From [110].

The recent, more involved, investigations along this line are, nevertheless, mainly based on the work of Fisher [188] on the condensation of water droplets in a vapour. The same techniques have actually been used in percolation and lattice gas model calculations (section 8.3.4). They are based on finite-size scaling [332]. We illustrate it here, following the experimental analysis performed by the EoS collaboration [174, 206].

According to Fisher's model, the mass distribution of the nuclear droplets reads:

$$n_A(\epsilon) = q_0 A^{-\tau} f(z) g(A, \Delta\mu, m) \tag{8.34}$$

in which the scaling variable is $z = \epsilon A^\tau$, $\epsilon = (m - m_c)/m$ (m is the fragment multiplicity and m_c the fragment multiplicity at the critical point). The function $f(z)$ (the so-called scaling function) is related to the surface free energy of fragments. As surface tension vanishes at the critical point, we have: $f(z = 0) = 1$. Finally, the factor q_0 in equation (8.34) is only a function of τ: $q_0 = 1/\zeta(\tau - 1)$ and $\Delta\mu = \mu_L - \mu_V$ where μ_L and μ_V are the chemical potentials of, respectively, the liquid and the gas phase. Note that, here, the multiplicity is used as the control parameter but the temperature can also be used in the same way (see for instance [227]).

The scaling function $g(A)$ entering equation (8.34) is related to the bulk free energy to produce fragments:

$$g(A) = e^{-\Delta\mu/T}. \tag{8.35}$$

At the critical point ($m = m_c$) or along the coexistence line, the two chemical potentials are equal so that the order parameter $\mu_L - \mu_V$ vanishes, thus leading (according to equation (8.34)) to a power-law mass distribution. Such power-law behaviours have been observed in various experiments [174, 206, 309] and a long time ago in proton-nucleus collisions in the GeV range [187, 316]. The issue is then to extract the two critical parameters τ and σ, which are often used to define two other critical parameters:

$$\beta = \frac{\tau - 2}{\sigma} \quad \text{and} \quad \gamma = \frac{3 - \tau}{\sigma}.$$

From an experimental point of view, the first step of the analysis is to find m_c. To this end, the charge distribution associated with a given multiplicity m is fitted with a power-law function. The distribution for which a sharp minimum is observed in the χ^2 provides the critical multiplicity m_c (see figure 8.8, middle panel). The associated exponent of the power-law distribution is then the critical parameter τ. Knowing m_c and τ and assuming that the system is always at the coexistence line, it is then possible to obtain the scaling function $f(z)$. An analysis of $f(z)$ then gives the value of the second critical parameter σ.

The results of several experiments as well as the predictions of different models concerning these critical exponents are shown in table 8.1. From a comparison between the models and the data, it is hard to conclude which model is best suited to describe the data because all of them give rather similar predictions.

These results could appear very promising. Let us, however, be cautious. It is not clear that a power-law distribution of fragment atomic numbers is a reliable signature of a second-order phase transition since it has been shown that such a distribution can be reproduced in simple simulations under non-critical conditions [367]. Thus, the search for a phase transition in nuclear fragmentation based on the study of nuclear partitions suffers from the same problems as the

Table 8.1. Critical exponents obtained for various systems. Theory: percolation [439], lattice gas [51], statistical multifragmentation model (SMM calculations [144] removing fission with prescription of [151]). Data: Au+emulsion at 1 GeV/u [109], Au+C at 1 GeV/u [206] (see also [174] for a re-analysis of the data which leads to the same results), Au+Au 35 MeV/u [144]. Adapted from [112].

System	τ	β	γ	$1 + \frac{\beta}{\gamma}$
Percolation	2.20	0.45	1.76	1.25
Lattice gas	2.21	0.33	1.24	1.27
Statistical multifragmentation model	2.168 ± 0.002	—	—	1.28 ± 0.002
Au+emulsion	2.27 ± 0.1	—	—	1.2 ± 0.1
Au+C	2.14 ± 0.06	0.29 ± 0.2	1.4 ± 0.1	1.21 ± 0.1
Au+Au	2.12 ± 0.02	—	—	1.28 ± 0.031

search for a liquid–gas phase transition in nuclear thermometry. There is no clear consensus on the experimental signals so that many different contradictory interpretations have been proposed. Most of the theoretical predictions are based on oversimplified hypotheses about the system under consideration. In particular, apart from the question of thermal equilibrium which may or may not be achieved, there is a problem with the fact that the system is open and can exchange matter and energy with the vacuum. This obviously raises the question of the importance of dynamical effects in the decay modes of hot nuclei and, in particular, in nuclear fragmentation. Here also, it is necessary to go further to be more conclusive. This may be performed by looking at the charge correlations between various fragments of the same events. This question is addressed at the end of the next section.

8.4.4 Microscopic dynamical description of nuclear fragmentation

In the previous sections, signatures of a liquid–gas transition have been discussed in the context of nuclear thermodynamics. By nature, these approaches do not address the question of the formation of the fragments. The only way to deal with such an issue is to 'follow in time' the whole process, whence the necessity of using dynamical approaches. The work of [356] has already been mentioned in section 8.2.3. Here, we would like to report on the confrontation of a stochastic transport model (the so-called BOB for Brownian-one-body approach) [224] with data obtained by the INDRA collaboration [403]. Two systems were considered (Xe+Sn at 32 MeV/u and Gd+U at 36 MeV/u) for which the most dissipative central collisions lead to the same excitation energy per nucleon (close to 8 MeV/u) but, of course, not to the same mass number for the fused system.

A first experimental result is the strong similarity between the atomic number

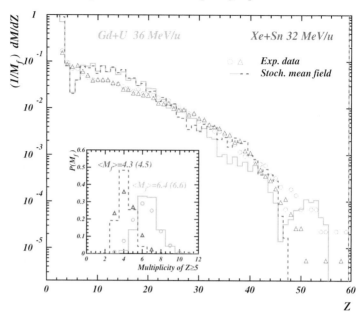

Figure 8.18. Charge and multiplicity distributions in central 32 MeV/u Xe+Sn (triangles) and 36 MeV/u Gd+U (circles) collisions. The histograms (broken: Xe+Sn, dotted: Gd+U) are the predictions of a dynamical simulation based on semi-classical transport theory with an additional stochastic term. The insert shows the corresponding fragment multiplicity M_f with the same symbols as in the main figure. From [403].

distributions of the two systems (see figure 8.18). Another interesting point is the scaling of the observed fragment multiplicity distribution with the total mass of each system (see insert in figure 8.18). It is worth noting that such trends can be interpreted in the framework of the statistical models [89]. However, these features can also be reasonably reproduced by a dynamical calculation (histograms in figure 8.18). A detailed analysis [191] of the time evolution of the system in the simulation shows that the system enters the spinodale region at finite temperature ($\rho_s \sim 0.4\rho_0$ for $T = 4$ MeV) after a maximum compression close to $1.25\rho_0$ at $t \sim 40$ fm/c. The fragment multiplicity reaches its asymptotic value for $t \sim 250$ fm/c after the beginning of the reaction. At this time, the system is deep inside the spinodale region and the disassembly of the system is driven by a mechanical instability (spinodale decomposition, section 8.2.1.3). But due to this rather long time, it is possible to understand why statistical assumptions are also able to reproduce the global features of the data [89].

Other dynamical approaches based on molecular dynamics models (QMD)

[339] advocate a faster, non-equilibrated process, triggered earlier in the reaction. The physical scenario that can be traced back from such calculations is the following. Free particles constituting the gas phase are produced in hard nucleon–nucleon collisions at the very beginning of the reaction, while the liquid phase is hardly excited and is constituted by those nucleons which only suffered soft 'secondary' collisions. Consequently, there is no strong mixing between the nucleons of the two partners of the reaction as required, for instance, in a statistical treatment. Transparency is hence predicted, even for the most central collisions. The kinetic energy of the fragments is provided, to a large extent, by the internal Fermi motion of the nucleons inside the medium. It is predicted that a strong memory effect of the entrance channel will be observed in the data. Results obtained at relativistic energies by the FOPI collaboration [380, 395] seem to show that full equilibration has indeed not been obtained even for the most central collisions, but similar experiments have not been performed below 100 MeV/u and they are clearly needed. The principle of such experiments is to induce collisions between nuclei with quite different N/Z ratios. The isotopic composition of IMF detected at forward or backward angles should keep the memory of the incident isospins of nuclei if chemical equilibrium is not achieved during the collision, i.e. if some transparency or rebound takes place. Such experiments will be undertaken in the near future but they need to identify, both in mass and charge, the outgoing products in any direction.

For the moment, recent data are in favour of a scenario in which multifragmentation results from a spinodale decomposition. They rely on correlation studies between fragments emitted in a single event. Stochastic statistical theories described at the beginning of this section predict that the size of the outgoing fragments should be strongly correlated if they result from a spinodale decomposition scenario. Such a behaviour has indeed been observed in central collisions [82]. A further coherence with the data is that the signal disappears if one selects less violent collisions for which the spinodale decomposition is not expected. This result has, however, to be confirmed by performing crossed comparisons between data and calculations for various systems and impact parameter selections.

8.5 Conclusion of the chapter

In this chapter we have discussed the main features of nuclear fragmentation from both experimental and theoretical points of view. Nuclear fragmentation studies are mainly motivated by the search for a liquid–gas phase transition in nuclear matter, which is a natural consequence of the structure of the nucleon–nucleon force. The fact that such a transition is sought for in finite transient systems makes it much more difficult. It should, however, be stressed that this is a unique situation in physical systems.

Experimentally, it is worth noting that such studies push the experimental and analysis techniques to their limits by requiring high granularity and high quality detectors. Although this represents a huge experimental effort, the quality of the obtained data is certainly rewarding. As far as the data are concerned, the characterization of nuclear fragmentation in terms of time and energy scales has been discussed. The transition (around $\epsilon \sim 3$ MeV/u) from a rather slow sequential process towards a fast simultaneous mechanism has been shown in conjunction with the evolution of the charge distributions towards symmetric partitions.

A collective motion has been evident when the incident energy reaches the Fermi energy. However, this radial motion remains rather moderate and could thus explain, to a certain extent, the dominance of phase space. This last point has been discussed both from the point of view of reducibility and thermal scaling and also through a comparison of the data with the multifragmentation models developed so far. A direct signature of a liquid–gas phase transition in hot nuclei is presumably provided by the structure of the caloric curves. Experimentally, it turns out that several such curves have been obtained due to the fact that the different methods of nuclear thermometry lead to different results. Therefore, the theoretical interpretation of the caloric cuves is still, to a large extent, not fully settled.

Theoretically, the use of generic approaches such as percolation models or lattice-gas models has proven to be extremely useful for our understanding of finite-size effects in phase transitions. Due to their simplicity, their link with the data remains questionable, mainly because such models, as well as the multifragmentation statistical models, consider idealized situations which may not be achieved in nuclear collisions. This leads us to the everlasting question of the role of the dynamics in the nuclear disassembly process. Very recently fully dynamical calculations have become available: such approaches are very ambitious since they treat the whole process from the very beginning up to times for which it is hoped that no strong dynamical effects will alter the process. First results are very promising and there is no doubt that such studies will be pursued in the future as computer capabilities now allow us to simulate thousands of collisions in a reasonable computer time.

Chapter 9

Epilogue

To conclude such a topic as the one we have been addressing in this book is quite a hard task. It is probably the hardest part of the job but it is also the one which makes the whole enterprise worth the effort, as it is the part which will allow us, and hopefully the reader, to put all the achievements into a wider perspective, both in time (we think here of the future of our field) and, to some extent, in space (by which we mean other fields of physics).

Fortunately we were cautious—we did not put any question in the title of the book which could have necessitated a definite answer. We have, nevertheless, tried to put forward the, still numerous, open questions in the field. As far as possible, we have tried to answer some of the more specific questions but we have also left several others open. This is, in our opinion, a sign of good health for the field, as these many open questions are presumably the best trigger for efficient and successful forthcoming research.

In the next few pages we shall, nevertheless, try to summarize some of our findings in terms of achievements, open questions and new emerging questions. We shall return first to a question of principle in the interest in heavy-ion collisions. Next we shall discuss the major experimental findings in terms of basic physical quantities and questions, such as time, energy and dynamics versus thermodynamics. We shall also summarize our theoretical understanding of the physics of nucleonic collisions and try to point out the possible (necessary) lines of development. The future will be addressed in the ensuing sections, mainly *in* the field, in particular, in terms of the opening in the physics of isospin. Finally, we shall superficially discuss the future in terms of openings in other fields of physics which are, more or less, loosely connected with heavy-ion collisions.

9.1 Why bombard nuclei against one another?

Heavy-ion collisions have been a central topic, and, to a large extent, the essence of this book. We presented them in the introductory chapter as a powerful means with which to explore the properties of nuclei. We have seen that they indeed

267

allow us a better understanding of some dynamical properties of nuclei. We also had in mind the possibility of exploring the phase diagram of nuclear matter and possible phase transitions. We have indeed made some steps along this line, although caution is necessary here. We had, in fact, to face both dynamical and thermodynamical questions. The links between the two aspects are both trivial and complex and, for sure, not fully understood in the context of heavy-ion collisions. However, we do not aim to discuss these achievements here in detail. We postpone that to section 9.2. Here we would like to focus on what is more a question of principle concerning, say, the feasibility of such a programme. In other words, are heavy-ion collisions in the nucleonic regime a proper tool with which to investigate the type of questions we had originally in mind or even for questions we might have overlooked up to here? Roughly speaking, such a question amounts to quantifying, or at least listing, the pros and cons of heavy-ion collisions in the nucleonic regime.

Nuclei are exceptional extended objects, although finite, in the sense of purity. As compared to many other physical situations (think, for example, of defects in crystals), experimenting on nuclei has the fundamental, satisfying and useful, advantage of allowing us to deal with well-defined objects. Very much is known about a given nuclear species, from gross properties like the binding energy or radius to very detailed ones spectroscopic in nature. This knowledge is preserved in heavy-ion collisions. The nature and energy of the projectile are perfectly known as is the nature of the target. In addition one is able to control the beam parameters to ensure that only one collision takes place at a time and is recorded. In turn, the new generation of 4π detectors has allowed us to measure most of the products of a collision, thus providing an incredibly detailed picture of the collision patterns, hardly attainable in comparable physical situations (think, for example, of collisions between complex molecules). In this respect heavy-ion collisions thus constitute a particularly favoured field. Note, however, that such a good knowledge of the initial conditions is not fully a gift of nature as one, for example, remains unable to control the impact parameter of the reactions. The need for modern 4π detectors is, to a large extent, due to this fact.

Still, heavy-ion collisions are not an easy tool to handle. The situations encountered are violently dynamical, which means that a proper understanding presumably requires a high degree of sophisticated modelling. In this respect a comparison of results obtained with light projectiles, like protons, deuterons, or antiprotons is quite telling, as it allows us to disentangle the various aspects at work (dynamical versus thermodynamical questions in particular). The complexity of the collisions itself reflects the difficulty in finding the most suitable observables to characterize a physical situation. Looking for the relevant variables has been, and remains, a basic quest in the field. It is interesting to note that the problem is actually both theoretical and experimental, as the necessary microscopic models used suffer from the same defects as the experimental data, namely the fact that one is facing too much unstructured information at once! But pessimism should presumably not win here. Over the years a wealth of experience has been gathered on these questions and, although some situations remain beyond simple analysis (the typical example is here multifragmentation), significant steps have been made in many other directions and should be acknowledged. For instance, the identification of the impact parameter (or at least the degree of violence of the collision) is more and more acceptable. Here we hit another inherent difficulty of this physics connected to the highly indirect nature of the measurements. It is physically impossible to measure directly distances of order 10^{-15} m! Of course this situation is not specific to nuclear collisions. There are many other fields of physics in which direct measurements are impossible but very often a few well-defined variables which provide a one-to-one clue to the sought after quantity exists. Most of the time this is not the case for heavy-ion collisions in the nucleonic regime. A good aspect of this is that one has thus to find clever and robust variables, which can even be exported to other situations. Altogether, although the situation is not always forgiving, years of patient work have shown that most of the intrinsic difficulties could be circumvented or, at worst, greatly reduced. To learn physics from heavy-ion collisions thus constitutes a plausible goal, which should hopefully become more and more real with time, particularly if a proper theoretical framework is able to accompany the experimental developments. Let us now discuss these aspects more specifically.

9.2   Nuclear collisions and the relevant observables

It is instructive to trace back (even in a schematic view) the evolution of the study of nuclear collisions in the last decades. The discovery of new phenomena and the emergence of new theoretical concepts are obviously linked with the development of new facilities and new experimental techniques.

Several decades ago, the use of light probes (say from nucleons up to α's) to induce nuclear collisions opened up the possibility of producing 'tepid' nuclei. The modest amount of excitation energy brought to the system enabled the use of statistical approaches pioneered by the invention of the concept of the compound nucleus by Niels Bohr. Nuclear level densities at moderate temperatures and emission barriers were then measured by studying evaporation processes. Collective phenomena such as giant resonances or fission were also explored. However, it was already recognized at that time that non-equilibrated processes were present in such collisions in the form of fast particle emission.

Although pioneering studies were performed in the 1950s, it was not until the early 1970s with the advent of powerful heavy-ion accelerators that the study of heavy-ion reactions triggered the possibility of exploring matter at higher excitation energies [415]. Indeed, the discovery of deep inelastic or highly damped reactions as well as the study of complete fusion lead to the conclusion that large amounts of excitation energy could be deposited in nuclei, which were then called 'hot nuclei'. However, the exit channels of such reactions remained rather simple: the two partners of the reaction retained, to a large extent, their identities although an important mass diffusion process (related to the dissipation inside the system) could be observed.

The timescales involved in such collisions allow the internal degrees of freedom (which evolve on a short timescale) and a few selected collective degrees of freedom slowly evolving in macroscopic potentials to be separated. It is thus possible to introduce a description based on the Fokker–Planck equation and thus to shed light on the mechanism of energy dissipation in nuclear collisions. Nuclear friction was introduced at that time with help of phenomenological theories based on the wall and window formulae.

The advent, in the early 1980s, of nuclear facilities accelerating heavy-ions at a few tenths of MeV/u opened up the possibility of reaching the very limits of the existence of nuclei. In nearly symmetrical collisions, part of the incident energy can be transferred into heat whose magnitude may be comparable with the total binding energy of the sytem. This is why the complete disintegration of nuclear edifices into nucleons and light composites has been observed in this domain.

The price to pay to characterize such complicated processes properly is a huge experimental effort to detect and identify the many products emitted in the most dissipative collisions. From this point of view, the incredible progress in detection technology is certainly one of the main achievements of the last decade. From a technical point of view, there is no possible comparison between a typical experiment performed in the late 1980s and one in the late 1990s.

Conceptually, the description of such very violent reactions requires new ideas because of the shortening of the collective timescales. A microscopic description using nucleons as degrees of freedom is then a necessity, hence the development of microscopic nuclear transport theory described in detail in this book.

The evolution of nuclear dissipation from a slow process at low incident energy to a fast and violent mechanism in the Fermi energy range is testified by at least three new features that have been discussed extensively in this book:

- the increase of particle pre-equilibrium emission and their related collective motion (sidewards flow and squeeze-out) on the one hand and, on the other hand, the associated sub-threshold production of particles (mainly pions but also kaons) as well as high-energy γ-rays;
- the formation of neck-like stuctures and the disappearance of complete fusion;
- the advent of new decay modes from sequential fragmentation up to vaporization via multifragmentation.

What have we learned by studying these new phenomena?

The study of fast processes and particle production has proven to provide a direct link with the properties of nuclear matter. The understanding of sidewards flow and squeeze-out requires, in the framework of transport models, a soft equation of state with momentum-dependent forces. This is probably one of the main achievements in the field in the last few years. However, we have seen that a consistent microscopic theory that could account for the production of composite particles in the early instances of the reaction is still not available although recent developments in molecular dynamics calculations are very promising (see section 9.3). The understanding of very rare processes such as pion or kaon production also remains, to a large extent, an open problem.

New reaction mechanisms have been discovered and explored. The characterization of neck-like structures and more generally the experimental evidence for strong deviations with respect to low-energy reactions is a major finding of the last decade. This represents a true challenge for the theory in the sense that it addresses its capability to reproduce rather complicated spacetime configurations. Such configurations depend, to a large extent, on a proper description of the transport properties of highly non-equilibrated matter. In particular, a correct treatment of dissipation through one-body and two-body dissipation mechanisms, and a proper evaluation of the corresponding macroscopic quantity (the nuclear viscosity) still remain problematic.

The study of the decay modes of hot nuclei is strongly motivated by the search for the liquid–gas phase transition of nuclear matter as suggested by the structure of the nucleon–nucleon interaction. We have seen that this programme represents a difficult task. The main reason for this is that it requires a control and an understanding of all the processes mentioned earlier, namely the characterization of non-equilibrium emission and a knowledge of the spacetime structure of the reaction on an event-by-event basis. Therefore, the measurement of the relevant variables turning the transition from a liquid-like state to a gas-like state such as the nuclear temperature, the excitation energy and the volume in which the system decays represents a formidable challenge for physicists.

However, the strong evolution from low-energy (evaporation and fission) towards high-energy decay processes (fragmentation and vaporization) suggests that such a transition has indeed been observed in nuclear collisions in the nucleonic regime. However, the obvious question here is whether one can relate such phenomena with the thermodynamical properties of nuclear matter. It is not impossible that the transition may occur out-of-equilibrium. This could be the reason why the usual behaviour (well known, for instance, for macroscopic real fluids) of a constant temperature associated with liquid–gas phase coexistence has not been unambiguously observed. In other words, the 'plateau' of the caloric curve is still disputed. This is so because such a plateau is expected only for well-defined thermodynamical conditions (for instance constant pressure). Such conditions are obviously not fulfiled in heavy-ion collisions in which dynamical effects play a determinant role. It is hence necessary to explore new signatures of phase transition. First attempts have been discussed in sections 8.4.3.4 and 8.4.3.5.

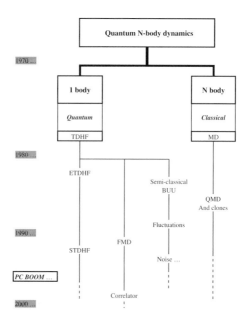

Figure 9.1. The various steps of the development of microscopic transport theories in nuclear physics

9.3   A consistent theory at hand?

As already mentioned, theory has a crucial role to play in this field because of the intrinsic complexity of the situations under study. Although many achievements have been attained over the years, in particular, with help of statistical and phenomenological dynamical models, a fully accepted and satisfying paradigm is not yet at work. And this deficiency could cause prejudice in this field. Before trying to envisage possible directions of investigation in this respect, let us briefly review the situation, again taking a historical perspective. To understand the collision of two nuclei means to solve a quantal many-body problem, which is known to be impossible as such, as soon as the total number of constituents becomes larger than four. One has thus to rely on approximate methods. Along the years, in the case of nuclear collisions, two lines of approximations have been developed, in which either the many-body or the quantal nature of the problem is lost.

In the first class of approximations one relies on the mean field, the power of which has been well established to describe ground-state nuclei. The mean field was first made time-dependent and extended to include two-body effects not accounted in the average mean field. During this process the original quantal problem has been further simplified to its semi-classical counterpart, and thus BUU now represents a typical tool of investigation. Still, as we know, one has to go even beyond BUU to account for fluctuations, which were proposed within the Boltzmann–Langevin approach, but little used in realistic cases. Nevertheless, the use of these kinetic equations (or their stochastic extensions) remains a matter of theoretical debate as a proper formal justification is not straigthforward, not to mention technical difficulties. The semi-classical approximation here bears a sizeable fraction of the responsibility as such approximations are only marginally valid for the collisions under study and remain hard to implement properly. As a consequence, these models have, to some extent, turned more and more towards phenomenology, and they have never succeeded in providing a truly reliable theoretical framework.

It is to be noted that the field suffered, particularly during the 1980s, from the inadequacy between computational capabilities and the problems to be described. Significant efforts were made in the early eighties to propose well-founded extensions of the mean field for describing nuclear collisions in the Fermi energy domain. But most of these attempts remained merely formal because of the impossibility of performing realistic calculations to test these theories. The great strength of BUU arose precisely from its numerical simplicity which allowed such realistic, although approximate, calculations. The situation is now quite different. The 'PC-boom' of the mid 1990s has provided software capable of testing these 'old' theories. It will be of great benefit to the field and, more generally speaking, to the theory of the many-body problem, to explore these directions properly.

In parallel to the development of the quantal mean field, the possibility of describing nuclear collisions at a purely classical level, by means of molecular dynamics (MD) methods was explored. This line of investigation gives priority to the many-body nature of the problem, at the price of losing quantal features. MD is well established in many fields of physics in which 'elementary' particles are either neutral atoms or inert molecules. In the nuclear case its justification is not so straightforward. It probably makes sense at very high beam energy, where quantal effects are indeed washed out, as was proved by the longstanding successes of cascade models. But in the nucleonic regime, the situation is much less forgiving in this respect. Early investigations of MD actually tried to account for the Pauli principle by means of a pseudo interaction but these attempts finally fell by the wayside. In turn, the mid 1980s saw the appearance of the so-called QMD which soon became a standard in the field, in spite of its intrinsic defects. The many calculations performed since then have shown several successes and failures, which are difficult to attribute to a clear theoretical aspect. Here too, phenomenology has probably also reached its limits.

In view of the structural defects of QMD, particularly with respect to the Pauli principle, since about 1990 an anti-symmetrized version of MD was developed. Looking at it in more detail in fact shows that the original FMD (or AMD) are nothing more than degraded versions of time-dependent mean-field theories. Thus they clearly point out again the unavoidable robustness of one-body theories in nuclear dynamics. Looking at it over the years, it turns out that FMD remains one of the few theoretical attempts in the field, which, while fulfilling basic nuclear requirements such as the Pauli principle, forming a basis for theoretical and computational developments, beyond the mean field, in particular, with the recent introduction of the correlator operator. The other promising direction is presumably provided by stochatic TDHF. Still, one should realize here that we are mainly back to, or a few steps forwards from the developments of the early 1980s around extended mean-field theories. This brings us back to our earlier conclusions, on the virtues of these investigations that were buried too soon. Thus, if there is one conclusion to be drawn here, it is certainly that the field should start to explore more systematically the many-body problem around the basis provided by the time-dependent mean field. This is probably the only possible way, outside exhausted phenomenology, to reach a coherent picture of nuclear dynamics in a wide range of beam energies in the nucleonic regime. A proper theoretical account of isospin also presumably relies on such well-founded theories, rather than on back-of-an-envelope models. Finally, it is important to note that such developments would also probably offer valuable openings towards other fields of physics in which strongly dynamical situations are also encountered.

9.4   Some future directions

9.4.1   Nuclear collisions in a 'large' N/Z range

From an experimental point of view, the advent of facilities capable of producing and accelerating radioactive beams with sizeable intensities in a large domain of the neutron-to-proton ratio (N/Z) is a revolution for the field of nuclear structure [333, 427, 458]. However, it is also good news for the study of excited nuclear matter.

First, collisions with radioactive beams can be used to study the degree of equilibration reached in the course of the reaction (as already demonstrated in [380]) but, in the present case, on a larger scale. To this end, during the interaction of two nuclei with initially very different N/Z values, the evolution of the corresponding 'isospin' degree of freedom is followed in the emitted products. According to the amount of stopping, the detected species should have different N/Z values depending on their emission angle or, equivalently, on their rapidity. Schematically, if nuclei are transparent or conversely if a rebound takes place in violent collisions, one expects to detect products which retain a memory of the initial N/Z asymmetry while in the case of complete mixing, (chemical) equilibrium is achieved. Then, similar N/Z values should be observed whatever the fragments and the detection angles. Second, the study of hot, equilibrated matter at various densities and with different N/Z ratios allows new regions of the phase diagram to be explored. For instance, it is expected that on increasing the N/Z ratio of the whole system, most neutrons should contribute to the gas phase rather than to the liquid phase (fragments).

From an experimental point of view, the use of multidetectors with good mass number resolution for intermediate mass fragments would certainly be a major improvement. For this purpose, it is necessary to detect and identify neutrons and fragments. These should be identified both in charge and mass which is not presently the case with the available detectors. It will hence be necessary to improve the existing 4π detectors in order to achieve both complete detection (large angular coverage, low thresholds) and complete charge *and* mass identification. The coupling of such a device with machines delivering new radioactive beams with sufficient intensities would allow us to pursue this new exploration of the nuclear-matter phase diagram.

9.4.2 Beyond specificity

Nowadays, interdisciplinarity has become a key word in the scientific community. Hence, we may wonder what could be the contribution of the study of nuclear collisions to other fields of physics. Here, we briefly describe two features that we think can cross-fertilize different fields of physics and in which nuclear collisions in the nucleonic regime can play a role.

9.4.2.1 'Species' production: what is common to hadronization and clusterization?

The production of species (we use this term here in a very generic manner: this can not only involve mesons, baryons, etc but also composite particles up to fragments) is the key process in which to learn about collision mechanisms. We should remember that this is true over a very large range of energies: from a few dozen MeV/u incident energy corresponding to the low-energy phenomena in the nucleonic regime discussed here up to several hundreds GeV e^+e^- collisions leading to the production of W's and Z_0's.

A striking and common feature of species production in such extremely different physical conditions is the incredible success met by the so-called thermal models. For instance, the hadronization process in e^+e^- collisions can be nicely described with help of Hagedorn's statistical bootstrap model [38]. Similar approaches have been used to reproduce particle production in ultra-relativistic nuclear reactions (see [438]) at the CERN-SPS (sulfur projectiles at 200 GeV/u) [95] and at BNL-AGS (silicon projectiles at 14.6 GeV/u) [94], in the relativistic regime at GSI-SIS (nickel projectiles between 0.8 and 2 GeV/u) [16] and (as discussed in section 7.3.2) in the nucleonic regime (argon projectiles at 95 MeV/u) [81]. Of course, in this last case, hadronization (coalescence of partons to produce hadrons) is replaced by clusterization (coalescence of nucleons to produce composites). However, it is amazing that both mechanisms can be explained with the help of simple phase-space arguments (namely the masses and spins of the considered species, the temperature and the 'freeze-out' volume).

9.4.2.2   Fragmentation: the essence of complexity?

Fragmentation is a process common to many physical systems. Fragmentation properties have been studied in a large variety of broken objects ranging from meteorites, terrestrial rocks, asteroids ([371] and references therein) up to more common structures such as plates [260] or gypsum structures [347]. All these systems exhibit universal scaling properties: in particular, fragment-size distributions show a power-law bahaviour. We have seen similar behaviour in section 8.4.3.5 in the nuclear context.

In a recent experiment, power-law behaviour has also been observed in the fragmentation of hydrogen-ion clusters (H_{25}^+) [179] colliding with carbon-60 fullerene molecules. A comparison in reduced units between the data from clusters and atomic nuclei is shown in [71]. A quite impressive similarity is obtained between the two systems although the physics involved in the two disassembly processes is very different. We believe that such comparisons are extremely useful for a full understanding of the (perhaps) underlying universal properties of fragmenting complex systems.

Chapter 10

Appendix

10.1 Units

- *Basic units*
  Energies are expressed in MeV, distances in fm and times in fm/c. This unit system is well suited to the physical situations considered here.
    - 1 MeV = 10^6 eV $\simeq 1.6 \times 10^{-13}$ J
    - 1 fm = 10^{-15} m
    - 1 fm/c $\simeq 10^{-15}$ m/$(3 \times 10^8$ m s$^{-1}) \simeq 3.3 \times 10^{-24}$ s
- *Fundamental constants*
    - $\hbar c \simeq 197$ MeV · fm
    - $\hbar^2/2m \simeq 20.73$ MeV · fm^2 (for a nucleon of mass $mc^2 \sim 940$ MeV)
    - $e^2/(4\pi\epsilon_0) = 1.44$ MeV · fm
- *Auxiliary units*
    - nucleon density $[\rho] \to$ fm^{-3}
    - temperature $[T] \to kT$ in MeV (with Boltzmann constant $k = 1$)
    - pressure $[P] \to$ MeV · fm^{-3}
    - cross-section $[\sigma] \to 1$ barn $= 100$ fm^2

10.2 Notation and conventions

Throughout this book we use the following notations:

- vectors and operators are denoted with bold characters (for example the Hamiltonian operator is denoted $\boldsymbol{H}$, and position $\boldsymbol{r}$)
- in formal developments we usually take $\hbar = 1$
- Boltzmann constant $k = 1$
- we use standard Dirac notation for kets, $|\Psi\rangle$; the corresponding wavefunction reads $\Psi(\boldsymbol{r}) = \langle r|\Psi\rangle$

Table 10.1. Notation.

Quantity	Notation
A-body antisymmetrization operator	$\mathcal{A}_{1\ldots A}$
A-body density matrix	$\boldsymbol{\rho}_{1\ldots A}$
A-body Hamiltonian	$H = \sum_{i=1}^{A} \boldsymbol{K}_i + \sum_{i<j} \boldsymbol{V}_{ij}$
A-body ket	$\lvert\Psi(1,\ldots,A)\rangle$
Beam energy per nucleon (lab. frame, MeV)	E_{lab}/u
Excitation energy (MeV)	E^*
Fermi energy (MeV)	ϵ_{F}
Incompressibility modulus (MeV)	K_∞
Kinetic energy operator	$\boldsymbol{K}_i$
Level density parameter (MeV^{-1})	a
Mean-field (one-body) potential	$\boldsymbol{U}_i$
Mean free path (fm)	λ
Neutron	n
Nucleon	N
Number of neutrons in a nucleus	N
Number of protons in a nucleus	Z
Number of nucleons in a nucleus	A
One-body density matrix in r space (fm^{-3})	$\rho(\boldsymbol{r}_1,\boldsymbol{r}_1') = \langle \boldsymbol{r}_1 \lvert \rho \rvert \boldsymbol{r}_1' \rangle$
One-body density of matter (fm^{-3})	$\varrho(\boldsymbol{r})$
One-body distribution function (phase space)	$f(\boldsymbol{r},\boldsymbol{p},t)$
One-body Hamiltonian	$\boldsymbol{h}_i = \boldsymbol{K}_i + \boldsymbol{U}_i$
One-body ket	$\lvert\phi(i)\rangle$
Pion	π
Proton	p
Two-body antisymmetrization operator	$\mathcal{A}_{12}$
Two-body density matrix	$\boldsymbol{\rho}_{12}$
Two-body interaction	$\boldsymbol{V}_{ij}$

10.3 Some basic relations

10.3.1 Properties of nuclei

10.3.1.1 *The nucleon–nucleon interaction*

- average nucleon–nucleon cross-section $\sigma \sim 40\,\mathrm{mb} \simeq 4\,\mathrm{fm}^2$
- typical range of the nucleon–nucleon interaction $d \sim 1\,\mathrm{fm}$
- hard core radius $r_{\mathrm{HC}} \sim 0.6\,\mathrm{fm}$
- typical mean free path in ground-state nuclei $\lambda \simeq 5\text{–}10\,\mathrm{fm}$

10.3.1.2   Ground-state nuclei

- *Gross properties*
  - nuclear radius $R \simeq r_0 A^{1/3}$ with $r_0 \simeq 1.12$ fm so that $R \sim 2\text{--}8$ fm
  - surface width $\sim 1\text{--}2$ fm
  - typical nuclear binding energy $E/A \simeq -8$ MeV
  - typical de Broglie wavelength in ground-state nuclei $\lambda_\mathrm{B} \simeq 5$ fm
- *Nuclear potential well*
  - typical nuclear potential depth $V \simeq -50$ MeV
  - typical nuclear chemical potential $\mu \simeq -8$ MeV
  - typical Fermi energy $\epsilon_\mathrm{F} \simeq 40$ MeV
- *Fermi gas properties (zero temperature, spin–isospin degenerate)*
  - Density

$$\rho = \frac{2}{3\pi^2} k_\mathrm{F}^3$$

  - Fermi energy

$$\epsilon_\mathrm{F} = \frac{\hbar^2}{2m} k_\mathrm{F}^2$$

  - values for saturation density $\rho = 0.17$ fm^{-3}

$$k_\mathrm{F} \simeq 1.36 \text{ fm}^{-1} \qquad \epsilon_\mathrm{F} \simeq 38 \text{ MeV}.$$

10.3.1.3   Thermodynamics of nuclei

- relation between thermal excitation energy and temperature (low-temperature limit)
$$E^* \simeq aT^2$$
- typical value of level density parameter (from experiments, overlooking shell effects) $a \simeq A/8$
- relation between entropy and temperature (low temperature limit)

$$S \simeq 2aT.$$

10.3.1.4   Low-energy dynamics

- *Giant resonances*
  - parametrization of frequencies

$$\hbar\omega_{l,s} = \kappa_{l,s} A^{-1/3}$$

for $l = 0, 1$ (Steinwedel–Jensen model) and $l = 2, 3, 4$ (isoscalar $s = 0$, isovector $s = 1$) or $\hbar\omega_{l,s} = \kappa_{l,s} A^{-1/6}$ for $l = 1$ and $s = 1$ (Goldhaber–Teller model)

Table 10.2. Some giant resonances energies and widths.

Multipolarity l	s	Energy (in $A^{-1/3}$) MeV	Width (in MeV)
0	0	80	2–4
1	1	78	4–8
2	0	64	2–8
3	0	30 and 120	

- typical giant resonance period $\tau_{\mathrm{GR}} \sim 50\text{--}150$ fm/c
- *Fission*
  - fissility parameter

$$x = \frac{E_{\mathrm{Coul}}(\text{sphere})}{2E_{\mathrm{surf}}(\text{sphere})} = \frac{Z^2/A}{(Z^2/A)_{\mathrm{crit}}}$$

  with $(Z^2/A)_{\mathrm{crit}} \simeq 50$
  - typical fission barrier height:
    heavy nuclei $B_{\mathrm{F}} \simeq 5$ MeV, light nuclei $B_{\mathrm{F}} \simeq 40$ MeV
  - typical fission time $\tau_{\mathrm{F}} \simeq 10^{-21}\text{--}10^{-20}$ s

10.3.2   The nucleonic equation of state

- *Saturation point (symmetric nuclear matter)*
  - saturation density $\rho_0 \simeq 0.17$ fm^{-3}
  - saturation energy $E/A_0 \simeq -16$ MeV
  - incompressibility modulus $K_\infty \simeq 220$ MeV
  - parabolic form of the nuclear matter equation of state in the vicinity of saturation point

$$E/A(\rho) \simeq E/A_0 + \frac{K_\infty}{18\rho_0^2}(\rho - \rho_0)^2$$

- *Thermodynamical properties (symmetric nuclear matter)*
  - typical range of nucleonic physics:

$$0 \le \rho \lesssim 1.5\rho_0 \quad \text{and} \quad 0 \le T \lesssim 20 \text{ MeV}$$

  - spinodale density (zero temperature) $\rho_s \simeq 2/3\rho_0 \simeq 0.10$ fm^{-3}
  - critical density (liquid–gas transition) $\rho_c \simeq 1/3\rho_0 \simeq 0.06\text{--}0.07$ fm^{-3}
  - critical temperature (liquid–gas transition) $T_c \simeq 16\text{--}18$ MeV

10.3.3 Kinematics and cross-sections

- *Available centre-of-mass energy*

Nuclear collisions in the Fermi energy range consist of bombarding a projectile on a fixed target nucleus. Hence part of the incident energy is spent as recoil energy of the centre-of-mass. Thus, the available centre-of-mass (CM) energy, i.e. the energy available for the collision E_{CM} is only a fraction of the laboratory energy of the projectile:

$$E_{CM} = E_{lab} \frac{A_t}{A_p + A_t}. \qquad (10.1)$$

This expression is a non-relativistic approximation; E_{lab} is the beam energy, $A_p(A_t)$ the projectile (target) mass number. The centre-of-mass energy per nucleon writes as:

$$\epsilon_{CM} = \frac{E_{CM}}{A_p + A_t} = \frac{E_{lab}}{A_p} \frac{A_p A_t}{(A_p + A_t)^2} = \epsilon_{lab} \frac{A_p A_t}{(A_p + A_t)^2}. \qquad (10.2)$$

The corresponding centre-of-mass velocity v_{CM} can be written as

$$v_{CM} = v_{lab} \frac{A_p}{A_p + A_t} \qquad (10.3)$$

where v_{lab} is the beam velocity in the laboratory frame. In the non-relativistic approximatiom, the conversion from the centre-of-mass frame to the laboratory frame is realized using the law of vectorial addition of the velocities: $v_{lab} = v^* + v_{CM}$ where v^* is the velocity of a given particle in the centre-of-mass frame, v_{lab} the corresponding velocity in the laboratory frame.

- *Direct and reverse kinematics*

The expression 'direct' ('inverse') kinematics is used when the projectile mass is lower (larger) than the target mass. From equation (10.2), the CM energy per nucleon has the same value for direct and reverse kinematics for a fixed incident energy per nucleon, ϵ_{lab}.

- *Invariant cross-sections*

The quantity $E\frac{d^3\sigma}{d^3p}$ (where E is the total energy of the particle) is Lorentz invariant. In the non-relativistic case, $E \simeq m$ where m is the mass of the particle, thus, $\frac{d^3\sigma}{d^3v}$ is Galilean invariant. From $d^3v = d\Omega\, v_\perp\, dv_\perp\, dv_\parallel$, isocontours in the plane $v_\perp - v_\parallel$ can be drawn as in section 4.2.1. From $E^2 = p^2 + m^2$, we have $E\,dE = p\,dp$ and thus $\frac{1}{p}\frac{d^2\sigma}{dE\,d\Omega}$ is an invariant that can be used to plot particles' kinetic energy distributions whatever the considered frame.

- *Rapidity*

Instead of $v_{\parallel}$, the rapidity y is sometimes used. This is defined by $y = \frac{1}{2}\ln[\frac{E+p_{\parallel}}{E-p_{\parallel}}]$ where $p_{\parallel}$ is the linear momentum parallel to the beam. From $\beta_{\parallel} = \frac{p_{\parallel}}{E}$, one gets: $y = \frac{1}{2}\ln[\frac{1+\beta_{\parallel}}{1-\beta_{\parallel}}]$.

In the non-relativistic approximation, one obtains: $y \simeq \beta_{\parallel}$. The rapidity is used instead of the velocity in the relativistic domain because of the convenient following additive property: $y_{\mathrm{lab}} = y^* + y_{\mathrm{CM}}$ in which y^* is the rapidity of the considered particle in the centre-of-mass frame, y_{lab} the rapidity in the laboratory frame and y_{CM} is the rapidity of the centre-of-mass in the laboratory frame. Of course, this property (which is equivalent to the additive property of velocities in the non-relativistic approximation) is true whatever the frames considered.

10.4   Abbreviations and acronyms

Table 10.3. Acronyms.

Acronym	Explicit name
AMD	Antisymmetrized molecular dynamics
BL	Boltzmann–Langevin
BNV	Boltzmann–Nordheim–Vlasov
BUU	Boltzmann–Uehling–Uhlenbeck
DIC	Deep inelastic collision
ECR	Electron cyclotron radiation
EES	Expanding emitting source
EoS	Equation of state
ER	Evaporation residue
FMD	Fermionic molecular dynamics
GDR	Giant dipole resonance
GR	Giant resonance
IMF	Intermediate mass fragment
INC	Intra-nuclear cascade
IQMD	Isospin quantum molecular dynamics
LCP	Light-charged particles
LDM	Liquid drop model
LV	Landau–Vlasov

Table 10.3. (Continued)

Acronym	Explicit name
MD	Molecular dynamics
NFD	Nuclear fluid dynamics
OBEP	One boson exchange potential
OPEP	One pion exchange potential
PLF	Projectile-like fragment
QGP	Quark gluon plasma
QMD	Quantum molecular dynamics
QP	Quasi-projectile
QSM	Quantum statistical model
QT	Quasi-Target
RQMD	Relativistic quantum molecular dynamics
SMM	Statistical multifragmentation model
STDHF	Stochastic time-dependent Hartree–Fock
TDHF	Time-dependent Hartree–Fock
TKE	Total kinetic energy
TLF	Target-like fragment
ToF	Time of flight
TPC	Time projection chamber
VUU	Vlasov–Uehling–Uhlenbeck

References

[1] Abe Y *et al* 1996 *Phys. Rep.* **275** 49
[2] Abel W R, Anderson A C and Wheatley J C 1966 *Phys. Rev. Lett.* **17** 74
[3] Aichelin J and Stöcker H 1986 *Phys. Lett.* B **176** 14
     Aichelin J 1986 *Phys. Rev.* C **33** 537
[4] Aichelin J *et al* 1988 *Phys. Rev.* C **37** 2451
[5] Aichelin J 1991 *Phys. Rep.* **202** 233
[6] Aiello S *et al* 1995 *Nucl. Phys.* A **583** 461c
[7] Alamanos N *et al* 1985 *Phys. Lett.* B **151** 100
[8] Alard J P *et al* 1987 *Nucl. Instrum. Methods* **261** 379
[9] Alard J P *et al* 1992 *Phys. Rev. Lett.* **69** 889
[10] Alard J P *et al* 1994 *Proc. XXXII Winter Meeting (Bormio)* ed I Iori
[11] Albergo S *et al* 1985 *Nuovo Cimento* A **89** 1
[12] Alhassid Y, Zingmann J and Levit S 1987 *Nucl. Phys.* A **469** 205
[13] Alton G D 1995 *Int. Conf. on Cyclotrons (Cape Town)*
[14] Anastasio M R *et al* 1983 *Phys. Rep.* **100** 327
[15] Ardouin D 1997 *Int. J. Mod. Phys.* E **6** 391
[16] Averbeck R *et al* 1997 *GSI Scientific Report* 97-1, p 58
[17] Ayik S and Grégoire Ch 1988 *Phys. Lett.* B **212** 269
[18] Ayik S and Grégoire Ch 1990 *Nucl. Phys.* A **513** 187
[19] Bacri C O, Chomaz Ph and Vautherin D 1988 *Phys. Rev. Lett.* **61** 1569
[20] Bacri C O *et al* 1995 *Phys. Lett.* B **353** 27
[21] Badala A *et al* 1995 *Phys. Rev. Lett.* **74** 4779
[22] Baden A *et al* 1982 *Nucl. Instrum. Methods* **203** 189
[23] Baldo M (ed) 1999 *Nuclear Methods and the Nuclear Equation of State (Int. Rev. Of Nucl. Physics, Vol 8)* (Singapore: World Scientific)
[24] Baldwin S P *et al* 1995 *Phys. Rev. Lett.* **74** 1299
[25] Balescu R 1975 *Nonequilibrium Statistical Mechanics* (New York: Wiley)
[26] Balian R and Vénéroni M 1981 *Ann. Phys., NY* **135** 270
[27] Balian R, Alhassid Y and Reinhardt H 1986 *Phys. Rep.* **131** 1
[28] Balian R 1991 *From Microphysics to Macrophysics* (Berlin: Springer)
[29] Balian R 1996 *Ann. Phys., Paris* **21** 437
[30] Bao-An Li *et al* 1997 *Phys. Rev. Lett.* **76** 4492
[31] Bartel J *et al* 1982 *Nucl. Phys.* A **386** 79
[32] Bauer M *et al* 1982 *J. Phys. G: Nucl. Phys.* **8** 525
[33] Bauer W *et al* 1985 *Phys. Lett.* B **150** 536
[34] Bauer W 1988 *Phys. Rev. Lett.* **61** 2534

[35] Bauer W, Gelbke K and Pratt S 1992 *Ann. Rev. Nucl. Sci.* **42** 77
[36] Bauer W and Pratt S 1999 *Phys. Rev.* C **59** 2695
[37] Baym G and Pethick C 1976 *The Physics of Liquid and Solid Helium* Part II, ed
       Bennemann and Ketterson (New York: Wiley)
[38] Becattini F 1996 *Z. Phys.* C **69** 485
[39] Behkami A N and Najafi S I 1980 *J. Physique* **6** 685
[40] Belkacem M, Suraud E and Ayik S 1993 *Phys. Rev.* C **47** R16
[41] Benenson W, Morissey D and Friedman W 1994 *Ann. Rev. Nucl. Sci.* **44** 27
[42] 1989 *Int. Conf. on Fifty Years of Nuclear Fission* (Berlin)
       1989 *Nucl. Phys.* A **502** 1c
[43] Bertsch G F and Siemens Ph 1983 *Phys. Lett.* B **126** 9
[44] Bertsch G F, Lynch W G and Tsang B M 1987 *Phys. Lett.* B **189** 384
[45] Bertsch G and Das Gupta S 1988 *Phys. Rep.* **160** 189
[46] Besprosvany J and Levit S 1988 *Phys. Lett.* B **217** 1
[47] Bethe H A 1937 *Rev. Mod. Phys.* **9** 69
[48] Bethe H A 1971 *Ann. Rev. Nucl. Sci.* **21** 93
[49] Bethe H A 1988 *Ann. Rev. Nucl. Part. Sci.* **38** 1
[50] Binder K and Stauffer D 1976 *Adv. Phys.* **25** 343
[51] Binney J J *et al* 1992 *The Theory of Critical Phenomena* (Oxford: Clarendon)
[52] Bixon M and Zwanzig R 1967 *Phys. Rev.* **187** 267
[53] Bizard G *et al* 1986 *Nucl. Instrum. Methods* A **244** 489
[54] Bizard G *et al* 1991 *Nuclear Physics News* **1** 15
[55] Bizard G *et al* 1993 *Phys. Lett.* B **302** 162
[56] Blaich Th *et al* 1992 *Nucl. Instrum. Methods* A **314** 136
[57] Blaizot J P 1980 *Phys. Rep.* **64** 171
[58] Blaizot J P 1999 *Nucl. Phys.* A **649** 61c
[59] Blann M 1975 *Ann. Rev. Nucl Sci.* **25** 123
[60] Blann M and Chadwick M B 1998 *Phys. Rev.* C **57** 233
[61] Blin A H *et al* 1986 *Phys. Lett.* B **182** 239
[62] Blocki J *et al* 1977 *Ann. Phys., NY* **105** 427
[63] Bocage F *et al* 1999 *Proc. XXXVI Int. Winter Meeting (Bormio)* ed I Iori
[64] Bocage F *et al Nucl. Phys.* at press
[65] Bohigas O 1979 *Phys. Rep.* **51** 267
[66] Bohr N 1936 *Nature* **137** 351
[67] Bohr N and Wheeler J A 1939 *Phys. Rev.* **56** 426
[68] Bohr A and Mottelson B 1969 *Nuclear Structure* vol I (New York: Benjamin)
[69] Bohr A and Mottelson B 1975 *Nuclear Structure* vol II (New York: Benjamin)
[70] Bonasera A, Gulminelli F and Molitoris J 1994 *Phys. Rep.* **243** 1
[71] Bonasera A 1999 *Physics World* **12** 20
[72] Bonche P, Koonin S and Negele W G 1976 *Phys. Rev.* C **13** 1010
[73] Bonche P, Levit S and Vautherin D 1984 *Nucl. Phys.* A **427** 278
[74] Bonche P, Levit S and Vautherin D 1985 *Nucl. Phys.* A **436** 265
[75] Bonche P 1985 *Proc. 'Ecole Joliot Curie'*
[76] Bonche P and Levit S 1985 *Nucl. Phys.* A **437** 426
[77] Bondorf J P 1978 *Nucl. Phys.* A **296** 320
[78] Bondorf J P *et al* 1995 *Phys. Rep.* **257**
[79] Bondorf J P *et al* 1995 *Phys. Lett.* B **359** 261
[80] Borderie B *et al* 1996 *Phys. Lett.* B **388** 224

[81]   Borderie B *et al* 1999 *Eur. Phys. J.* A **6** 197

[82]   Borderie B *et al* 2000 *Proc. XXXVIII Int. Winter Meeting (Bormio)* ed I Iori and A Moroni

[83]   Bortignon P F and Dasso C H 1987 *Phys. Lett.* B **189** 381

[84]   Botermans W and Malfliet R 1986 *Phys. Lett.* B **171** 22

[85]   Botermans W *et al* 1987 *J. Physique Coll.* C **2** 287

[86]   Botvina A S and Gross D H E 1995 *Phys. Lett.* B **344** 6

[87]   Bougault R *et al* 1987 *Nucl. Instrum. Methods* A **259** 473

[88]   Bougault R *et al* 1995 *Nucl. Phys.* A **587** 499

[89]   Bougault R *et al* 1999 *Proc. XXVII Winter Meeting (Hirshegg)* ed H Feldmeier *et al*

[90]   Bougault R *et al* 2000 *Proc. XXXVIII Int. Winter Meeting (Bormio)* ed I Iori

[91]   Brack M 1985 *et al Phys. Rep.* **123** 275

[92]   Brack M and Bhaduri R K 1997 *Semi Classical Physics* (New York: Addison-Wesley)

[93]   Braun-Munzinger P and Stachel J 1987 *Ann. Rev. Nuc. Part. Sci.* **37** 97

[94]   Braun-Munzinger P *et al* 1995 *Phys. Lett.* B **344** 43

[95]   Braun-Munzinger P *et al* 1996 *Phys. Lett.* B **365** 1

[96]   Bresson S 1993 *PhD Thesis, Preprint* Ganil T 93-02

[97]   Brink D M 1985 *Semi-Classical Methods in Nucleus–Nucleus Scattering* (Cambridge: Cambridge University Press)

[98]   Brown G E and Jackson A D 1976 *The Nucleon–Nucleon Interaction* (Amsterdam: North-Holland)

[99]   Brown G E (ed) 1988 Theory of supernovae *Phys. Rep.* **163** 171

[100] Brown G E 1988 *Z. Phys.* C **38** 291

[101] Brückner K A, Levinson C A and Mahmoud H M 1954 *Phys. Rev.* **95** 217

[102] Brückner K A and Levinson C A 1955 *Phys. Rev.* **97** 1344

[103] Brückner K A 1955 *Phys. Rev.* **97** 1353

[104] Burgio G F, Chomaz Ph and Randrup J 1991 *Nucl. Phys.* A **529** 57

[105] Burgio G F, Chomaz Ph and Randrup J 1992 *Phys. Rev. Lett.* **69** 885

[106] Buta A *et al* 1995 *Nucl. Phys.* A **584** 397

[107] Cagin T and Pettitt M 1991 *Mol. Phys.* **5** 72

[108] Callaway D J E, Wilets L and Yariv Y 1979 *Nucl. Phys.* A **327** 250

[109] Campi X 1986 *J. Phys. A: Math. Gen.* **19** L917

[110] Campi X 1988 *Phys. Lett.* B **208** 351

[111] Campi X *et al* 1996 *Phys. Lett.* B **385** 1

[112] Campi X and Krivine H 1996 *Trends in Nuclear Physics, 100 Years Later (Les Houches, Session LXVI)* ed H Nifenecker *et al*

[113] Campi X and Krivine H 1997 *Nucl. Phys.* A **620** 46

[114] *Proc. 14th Int. Conf. on Cyclotrons (Cape Town)*

[115] Car R and Parinello M 1985 *Phys. Rev. Lett.* **55** 2471

[116] Casandjian J M *et al* 1998 *Phys. Lett.* B **430** 43

[117] Casini G *et al* 1991 *Phys. Rev. Lett.* **67** 3364

[118] Cassing W and Mosel U 1990 *Prog. Part. Nucl. Phys.* **25** 235

[119] Cassing W *et al* 1990 *Phys. Rep.* **188** 363

[120] Charity R J *et al* 1990 *Nucl. Phys.* A **511** 59

[121] Charity R J *et al* 1991 *Z. Phys.* A **341** 53

[122] Chandler D 1987 *Introduction to Modern Statistical Mechanics* (New York: Oxford University Press)

[123] Chomaz Ph, Burgio G F and Randrup J 1991 *Phys. Lett.* B **254** 340
[124] Chomaz Ph 1996 *Ann. Phys., Paris* **21** 669
[125] Chomaz Ph and Gulminelli F 1999 *Nucl. Phys.* A **647** 153
[126] Chomaz Ph, Gulminelli F and Duflot V *Phys. Rev. Lett.* submitted
[127] Coester F *et al* 1970 *Phys. Rev.* C **1** 769
[128] Colin E *et al* 1998 *Phys. Rev.* C **57** R1032
[129] Colonna M *et al* 1992 *Prog. Part. Nucl. Phys.* **30** 17
[130] Colonna M *et al* 1995 *Nucl. Phys.* A **583** 525c
[131] Colonna N *et al* 1995 *Phys. Rev. Lett.* **75** 4190
[132] Cooperstein J 1988 *Phys. Rev.* C **37** 786
[133] Cortès E, West B J and Lindenberg K 1985 *J. Chem. Phys.* **82** 2708
[134] Cottingham W A and Greenwood D A *An Introduction to Nuclear Physics* (Cambridge: Cambridge University Press)
[135] Crema E *et al* 1991 *Phys. Lett.* B **258** 266
[136] Csernai L P and Barz H W 1981 *Z. Phys.* A **296** 173
[137] Cugnon J, Mizutani T and Vandermeulen J 1981 *Nucl. Phys.* A **352** 505
[138] Cugnon J 1984 *Phys. Lett.* B **135** 374
[139] Cugnon J 1986 *Ann. Phys., Paris* **11** 201
[140] Cugnon J, Deneye P and Lejeune A 1987 *Z. Phys.* A **328** 409
[141] Cugnon J 1996 *Ann. Phys., Paris* **21** 537
[142] Cussol D *et al* 1993 *Nucl. Phys.* A **561** 298
[143] D'Agostino M *et al* 1996 *Phys. Lett.* B **368** 259
[144] D'Agostino M *et al* 1999 *Nucl. Phys.* A **650** 329
[145] D'Agostino M *et al* 2000 *Phys. Lett.* B **473** 219
[146] Danielewicz P 1984 *Ann. Phys., NY* **152** 239
[147] Danielewicz P and Odyniec G 1985 *Phys. Lett.* B **157** 146
[148] Danielewicz P 1995 *Phys. Rev.* C **51** 716
[149] Das Gupta S and Lam C 1979 *Phys. Rev.* C **20** 1192
[150] Deak F *et al* 1995 *Phys. Rev.* C **52** 219
[151] De Angelis A R, Gross D H E and Heck R 1992 *Nucl. Phys.* A **537** 606
[152] de la Mota V *et al* 1992 *Phys. Rev.* C **46** 677
[153] Demoulins M *et al* 1990 *Phys. Lett.* B **241** 476
[154] Dempsey J F *et al* 1996 *Phys. Rev.* C **54** 1710
[155] deShalit A and Feschbach H 1974 *Theoretical Nuclear Physics. Nuclear Structure* (New York: Wiley–Interscience)
[156] Desesquelles P 1995 *Ann. Phys., Paris* **20** 1
[157] De Souza R T *et al* 1991 *Nucl. Instrum. Methods* A **295** 109
[158] De Souza R T *et al* 1991 *Phys. Lett.* B **268** 6
[159] Dirac P A M 1930 *Proc. Camb, Phil. Soc.* **26** 376
[160] Donangelo R *et al* 1989 *Phys. Lett.* B **219** 165
[161] Donangelo R and Souza S R 1997 *Phys. Rev.* C **56** 1504
[162] Donangelo R and Souza S R 1998 *Phys. Rev.* C **58** R2662
[163] Dorso C and Randrup J 1987 *Phys. Lett.* B **188** 287
[164] Dorso C and Randrup J 1988 *Phys. Lett.* B **215** 611
[165] Dorso C and Randrup J 1989 *Phys. Lett.* B **232** 29
[166] Dorso C and Randrup J 1993 *Phys. Lett.* B **301** 328
[167] Dorvaux O *et al* 1999 *Nucl. Phys.* A **651** 225
[168] Drain D *et al* 1989 *Nucl. Instrum. Methods* A **281** 528

[169] Drees A 1998 *Nucl. Phys.* A **630** 449c
[170] Durand D 1992 *Nucl. Phys.* A **451** 266
[171] Durand D 1998 *Nucl. Phys.* A **630** 52c
[172] Durand D 1999 *Nucl. Phys.* A **654** 273c
[173] Egido J L *et al* 1993 *J. Phys. G: Nucl. Part. Phys.* **19** 1
[174] Elliott J B *et al* 1996 *Phys. Lett.* B **382** 35
[175] Ericson T 1960 *Adv. Phys.* **9** 425
[176] Ericson T 1963 *Ann. Phys., NY* **23** 390
[177] Ericson T and Mayer-Tuckuck T 1966 *Ann. Rev. Nucl. Sci.* **16** 183
[178] Fai G and Randrup J 1986 *Comput. Phys. Commun.* **42** 385
[179] Farizon B *et al* 1998 *Phys. Rev. Lett.* **81** 4108
[180] Fatyga M *et al* 1985 *Phys. Rev. Lett.* **55** 1376
[181] Feldmeier H 1990 *Nucl. Phys.* A **515** 147
[182] Feldmeier H, Bieler K and Schnack J 1995 *Nucl. Phys.* A **586** 493
[183] Feldmeier H and Schnack J 1997 *Prog. Nucl. Part. Phys.* **39** 1
[184] Feldmeier H *et al* 1999 GSI-Nachrichten
[185] Feshbach H 1992 *Theoretical Nuclear Physics. Nuclear Reactions* (New York: Wiley)
[186] Fetter A L and Walecka J D 1971 *Quantum Theory of Many Body Systems* (New York: McGraw-Hill)
[187] Finn J E *et al* 1982 *Phys. Rev. Lett.* **49** 1321
[188] Fisher M E 1967 *Rep. Prog. Phys.* **67** 615
[189] Fong P 1956 *Phys. Rev.* **102** 434
[190] Fox G C and Wolfram S 1978 *Phys. Rev. Lett.* **41** 1581
[191] Frankland J D 1998 *Proc. XXXVI Int. Winter Meeting (Bormio)*
[192] Friedman W and Pandharipande V R 1981 *Nucl. Phys.* A **361** 502
[193] Friedman W and Lynch W G 1983 *Phys. Rev.* C **28** 16
[194] Friedman W 1990 *Phys. Rev.* C **42** 667
[195] Fröbrich P and Gontchar I I 1998 *Phys. Rep.* **292** 131
[196] Frois B and Papanicolas C N 1987 *Ann. Rev. Nuc. Part. Sci.* **37** 133
[197] Gaarhoje J J 1992 *Ann. Rev. Nucl. Part. Sci.* **42** 483
[198] Gaddioli E and Hodgson P E 1991 *Pre-Equilibrium Nuclear Reactions* (New York: Oxford University Press)
[199] Gale C *et al* 1987 *Phys. Rev.* C **35** 1666
[200] Galin J and Jahnke U 1994 *J. Phys. G: Nucl. Part. Phys.* **20** 1105
[201] Gallardo M *et al* 1985 *Nucl. Phys.* A **443** 415
[202] Garcias F *et al* 1990 *Z. Phys.* A **337** 261
[203] Gardiner C W 1990 *Handbook of Stochastic Methods* (Berlin: Springer)
[204] Ghetti R 1997 *Phys. Rev.* C **56** 1651
[205] Gibson B F and MacKellar B H J 1988 *Few Body Syst.* **3** 143
[206] Gilkes M L *et al* 1994 *Phys. Rev. Lett.* **73** 1590
[207] Glendenning N 1988 *Phys. Rev.* C **37** 2733
[208] Gobbi A *et al* 1993 *Nucl. Instrum. Methods* **324** 156
[209] Goeke K and Reinhard P G (ed) 1992 *TDHF and Beyond (Lecture Notes in Physics, vol 171)* (Berlin: Springer)
[210] Gogny D 1975 *Nucl. Phys.* A **237** 399
[211] Goldenbaum F *et al* 1996 *Phys. Rev. Lett.* **77** 1230
[212] Goldenbaum F *et al* 1999 *Phys. Rev. Lett.* **82** 5012

[213] Goldstone J 1957 *Proc. R. Soc.* A **239** 267
[214] Gossiaux P B *et al* 1995 *Phys. Rev.* C **51** 3357
[215] Gossiaux P B and Aichelin J 1997 *Phys. Rev.* C **56** 2109
[216] Gosset J, Kapusta J I and Westfall G 1978 *Phys. Rev.* C **18** 844
[217] Grangé P and Weidenmüller H A 1980 *Phys. Lett.* B **96** 26
      Grangé P, Pauli H C and Weidenmüller H A 1980 *Z. Phys.* A **296** 107
[218] Grangé P 1984 *Nucl. Phys.* A **428** 37c
[219] Greiner C, Wagner K and Reinhard P-G 1994 *Phys. Rev.* C **49** 1693
[220] 1996 Proc. Groningen Conf. on Giant Resonances *Nucl. Phys.* A **599** 1c
[221] Gross D H E 1990 *Rep. Prog. Phys.* **59** 605
[222] Gross D H E *et al* 1997 *Phys. Rep.* **279** 119
[223] Guarnera A *et al* 1996 *Phys. Lett.* B **373** 267
[224] Guarnera A *et al* 1997 *Phys. Lett.* B **403** 191
[225] Guet C *et al* 1988 *Phys. Lett.* B **205** 427
[226] Gulminelli F and Durand D 1997 *Nucl. Phys.* A **615** 117
[227] Gulminelli F and Chomaz Ph 1999 *Phys. Rev. Lett.* **82** 1402
      Gulminelli F and Chomaz Ph 1999 *Phys. Lett.* B **447** 221
[228] Gutbrod H H *et al* 1989 *Rep. Prog. Phys.* **52** 1267
      Gutbrod H H *et al* 1989 *Phys. Lett.* B **216** 267
[229] Gutbrod H H *et al* 1990 *Phys. Rev.* C **42** 640
[230] Haddad F *et al* 1996 *Phys. Rev.* C **53** 1437
[231] Hagel K *et al* 1988 *Nucl. Phys.* A **486** 429
[232] Hahn D and Stöcker H 1988 *Nucl. Phys.* A **476** 719
[233] Hanbury-Brown R and Twiss R Q 1954 *Phil. Mag.* **45** 663
[234] Hansen J P and McDonald I R 1976 *Theory of Simple Liquids* (London: Academic)
[235] Hartnack C *et al* 1989 *Nucl. Phys.* A **495** 303
[236] Hartnack C *et al* 1998 *Eur. J. Phys.* A **1** 151
[237] Hassani S and Grangé P 1984 *Phys. Lett.* B **137** 281
[238] Hasse R W and Schuck P 1986 *Phys. Lett.* B **179** 313
[239] Hauger J A *et al* 1998 *Phys. Rev.* C **57** 764
[240] Haxel O, Jensen J H D and Suess H E 1949 *Phys. Rev.* **75** 1354
[241] Heiselberg H, Pethick Ch J and Ravenhall D G 1988 *Phys. Rev. Lett.* **61** 818
[242] Hilscher D and Rossner H 1992 *Ann. Phys., Paris* **17** 471
[243] Hinde D J *et al* 1986 *Nucl. Phys.* A **452** 550
[244] Hinde D J *et al* 1992 *Phys. Rev.* C **45** 1229
[245] Holub E *et al* 1986 *Phys. Rev.* C **33** 143
[246] Hongfei Xi *et al* 1999 *Phys. Rev.* C **59** 1567
[247] Hoover W G 1985 *Phys. Rev.* A **31** 1685
[248] Hoover W G 1991 *Computational Statistical Mechanics* (Amsterdam: Elsevier)
[249] Huang K 1963 *Statistical Mechanics* (New York: Wiley)
[250] Hubele J *et al* 1991 *Z. Phys.* A **340** 263
[251] Hüfner J and Knoll J 1977 *Nucl. Phys.* A **290** 460
[252] Husimi K 1940 *Prog. Phys. Math. Soc. Jap.* **22** 264
[253] Imme G *et al* 1996 *Preprint* GSI 96-30
[254] Iori I *et al* 1993 *Nucl. Instrum. Methods* A **325** 458
[255] Jaqaman H, Mekjian A Z and Zamick L 1983 *Phys. Rev.* C **27** 2782
[256] Jarzynski C and Bertsch G F 1996 *Phys. Rev.* C **53** 1028
[257] Johnston H *et al* 1996 *Phys. Lett.* B **371** 186

[258] Jouan D *et al* 1991 *Z. Phys.* A **340** 63
[259] Julien J *et al* 1991 *Phys. Lett.* B **264** 269
[260] Kadono T 1997 *Phys. Rev. Lett.* **78** 1444
[261] Kahana S H 1989 *Ann. Rev. Nucl. Part. Sci.* **39** 231
[262] Kirschbaum C L and Wilets L 1980 *Phys. Rev.* A **21** 834
[263] Klakow D, Toepffer C and Reinhard P G 1994 *Phys. Lett.* A **192** 55
[264] Knoll G F 1979 *Radiation Detection and Measurement* (New York: Wiley)
[265] Knoll J and Wu J 1988 *Nucl. Phys.* A **481** 173
[266] Köhler H S and Nilsson B S 1988 *Nucl. Phys.* A **477** 318
[267] Köhler H S 1989 *Nucl. Phys.* A **494** 281
[268] Köhler H S 1991 *Nucl. Phys.* A **529** 209
[269] Konopka J *et al* 1994 *Phys. Rev.* C **50** 2085
[270] Koonin S and Randrup J 1987 *Nucl. Phys.* A **474** 173
[271] Kotov A *et al* 1995 *Nucl. Phys.* A **583** 575c
[272] Kramers H A 1940 *Physica* VII **4** 284
[273] Kreutz P *et al* 1993 *Nucl. Phys.* A **556** 672
[274] Krivine H, Treiner J and Bohigas O 1980 *Nucl. Phys.* A **336** 155
[275] Kubo R, Toda M and Hashitsume N 1985 *Statistical Physics* vols 1 and 2 (Heidelberg: Springer)
[276] Kwiatkowski K *et al* 1995 *Nucl. Instrum. Methods* A **360** 571
[277] Landau L and Lifshitz E M 1959 *Quantum Mechanics* (London: Pergamon)
[278] Langevin P 1908 *C. R. Acad. Sci., Paris* **146** 530
[279] Lassaut M *et al* 1987 *Astron. Astrophys.* **183** L3
[280] Lecolley J F *et al* 1994 *Phys. Lett.* B **325** 317
[281] Lecolley J F *et al* 1995 *Phys. Lett.* B **354** 202
[282] Lecolley J F *et al* 1996 *Phys. Lett.* B **387** 460
[283] Le Fèvre A *et al* 1999 *Proc. XXVII Winter Meeting (Hirshegg)* ed H Feldmeier *et al*
[284] Lefort T *et al* 1999 *Proc. XXXVI Int. Winter Meeting (Bormio)* ed I Iori
[285] 1988 First topical meeting on giant resonance in heavy-ion collisions *Nucl. Phys.* A **482** 1c
[286] Legrain R *et al* 1999 *Phys. Rev.* C **59** 1464
[287] Lemmon R C *et al* 1999 *Phys. Lett.* B **446** 197
[288] Lenk R J, Schlagel T J and Pandharipande V R 1990 *Phys. Rev.* C **42** 372
[289] L'Eplattenier P *et al* 1995 *Ann. Phys., NY* **244** 426
[290] Leray S 1986 *J. Physique* **47** C4-275
[291] L'Hote D and Cugnon J 1983 *Nucl. Phys.* A **397** 519
[292] Lopez J and Randrup J 1990 *Nucl. Phys.* A **512** 345
[293] Lopez O *et al* 1993 *Phys. Lett.* B **315** 34
[294] Lott B *et al* 1992 *Phys. Rev. Lett.* **68** 3141
[295] Louvel M *et al* 1994 *Phys. Lett.* B **320** 221
[296] Lukasik J *et al* 1997 *Phys. Rev.* C **55** 1906
[297] Lynch W G 1987 *Ann. Rev. Nucl. Part. Sci.* **37** 493
[298] Lynen U *et al* 1998 *Nucl. Phys.* A **630** 176c
[299] Ma Y G *et al* 1997 *Phys. Lett.* B **390** 41
[300] Madelung E 1927 *Z. Phys.* **40** 322
[301] Mahaux C *et al* 1985 *Phys. Rep.* **120** 1
[302] Mahaux C and Sartor R 1989 *Nuclear Matter and Heavy-Ion Collisions (Les Houches, France) (NATO, ASI, B 205)* (New York: Plenum)

[303] Majka Z *et al* 1996 *Preprint* TAM 96-03
[304] Malfliet R 1992 *Nucl. Phys.* A **545** 3c
[305] Marie N 1995 *Thesis* Université de CAEN, unpublished
[306] Marie N *et al* 1996 *Phys. Lett.* B **391** 15
[307] Marques F M *et al* 1997 *Phys. Rep.* **284** 91
[308] Maskay A M *et al* to be published
[309] Mastinu P F *et al* 1996 *Phys. Rev. Lett.* **76** 2646
[310] Mayer M G 1949 *Phys. Rev.* **75** 1969
[311] Mayer M G and Jensen J H D 1955 *Elementary Theory of Nuclear Shell Structure* (New York: Wiley)
[312] Metag V 1998 *Nucl. Phys.* A **630** 1c
[313] Métivier V 1995 *PhD Thesis* Université de Caen
[314] Métivier V *et al* 2000 *Nucl. Phys.* A **672** 357
[315] Migneco E *et al* 1992 *Nucl. Instrum. Methods* A **314** 31
[316] Minich R *et al* 1982 *Phys. Lett.* B **118** 458
[317] Mishustin I N 1998 *Nucl. Phys.* A **630** 111c
[318] Molitoris J and Stöcker H 1985 *Phys. Lett.* B **162** 47
[319] Montoya C P *et al* 1994 *Phys. Rev. Lett.* **73** 3070
[320] Moretto L 1975 *Nucl. Phys.* A **427** 211
[321] Moretto L and Wozniak G 1984 *Ann. Rev. Nucl. Sci.* **34** 189
[322] Moretto L and Wozniak G 1993 *Ann. Rev. Nucl. Sci.* **43** 379
[323] Moretto L *et al* 1997 *Phys. Rep.* **287** 249
[324] Morjean M *et al* 1995 *Nucl. Phys.* A **591** 371
[325] Morjean M *et al* 1998 *Nucl. Phys.* A **630** 200c
[326] Morjean M Private communication and in preparation
[327] Morley K B *et al* 1995 *Phys. Lett.* B **355** 52
[328] Mosel U 1991 *Ann. Rev. Nucl. Part. Sci.* **41** 29
[329] Mottelson B 1996 *Trends in Nuclear Physics, 100 Years Later (Les Houches, Session LXVI)* ed H Nifenecker *et al*
[330] Müller E 1990 *J. Phys. G: Nucl. Part. Phys.* **16** 1571
[331] Mueller H and Serot B 1995 *Phys. Rev.* C **52** 2072
[332] Mueller W F *et al* 1996 *Preprint* GSI 96-29 First Catania Relativistic Ion Studies, Acicastello
[333] Mueller A C 1999 *Nucl. Phys.* A **654** 215c
[334] Myers W D and Swiatecki W J 1966 *Nucl. Phys.* **81** 1
[335] Myers W D 1978 *Nucl. Phys.* A **296** 177
[336] Nagamiya S 1988 *Nucl. Phys.* A **488** 3c
[337] Natowitz J *et al* 1992 *Nucl. Phys.* A **538** 263c
[338] Natowitz J *et al* 1995 *Phys. Rev.* C **52** R2322
[339] Nebauer R *et al* 1999 *Nucl. Phys.* A **650** 65
[340] Negele J W and Vautherin D 1972 *Phys. Rev.* C **5** 1472
[341] Negele J W 1982 *Rev. Mod. Phys.* **54** 912
[342] Nguyen A D *et al* 1998 *Proc. XXXV Int. Winter Meeting (Bormio)* ed I Iori
[343] Nix J R and Strottman D 1981 *Phys. Rev.* C **23** 2548
[344] Nordheim L W 1928 *Proc. R. Soc.* A **119** 689
[345] Nosé S 1984 *J. Chem. Phys.* **81** 511
[346] Novotny R *et al* 1991 *IEEE Trans. Nucl. Sci.* **38** 379
[347] Oddershede L, Dimon P and Bohr J 1993 *Phys. Rev. Lett.* **71** 3107

[348] Oeschler H *et al* 1999 *Proc. XXVII Winter Meeting (Hirshegg)* ed H Feldmeier *et al*
[349] Ohnishi A and Randrup J 1993 *Nucl. Phys.* A **565** 474
[350] Ohnishi A and Randrup J 1995 *Phys. Rev. Lett.* **75** 596
[351] Ono A *et al* 1992 *Phys. Rev. Lett.* **68** 2898
      Ono A *et al* 1992 *Prog. Theor. Phys.* **87** 1185
[352] Ono A and Horiuchi H 1995 *Phys. Rev.* C **51** 299
[353] Ono A and Horiuchi H 1996 *Phys. Rev.* C **53** 845
[354] Ono A and Horiuchi H 1996 *Phys. Rev.* C **53** 2344
[355] Ono A and Horiuchi H 1996 *Phys. Rev.* C **53** 2958
[356] Ono A *et al* 1998 *Nucl. Phys.* A **630** 148c
[357] Pak R *et al* 1997 *Phys. Rev. Lett.* **78** 1022
[358] Pan J *et al* 1998 *Phys. Rev. Lett.* **80** 1182
[359] Paul P 1998 *Nucl. Phys.* A **630** 92c
[360] Péghaire A *et al* 1990 *Nucl. Instrum. Methods* A **295** 365
[361] Peilert G *et al* 1989 *Phys. Rev.* C **39** 1402
[362] Péter J *et al* 1995 *Nucl. Phys.* A **593** 95
[363] Pethick Ch J and Ravenhall D G 1987 *Nucl. Phys.* A **471** 19c
[364] Pethick Ch J and Ravenhall D G 1996 *Trends in Nuclear Physics, 100 Years Later (Les Houches, Session LXVI)* ed H Nifenecker *et al*
[365] Phair L *et al* 1992 *Nucl. Phys.* A **548** 489
[366] Phair L *et al* 1995 *Phys. Rev. Lett.* **75** 213
[367] Phair L *et al* 1997 *Phys. Rev. Lett.* **79** 3538
[368] Pi M, Suraud E and Schuck P 1991 *Nucl. Phys.* A **524** 537
[369] Piattelli P *et al* 1996 *Nucl. Phys.* A **599** 63c
[370] Pienkowski L *et al* 2000 *Phys. Lett.* B **472** 15
[371] Ploszajczak M, Botet R and Srokowski T 1997 *Proc. IV TAPS Workshop (Mt St Odile, France)* ed Y Schutz and H Lohner (Éditions Frontières) p 123
[372] Pochodzalla J *et al* 1995 *Phys. Rev. Lett.* **75** 1040
[373] Polacco E C *et al* 1995 *Nucl. Phys.* A **583** 441
[374] Porile N T *et al* 1989 *Phys. Rev.* C **39** 1914
[375] Pouthas J *et al* 1995 *Nucl. Instrum. Methods* A **357** 418
[376] Preston M A and Bhaduri R K 1982 *Structure of the Nucleus* (London: Addison-Wesley)
[377] Puri R *et al* 1996 *Phys. Rev.* C **54** R28
[378] Quentin Ph and Flocard H 1978 *Ann. Rev. Nucl. Part. Sci.* **28** 523
[379] Raduta Al H and Raduta Ad R 1999 *Phys. Rev.* C **59** 323
[380] Rami F *et al* 2000 *Phys. Rev. Lett.* **84** 1120
[381] Randrup J and Koonin S 1981 *Nucl. Phys.* A **356** 223
[382] Randrup J and Remaud B 1990 *Nucl. Phys.* A **514** 339
[383] Randrup J 1993 *Comput. Phys. Commun.* **77** 153
[384] Ravenhall D G, Pethick C J and Wilson J R 1983 *Phys. Rev. Lett.* **50** 2066
[385] Ray S *et al* 1997 *Phys. Lett.* B **392** 7
[386] Reif F 1965 *Fundamentals of Statistical and Thermal Physics* (New York: McGraw-Hill)
[387] Reinhard P-G, Suraud E and Ayik S 1992 *Ann. Phys., NY* **213** 204
[388] Reinhard P-G and Suraud E 1992 *Ann. Phys., NY* **216** 98
[389] Reinhard P-G and Toepffer C 1994 *Int. J. Mod. Phys.* E **3** 435
[390] Reinhard P-G and Suraud E 1995 *Ann. Phys., NY* **239** 193

[391] Reinhard P-G and Suraud E 1995 *Ann. Phys., NY* **239** 216
[392] Reinhard P-G and Suraud E 1996 *Z. Phys.* A **355** 339
[393] Reisdorf W and Ritter H G 1997 *Ann. Rev. Nucl. Part. Sci.* **47** 663
[394] Reisdorf W 1998 *Nucl. Phys.* A **630** 15c
[395] Reisdorf W *et al* 1999 *Proc. XXVII Winter Meeting (Hirshegg)*
[396] Remaud B 1996 *Ann. Phys., Paris* **21** 503
[397] Richert J and Wagner P 1992 *Z. Phys.* A **341** 171
[398] Ring P and Schuck P 1980 *The Nuclear Many-Body Problem* (Heidelberg: Springer)
[399] Risken H 1984 *The Fokker–Planck Equation* (Berlin: Springer)
[400] Rivet M F *et al* 1993 *Proc. XXXI Int. Winter Meeting (Bormio)* ed I Iori
[401] Rivet M F *et al* 1996 *Phys. Lett.* B **388** 219
[402] Rivet M F *et al* 1997 *Proc. XXXV Int. Winter Meeting (Bormio)* ed I Iori
[403] Rivet M F *et al* 1998 *Phys. Lett.* B **430** 217
[404] Rivet M F Private Communication
[405] Rivet M F *et al* 1999 *Proc. XXVII Winter Meeting (Hirshegg)* ed H Feldmeier *et al*
[406] Rudolf G *et al* 1991 *Nucl. Instrum. Methods* A **307** 325
[407] Rudolf G *et al* 1993 *Phys. Lett.* B **307** 278
[408] Samadar J *et al* 1997 *Phys. Rev. Lett.* **79** 4962
[409] Samaddar S K and Richert J 1989 *Z. Phys.* A **332** 443
[410] Sauer G *et al* 1976 *Nucl. Phys.* A **264** 221
[411] Schlagel T J and Pandharipande V R 1987 *Phys. Rev.* C **36** 162
[412] Schmelzer J, Ropke G and Ludwig F P 1997 *Phys. Rev.* C **55** 1917
[413] Schmitt K J, Reinhard P-G and Toepffer Ch 1990 *Z. Phys.* A **336** 123
[414] Schnack J and Feldmeier H 1996 *Nucl. Phys.* A **601** 181
[415] Schröder W U and Huizenga J R 1984 *Treatise on Heavy Ion Science* vol 2, ed A Bromley (New York: Plenum)
[416] Schröder W U and Huizenga J R 1989 *Nucl. Phys.* A **502** 473c
[417] Schussler F *et al* 1995 *Nucl. Phys.* A **584** 704
[418] Schuttauf A *et al* 1996 *Nucl. Phys.* A **607** 457
[419] Schutz Y 1996 *Nucl. Phys.* A **597** 97c
[420] Schweber S S 1961 *Introduction to Relativistic Quantum Field Theory* (London: Harper)
[421] Sehn L and Wolter H H 1990 *Nucl. Phys.* A **519** 289c
[422] Serber R 1947 *Phys. Rev.* **72** 1114
[423] Serot B and Walecka J D 1986 *Advances in Nuclear Physics* vol 16, eds J W Negele and E Vogt (New York: Plenum)
[424] Shapiro S I and Teukolsky S A 1983 *Black Holes, White Dwarfs and Neutron Stars* (New York: Wiley)
[425] Shlomo S and Natowitz J B 1990 *Phys. Lett.* B **252** 187
[426] Siemens Ph and Jensen A 1987 *Elements of Nuclei* (Reading, MA: Addison-Wesley)
[427] Siemssen R H 1996 *Trends in Nuclear Physics, 100 Years Later (Les Houches, Session LXVI)* ed H Nifenecker *et al*
[428] Sierk A J and Nix J R 1980 *Phys. Rev.* C **21** 982
[429] Siwek A *et al* 1998 *Phys. Rev.* C **57** 2507
[430] Skulski W *et al* 1996 *Phys. Rev.* C **53** R2594
[431] Skyrme T H R 1956 *Phil. Mag.* **1** 1043
[432] Snider R F 1960 *J. Chem. Phys.* **32** 1051
[433] Snover K A 1986 *Ann. Rev. Nucl. Part. Sci.* **36** 545

[434] Sobotka L *et al* 1997 *Phys. Rev.* C **55** 2109
[435] Soff S *et al* 1995 *Phys. Rev.* C **51** 3320
[436] Sorge H, Stöcker H and Greiner W 1989 *Ann. Phys., NY* **192** 266
[437] Sprung D W L 1972 *Adv. Nucl. Phys.* **5** 225
[438] Stachel J 1999 *Nucl. Phys.* A **654** 119c
[439] Stauffer D and Aharony A 1994 *Introduction to Percolation Theory* (London: Taylor and Francis)
[440] Steckmeyer J C *et al* 1996 *Phys. Rev. Lett.* **76** 4895
[441] Stock R 1979 *Heavy Ion Collisions* vol I, ed R Bock (Amsterdam: North-Holland) p 607
[442] Stock R 1986 *Phys. Rep.* **135** 261
[443] Stöcker H, Maruhn J A and Greiner W 1979 *Z. Phys.* A **290** 297
[444] Stöcker H and Greiner W 1986 *Phys. Rep.* **137** 277
[445] Stracener D W *et al* 1990 *Nucl. Instrum. Methods* A **294** 485
[446] Suraud E, Schuck P and Hasse R W 1985 *Phys. Lett.* B **164** 212
[447] Suraud E 1987 *Nucl. Phys.* A **462** 109
[448] Suraud E, Barranco M and Treiner J 1988 *Nucl. Phys.* A **480** 29
[449] Suraud E, Grégoire Ch and Tamain B 1989 *Prog. Part. Nucl. Phys.* **23** 357
[450] Suraud E *et al* 1989 *Phys. Lett.* B **229** 359
[451] Suraud E *et al* 1990 *Nucl. Phys.* A **519** 171c
[452] Suraud E *et al* 1992 *Nucl. Phys.* A **542** 141
[453] Suraud E *et al* 1994 *Nucl. Phys.* A **580** 323
[454] Suraud E 1996 *Ann. Phys., Paris* **21** 461
[455] Suraud E and Reinhard P G 1998 *Cze. J. Phys.* **48** 862
[456] Takahashi K 1989 *Prog. Theor. Ph. Suppl.* **98** 109
[457] Tanaka K *et al* 1995 *Nucl. Phys.* A **583** 581c
[458] Tanihata I 1999 *Nucl. Phys.* A **654** 235c
[459] Ter Haar B and Malfliet R 1987 *Phys. Rep.* **149** 207
[460] Tilquin I *et al* 1995 *Nucl. Instrum. Methods* A **365** 446
[461] Tirel O 1998 *Thesis* Caen, *Preprint* GANIL T98-02
[462] Tohyama M and Suraud E 1992 *Nucl. Phys.* A **459** 461
[463] Toke J and Schröder U 1992 *Ann. Rev. Nucl. Part. Sci.* **42** 401
[464] Toke J *et al* 1995 *Nucl. Phys.* A **583** 519c
[465] Toke J *et al* 1995 *Phys. Rev. Lett.* **75** 2920
[466] Toke J *et al* 1996 *Phys. Rev. Lett.* **77** 3514
[467] Toke J *et al* 1997 *Phys. Rev.* C **56** R1683
[468] Trockel R *et al* 1987 *Phys. Rev. Lett.* **59** 2844
[469] Tsang B *et al* 1986 *Phys. Rev. Lett.* **57** 559
[470] Tsang B *et al* 1996 *Phys. Rev.* C **53** 1959
[471] Tsang B *et al* 1996 *Phys. Rev.* C **53** R1057
[472] Tsang B and Danielewicz P 1998 *Phys. Rev. Lett.* **80** 1178
[473] Uehling E A and Uhlenbeck G E 1933 *Phys. Rev.* **43** 552
[474] van der Ploeg H *et al* 1996 *Nucl. Phys.* A **599** 117c
[475] Van Kampen N G 1987 *Stochastic Processes in Physics and Chemistry* (Amsterdam: North-Holland)
[476] 1999 Proc. Topical Conf. on Giant Resonances *Nucl. Phys.* A **649** 1c
[477] Vautherin D and Brink D 1972 *Phys. Rev.* C **5** 626
[478] Vautherin D *et al* 1984 *Nucl. Phys.* A **422** 140

[479] Vautherin D *et al* 1987 *Phys. Lett.* B **191** 6
[480] Viola V E, Kwiatkowski K and Walker M 1985 *Phys. Rev.* C **31** 1550
[481] Viola V *et al* 1999 *Proc. XXVII Winter Meeting (Hirshegg)* ed H Feldmeier *et al*
[482] Vlasov A A 1938 *Zh. Eksp. Teor. Fiz.* **8** 291
[483] Wada T, Abe Y and Carjan N 1993 *Phys. Rev. Lett.* **70** 3538
[484] Waddington C J and Freier P S 1985 *Phys. Rev.* C **31** 888
[485] Walecka J D 1974 *Ann. Phys., NY* **83** 491
[486] Weisskopf V 1937 *Phys. Rev.* **52** 295
[487] Weidenmüller H A 1980 *Prog. Nucl. Part. Phys.* **3** 49
[488] Welke G *et al* 1989 *Phys. Rev.* C **40** 317
[489] Westfall G D *et al* 1976 *Phys. Rev. Lett.* **27** 1202
[490] Westfall G D 1998 *Nucl. Phys.* A **630** 27c
[491] Wieloch A and Durand D 1997 *Z. Phys.* **359** 345
[492] Wieloch A *et al* 1998 *Phys. Lett.* B **432** 29
[493] Wilets L *et al* 1977 *Nucl. Phys.* A **282** 341
[494] Wilets L, Yariv Y and Chestnut R 1978 *Nucl. Phys.* A **301** 359
[495] Wilczynski J 1973 *Phys. Lett.* B **47** 124
      Wilczynski J 1973 *Phys. Lett.* B **47** 484
[496] Wilson W K *et al* 1992 *Phys. Rev.* C **45** 738
[497] Williams R D and Koonin S E 1985 *Nucl. Phys.* **435** 844
[498] Wong C-y 1994 *Introduction to High-Energy Heavy Ion Collisions* (Singapore: World Scientific)
[499] Wu J R *et al* 1979 *Phys. Rev.* C **19** 698
[500] Yanez R *et al* 1999 *Phys. Rev. Lett.* **82** 3585
[501] Yariv Y and Fraenkel Z 1979 *Phys. Rev.* C **20** 2227
[502] Zhang F S and Suraud E 1995 *Phys. Rev.* C **51** 3201
[503] Zhu F *et al* 1992 *Phys. Lett.* B **282** 299
[504] Zhu F *et al* 1995 *Phys. Rev.* C **52** 784
[505] Zwanzig R 1973 *J. Stat. Phys.* **9** 215

Related references

1992 Dynamical fluctuations and correlations in nuclear collisions (Aussois) *Nucl. Phys.* A **545** 1c
1988 Nuclear dynamics at medium and high energies (Bad-Honnef) *Nucl. Phys.* A **495** 1c
Balescu R 1975 *Nonequilibrium Statistical Mechanics* (New York: Wiley)
Balian R 1991 *From Microphysics to Macrophysics* (Berlin: Springer)
Baym G and Pethick C 1976 *The Physics of Liquid and Solid Helium* Part II, ed Bennemann and Ketterson (New York: Wiley)
1989 *Int. Conf. on Fifty Years of Nuclear Fission* (Berlin)
1989 *Nucl. Phys.* A **502** 1c
Bertsch G F and Broglia R 1994 *Oscillations in Finite Quantum Systems* (Cambridge: Cambridge University Press)
Binney J J *et al* 1992 *The Theory of Critical Phenomena* (Oxford: Clarendon)
Bohr A and Mottelson B 1969 *Nuclear Structure* vol I (New York: Benjamin)
——1975 *Nuclear Structure* vol II (New York: Benjamin)

Brack M and Bhaduri R K 1997 *Semi Classical Physics* (New York: Addison-Wesley)

Brink D M 1985 *Semi-Classical Methods in Nucleus–Nucleus Scattering* (Cambridge: Cambridge University Press)

Brown G E and Jackson A D 1976 *The Nucleon–Nucleon Interaction* (Amsterdam: North-Holland)

Campi X (ed) 1993 *Fragmentation Phenomena (Les Houches, Spring)*

Proc. 14th Int. Conf. on Cyclotrons (Cape Town)

Chandler D 1987 *Introduction to Modern Statistical Mechanics* (New York: Oxford University Press)

Cottingham W A and Greenwood D A *An Introduction to Nuclear Physics* (Cambridge: Cambridge University Press)

deShalit A and Feschbach H 1974 *Theoretical Nuclear Physics. Nuclear Structure* (New York: Wiley–Interscience)

1990 Nuclear dynamics (Elba) *Nucl. Phys.* A **519** 1c

Feshbach H 1992 *Theoretical Nuclear Physics. Nuclear Reactions* (New York: Wiley)

Fetter A L and Walecka J D 1971 *Quantum Theory of Many Body Systems* (New York: McGraw-Hill)

Fröbrich P and Lipperheide R 1996 *Theory of Nuclear Reactions (Oxford Studies in Nuclear Physics, vol 18)* Oxford: Clarendon)

Gaddioli E and Hodgson P E 1991 *Pre-Equilibrium Nuclear Reactions* (New York: Oxford University Press)

Gardiner C W 1990 *Handbook of Stochastic Methods* (Berlin: Springer)

Hansen J P and McDonald I R 1976 *Theory of Simple Liquids* (London: Academic)

1999 *Proc. Int. Workshop XXVII on Gross Properties of Nuclei and Nuclear Excitations (Hirshegg)* ed H Feldmeier *et al*

Hoover W G 1991 *Computational Statistical Mechanics* (Amsterdam: Elsevier)

Huang K 1963 *Statistical Mechanics* (New York: Wiley)

Kadanoff L P and Baym G 1962 *Quantum Statistical Mechanics* (New York: Benjamin)

1992 Nucleus–nucleus IV (Kanazawa, 1991) *Nucl. Phys.* A **538** 1c

Knoll G F 1979 *Radiation Detection and Measurement* (New York: Wiley)

Kubo R, Toda M and Hashitsume N 1985 *Statistical Physics* vols 1 and 2 (Heidelberg: Springer)

Landau L and Lifshitz E M 1959 *Quantum Mechanics* (London: Pergamon)

1988 First topical meeting on giant resonance in heavy-ion collisions *Nucl. Phys.* A **482** 1c

Liboff R L 1990 *Kinetic Theory* (New York: Prentice-Hall)

Lifshitz E M and Pitaevski L P 1979 *Physical Kinetics* (New York: Pergamon)

Mahaux C and Sartor R 1989 *Nuclear Matter and Heavy-Ion Collisions (Les Houches, France) (NATO, ASI, B 205)* (New York: Plenum)

Mayer M G and Jensen J H D 1955 *Elementary Theory of Nuclear Shell Structure* (New York: Wiley)

1990 *The Nuclear Equation of State, (Peniscola, 1989) (NATO, ASI, 208)* (New York: Plenum)

Press W H *et al* 1986 *Numerical Recipes* (Cambridge: Cambridge University Press)

Preston M A and Bhaduri R K 1982 *Structure of the Bucleus* (London: Addison-Wesley)

Reif F 1965 *Fundamentals of Statistical and Thermal Physics* (New York: McGraw-Hill)

Ring P and Schuck P 1980 *The Nuclear Many-Body Problem* (Heidelberg: Springer)

Risken H 1984 *The Fokker–Planck Equation* (Berlin: Springer)

1988 Nucleus–nucleus III (Saint-Malo) *Nucl. Phys.* A **488** 1c

Schröder W U and Huizenga J R 1984 *Treatise on Heavy Ion Science* vol 2, ed A Bromley (New York: Plenum)

Schweber S S 1961 *Introduction to Relativistic Quantum Field Theory* (London: Harper)

Shapiro S I and Teukolsky S A 1983 *Black Holes, White Dwarfs and Neutron Stars* (New York: Wiley)

Siemens Ph and Jensen A 1987 *Elements of Nuclei* (Reading, MA: Addison-Wesley)

Stauffer D and Aharony A 1994 *Introduction to Percolation Theory* (London: Taylor and Francis)

Stock R 1979 *Heavy Ion Collisions* vol I, ed R Bock (Amsterdam: North-Holland) p 607

Thoenessen M *et al* (ed) 1998 Nucleus–nucleus VI (Gattlinburg, 1997) *Nucl. Phys.* A **630** 1c

Van Kampen N G 1987 *Stochastic Processes in Physics and Chemistry* (Amsterdam: North-Holland)

1999 Proc. Topical Conf. on Giant Resonances *Nucl. Phys.* A **649** 1c

Wong C-y 1994 *Introduction to High-Energy Heavy Ion Collisions* (Singapore: World Scientific)

Index